IUTAM Symposium on Nonlinear Stochastic Dynamics and Control

IUTAM BOOKSERIES

VOLUME 29

Aims and Scope of the Series

The IUTAM Bookseries publishes the proceedings of IUTAM symposia under the auspices of the IUTAM Board.

For other titles published in this series, go to
www.springer.com/series/7695

W.Q. Zhu, Y.K. Lin, and G.Q. Cai
Editors

IUTAM Symposium on Nonlinear Stochastic Dynamics and Control

Proceedings of the IUTAM Symposium held in Hangzhou, China, May 10–14, 2010

Springer

Editors

W.Q. Zhu
Zhejiang University
Zhe Da Road 38
310027 Hangzhou
People's Republic of China
E-mail: wqzhu@yahoo.com

Y.K. Lin
Florida Atlantic University
College of Engineering and
Computer Science
33431 Boca Raton Florida
USA

G.Q. Cai
Florida Atlantic University
College of Engineering and
Computer Science
33431 Boca Raton Florida
USA
E-mail: caig@fau.edu

ISSN 1875-3507
ISSN 1875-3493 (eBook)
ISBN 978-94-017-8184-8
ISBN 978-94-007-0732-0 (eBook)
DOI 10.1007/978-94-007-0732-0
Springer Dordrecht Heidelberg London New York

Cover design: VTEX,Vilnius

Printed on acid-free paper

Springer is part of Springer Science+Business Media (www.springer.com)

Preface

The history of stochastic dynamics may be traced back over 100 years ago to Einstein's paper on Brownian motion in 1905. Since then, its scope and depth have been broadened and enhanced by a wide variety of applications in science and engineering. From 1972 to 2002 ten symposia on stochastic dynamics were sponsored by International Union of Theoretical and Applied Mechanics (IUTAM). More recently, much progress on various aspects of nonlinear stochastic dynamics has been made and the research activities in the field have placed greater emphasis on systems with high nonlinearity, and on systems in which controllers play an important role.

This volume contains the papers presented at the IUTAM Symposium on Nonlinear Stochastic Dynamics and Control, held in Zhejiang University, China, May 10-14, 2010. The scientific committee appointed by the Bureau of IUTAM includes the following members:

W.Q. Zhu, Hangzhou, China (Chairman)
Y.K. Lin, Boca Raton,USA (Co-chairman)
A. Naess, Trondheim, Norway
W. Schiehlen, Stuttgart,Germany
G.I. Schüller, Innsbruck, Austria
K. Sobczyk, Warsaw, Poland
T.T. Soong, Buffalo, USA

A total of 44 active scientists from 13 countries (regions) accepted the invitation from the scientific committee, and 31 presentations, covering the following six groups of topics, were scheduled for the symposium:

1. Response and reliability of nonlinear stochastic systems. K. Sobczyk and P. Holobut gave a brief analysis of the entropy/information change in stochastic dynamical system with a special emphasis on the effect of the system parameters and intensities of random noise. V.V. Malanin and I.E. Poloskov proposed a scheme of study for systems with different forms of time aftereffect which has a transparent algorithm and can be simply combined with Monte Carlo method. W. Xu *et al* focused on the steady-state analysis of a class of nonlinear dynamical systems with multi-delayed feedbacks. J.Q. Sun presented a finite-dimensional Markov process approximation which opens a gate to various studies of stochastic dynamical systems with time delay. G.Q. Cai and Y.K. Lin studied the stochastic nonlinear behavior of two competing species of grass and woody vegetation. M.L. Deng and W.Q. Zhu investigated the stochastic energy transition of peptide-bond (PB) in enzyme-substrate-complex (ESC) by using stochastic averaging method and

Monte Carlo simulation. X.L. Jin and Z.L. Huang obtained the non-stationary probability densities of system responses for multi-degree-of-freedom nonlinear systems subject to stochastic parametric and external excitations. Y. Zeng and W.Q. Zhu proposed a stochastic averaging method for single-degree-of-freedom strongly nonlinear oscillators under Poisson white noise excitation using the so-called generalized harmonic functions. Cho W. S. To presented a novel approach for response analysis of MDOF nonlinear systems under non-Gaussian nonstationary random excitations using the stochastic central difference method, co-ordinate transformation and adaptive time scheme. M. Vasta and M. Di Paola obtained an approximate explicit response probability density function of a beam under external impulsive random Poisson excitation as approximate solution of integro-differential KF equation. G.K. Er and V.P. Iu proposed a method for determining the probabilistic stationary solutions for multi-dimensional nonlinear stochastic dynamic systems. S. Narayanan and P. Kumar developed an efficient numerical implementation for the path integral (PI) method based on non-Gaussian transition probability density function (PDF) and the Gauss-Legendre integration scheme. P.D. Spanos and I.A. Kougioumtzoglou proposed a novel harmonic wavelet-based statistical linearization approach for determining the evolutionary power spectrum (EPS) of the response of nonlinear oscillators subject to stochastic excitation. C.H. Loh *et al* developed methods for analyzing the seismic response data and the long-term static data of the Fei-tsui arch dam and thus setting an early warning threshold level for dam safety evaluation. A. Naess *et al* considered the first-passage type failure for systems subject to multiplicative and additive white noises excitations and obtained the numerical results using the path integration method.

2. Stability, bifurcation and chaos of nonlinear stochastic systems. M. F. Dimentberg *et al* found that temporal random variations of parameters in dynamic systems might “smear” classical neutral stability boundaries. The system’s response within such a “twilight zone” of marginal instability was found to be of an intermittent nature, with alternating periods of zero (or almost zero) response and rare short outbreaks. S.H. Li and X.B Liu investigated the *p*th moment Lyapunov exponent of a co-dimension two bifurcation system, that is on a three-dimensional centermanifold and excited parametrically by a white noise. T. Fang *et al* proposed a practical strategy for studying stochastic chaos based on orthogonal polynomial approximation and ergodic theorem. C.B. Gan studied the noisy scattering dynamics in the randomly-driven Henon-Heiles oscillator when the energy was large enough to permit particles to escape from the exits. The author paid special attention to the computation of the exit basins, which show a rich pattern of noisy fractal structures and the uncertainty dimensions of the fractal sets. N. Gaus and C. Proppe investigated the bifurcation behavior of the non-smooth mass on a belt system.

3. Resonance and synchronization of nonlinear stochastic systems. W. Wedig investigated stochastic resonances (SR), a phenomenon that nonlinear system synchronizes with noise to boost a resonant-like behavior, for road-vehicle systems and related bifurcation problems. Y.B. Yang and B.H. Xu presented a review on the parametric SR technique and its applications in signal processing and target detection in shallow water reverberation. X.J. Sun and Q.S. Lu discussed the synchronization behavior of a clustered neuronal network with additive noise to reveal the role played by the clustered structure of networks and the effects of the coupling strength and cluster number.

4. Control of nonlinear stochastic systems. Z.H. Liu and W.Q. Zhu studied stochastic optimal time-delay control and stabilization of quasi-integrable Hamiltonian systems by converting the original control problem into a stochastic optimal control problem without time-delay. P. Kaczynski and L. Socha solved the problem of quasi-optimal control for Duffing oscillator with parametric and external Gaussian and Poisson noise excitations, using an iterative procedure combining Gaussian statistical linearization and LQGP technique. V. Gattulli *et al* presented semi-active control strategies for asymmetric structures based on optimal sizing of an equivalent Kelvin-Voight model describing the constitutive behavior of semi-active magneto-rheological devices, when operating in passive modality with maximum achievable modal damping.

5. Modeling of stochastic dynamical systems and stochastic excitations. J. Li *et al* presented a review on modeling stochastic dynamic excitations, which is incorporated with a probability density evolution method for response analysis of nonlinear structures. G. I. Schuëller and B. Goller proposed model updating procedures to improve the match between experimental data and corresponding model output. X.Q. Wang *et al* proposed a nonparametric stochastic modeling technique to reduce the order of geometrically nonlinear structural models.

6. Structural health monitoring. B. F. Spencer *et al* proposed a RDT-based decentralized data aggregation approach for efficient data condensation and feature extraction, and verified experimentally the results. The performance of decentralized RDT was assessed in terms of accuracy of the estimated modal properties and efficiency in the wireless data communication. R. Zhang *et al* proposed a wave-based approach to model and analyze seismic building motion, providing additional perspective of seismic behavior of building structures, not clearly obtainable with the traditional vibration-based approach.

We wish to thank all participants of this IUTAM Symposium, and all organizers for their enthusiastic and valuable contributions to the Symposium. In addition, we gratefully acknowledge the financial supports from IUTAM and National Natural Science Foundation of China.

Symposium Chairmen
W.Q. Zhu
Y.K. Lin

Contents

List of Participants

Names

G.Q. Cai
Y.W. Chang
C.H. Chen
J.B. Chen
M.L. Deng
G.K. ER
T. Fang
C.B. Gan
V. Gattulli
B. Goller
N. Gaus
R.H. Huan
Z.L. Huang
P. Kaczyński
X.L. Leng
Y. Lei
H. Li
J. Li
Y.K. Lin
X.B. Liu
Z.H. Liu
C.H. Loh
Q.S. Lu
A. Naess
S. Narayanan
I.E. Poloskov
M.D. Paola
K. Sobczyk
L. Socha
T.T. Soong
C. Soize
P. Spanos
B.F. Spencer
J.Q. Sun
W.L. Sun
Cho W.S. To
M. Vasta
W.V. WEDIG
W. Xu
Y.B. Yang
Y. Zeng
H.Q. Zhang
R.C. Zhang
W.Q. Zhu

Welcome to IUTAM Symposium on Nonlinear Stochastic Dynamics and Control

Welcome to IUTAM Symposium on Nonlinear Stochastic Dynamics and Control
1
2
3
4
5
6
7
8
9
10
11
12
13
14
15
16
17
18
19
20
21
22
23
24
25
26
27
28
29
30
31
32
33
34
35
36
37
38
39
40
41

1. B.F. Spencer
2. C. Soize
3. A. Naess
4. Y.K. Lin
5. W. Yang
6. W.Q. Zhu
7. T.T. Soong
8. K. Sobczyk
9. P. Spanos
10. I.E. Poloskov
11. T. Fang
12. Q.S. Lu
13. G.Q. Cai
14. R. Zhang
15. W. Wedig
16. S. Narayanan
17. N. Gaus
18. G.K. Er
19. R.H. Huan
20. X.L. Leng
21. H.Q. Zhang
22. J.Q. Sun
23. X.B. Liu
24. J.B. Chen
25. B. Goller
26. L. Socha
27. Z.G. Huang
28. C.H. Chen
29. Y.W. Chang
30. Z.L. Huang
31. H.W. Rong
32. Y. Zeng
33. Z.H. Liu
34. P. Kaczyński
35. Y. Lei
36. J. Li
37. H. Li
38. W. Xu
39. C.B. Gan
40. Cho W.S. To
41. V. Gattulli

Part 1
Response and Reliability of Nonlinear Stochastic Systems

Nonlinear Stochastic Ecosystem of Two Competing Species

G.Q. Cai and Y.K. Lin

College of Engineering and Computer Science, Florida Atlantic University,
Boca Raton, FL 33431, USA

Abstract. A nonlinear stochastic model describing two competing species of grass and woody vegetation is proposed. Two types of stochastic processes are proposed to model the variation part of the stocking rate. One is Gaussian white-noise, and another is randomized sinusoidal process. The system nonlinear behaviors are investigated for the deterministic case and the cases with two different stochastic variations. Some key characteristics of the system are found to be different for cases with and without stochastic variations. With the stochastic variation, a single stable state in the deterministic system is diffused into a region of stable states, and a separatrix dividing the two attraction zones no longer exists. The system may follow different trajectories and lead to different outcomes, beginning from the same initial state. It is also found that the stationary probability of the system response depends on the initial conditions, a special phenomenon for the investigated nonlinear system. Although the white-noise process and the randomized sinusoidal process are quite different in nature, the qualitative behaviors of the system are similar. Furthermore, effects of the initial state and the intensity of the stochastic variations on the system behaviors are investigated.

Keywords: Nonlinear ecosystem, Stochastic variation, Probability density, Monte-Carlo simulation.

1 Introduction

In its natural state, a semi-arid savanna has a predominant grass cover, with scattered trees and shrubs. It is an ecosystem of two species competing for resources. Such a natural state is found in the south-western U.S.A, and in Africa, India and Australia. Sustainability, stability and resilience of such a state have generated considerable interest among ecologists. Early studies of the dynamics of such systems were presented by Ludwig et al. [3] and Walker et al. [7]. More recently, Ludwig et al. [4] presented a comprehensive investigation based on a modified version of the well-known model of Lotka [2] and Volterra [5,6] as follows

$$\begin{aligned} \dot{g} &= r_g\, g(1 - s - c_{gg}\, g - c_{wg}\, w) \\ \dot{w} &= r_w[a + w(1 - c_{gw}\, g - c_{ww}\, w)] \end{aligned} \tag{1}$$

W.Q. Zhu et al. (Eds.): Nonlinear Stochastic Dynamics & Control, IUTAM BOOK SERIES 29, pp. 3–12.
springerlink.com

where g and w are the densities of the grass and woody vegetation, respectively, r_g and r_w are their growth rates, s is the cattle stocking rate, c_{wg} and c_{gw} are the inter-species-competition coefficients, c_{gg} and c_{ww} are the intraspecies-competition coefficients, and a is a source term for the woody vegetation. Model (1) differs from the original Lotka-Volterra model by adding the intraspecies competition terms $-c_{gg}g^2$ and $-c_{ww}w^2$, a source term to the woody vegetation, and a cattle stocking term. Equation set (1) is nonlinear and deterministic; namely, all parameters in the set are assumed to be precisely known. Such assumption is idealistic since changes in the environment are always present, and in most cases, they cannot be predicted in advance.

In the present paper, the nonlinear qualitative behaviors are investigated first for the deterministic model (1). Then a stochastic model is proposed for the two competing species of grass and woody vegetation with the stocking rate varying randomly. Two stochastic processes, Gaussian white-noise process and randomized sinusoidal process, are used to model the variation of the stocking rate, respectively. Since the system is highly nonlinear, Monte Carlo type simulations are carried out to analyze the characteristics of the stochastic systems. The purpose of the paper is aimed at shedding some light on the profound effects of the nonlinearity and random variability on system behaviors.

2 Characteristics of Deterministic Nonlinear System

The equilibrium points of system (1) can be found by letting the right-hand-sides of (1) be zero, namely,

$$\begin{aligned} g(1-s-c_{gg}g-c_{wg}w) &= 0 \\ a+w(1-c_{gw}g-c_{ww}w) &= 0 \end{aligned} \tag{2}$$

Denoting

$$A = c_{wg}c_{gw} - c_{gg}c_{ww}, \quad B = c_{gw}(1-s) - c_{gg}, \quad C = ac_{gg} \tag{3}$$

The following three equilibrium centers exist if $B^2 - 4AC > 0$

Unstable center **S**: $w = \dfrac{1}{2A}(B+\sqrt{B^2-4AC})$, $g = \dfrac{1-s-c_{wg}w}{c_{gg}}$

Asymptotic stable center **W**: $w = \dfrac{1}{2c_{ww}}(1+\sqrt{1+4ac_{ww}})$, $g = 0$

Asymptotic stable center **G**: $w = \dfrac{1}{2A}(B-\sqrt{B^2-4AC})$, $g = \dfrac{1-s-c_{wg}w}{c_{gg}}$

On the other hand, if B^2 - $4AC = 0$, then system (1) has an unstable equilibrium center **S**, and an asymptotic stable equilibrium center **W**. Finally, if B^2 - $4AC < 0$, then there exists just an asymptotic stable equilibrium center **W**.

The case of three equilibrium centers is shown in Fig. 1 with $r_g = 1.5$, $r_w = 1$, $s = 0.3$, $a = 0.03$, $c_{gg} = 0.7$, $c_{wg} = 1$, $c_{gw} = 2$, and $c_{ww} = 1.03$, respectively. These values are non-dimensional and have been used in literatures. The stable equilibrium center **W** corresponds to the extinction of grass, while the stable equilibrium center **G** corresponds nearly to the opposite; namely, the grass density is high and the woody vegetation density is very low. Depending on the initial conditions, the system will approach definitely to one of the stable equilibrium centers. As shown in Fig. 1, a trajectory starting from an initial state of $g_0 = 0.1$ and $w_0 = 0.1$ will approach the stable center **G**, while one will end up at the stable center **W** if it begins from $g_0 = 0.1$ and $w_0 = 0.3$. A separatrix passes through center **S**, and divides the whole permissible domain into two regions of attraction.

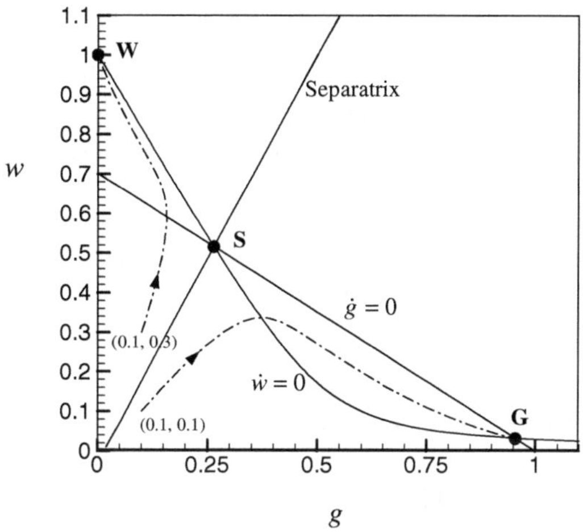

Fig. 1 Phase plane of deterministic system (1)

While Equation (1) catches some basic characteristics of the nonlinear ecosystem, it fails to account for variations in a real ecological environment. According to the model, the system would eventually approach either state **S** or state **W**, and the outcome would depend entirely on the initial populations of the grass and woody vegetation, respectively. Such type of behavior is obviously not realistic.

3 Stochastic Nonlinear Model

The present paper attempts to find out what will happen if certain parameters in the system are allowed to change randomly with time. Specifically, the stocking rate is allowed to vary randomly, namely, the parameter s in Equation (1) is now replaced by

$$s = s_0 + \xi(t) \tag{4}$$

where s_0 is a constant, and $\xi(t)$ is a stochastic process with zero mean value. The stochastic version of deterministic model (1) is then proposed as follows

$$\begin{aligned} \frac{d}{dt}G &= r_g G(1 - s_0 - c_{gg}G - c_{wg}W) + r_g G\xi(t) \\ \frac{d}{dt}W &= r_w[a + W(1 - c_{gw}G - c_{ww}W)] \end{aligned} \tag{5}$$

where $G(t)$ and $W(t)$ are now stochastic processes representing the uncertain grass density and woody vegetation density, respectively. The tradition of using a capital letter to represent a random variable or a stochastic process is adopted here.

The random variation of the stocking rate, $\xi(t)$, should be estimated from the observation data. Depending on different situations, two stochastic processes may be feasible to model $\xi(t)$. One is Gaussian white noise, describing fast changing variations; while another is randomized sinusoidal process, allowing random variations embedded in periodic processes.

Due to the strong system nonlinearity, analytical solution is presently not available for (5); thus, a Monte Carlo type simulation study has been carried out with the same parameters as those in Fig. 1, i.e. $r_g = 1.5$, $r_w = 1$, $a = 0.03$, $c_{gg} = 0.7$, $c_{wg} = 1$, $c_{gw} = 2$, $c_{ww} = 1.03$, and $s_0 = 0.3$.

3.1 Gaussian White-Noise Model

$\xi(t)$ is modeled as a Gaussian white noise with an intensity D, i.e.,

$$E[\xi(t)\xi(t+\tau)] = D\delta(\tau) \tag{6}$$

Equation (6) indicates that the correlation time is very short, namely, the change of $\xi(t)$ is very fast. Due to the random variability in the stocking rate, the behavior of the stochastic system (5) is substantially different from that of its deterministic counterpart. Fig. 2 shows two trajectories of the stochastic system (5) beginning from $g_0 = 0.2$ and $w_0 = 0.1$, and corresponding to two different noise intensities $D = 0.01$ and 0.05 respectively. The starting point (0.2, 01) of the trajectory is not included in the figure to permit using a larger scale, thus a clearer display of the trajectory near point **G**. It is seen that the steady state of the system can no longer be represented by a single point; instead, it is represented by a region around **G**, which is in better agreement with what is expected of a real ecosystem. Also shown in the figure, the larger the noise intensity is, the more diffusive of the system state is.

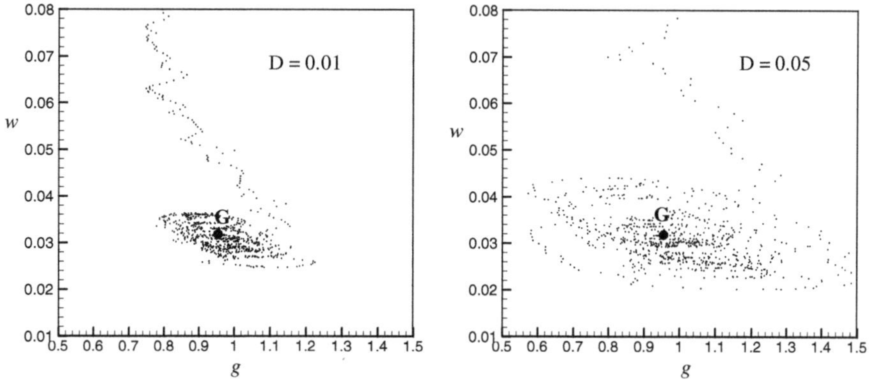

Fig. 2 Trajectories of stochastic system (6) with white noise variation for two cases with different noise intensities

It is noted that, with a random variation in the cattle stocking rate, which affects the grass consumption, the stable equilibrium point **G** is transformed into a region of stable steady states, while **W** remains a single stable equilibrium point. If random variation is also included in the source term of the woody vegetation, then point **W** will also be diffused into a region of stable steady states.

Another significant change of the system characteristics is that a distinctive separatrix of the system no longer exists, nor does an explicit boundary between two domains of attraction. Fig. 3 shows two different trajectories of the same system (5) with a noise intensity $D = 0.01$ and beginning from the same initial state of $g_0 = 0.06$ and $w_0 = 0.1$. One trajectory approaches the stable center **W**, and another ends up in a region around **G**. It is noted that the initial point (0.06, 01) is in the attraction zone leading to stable center **W** for the deterministic system. For the stochastic system, however, two possible outcomes exist. Therefore, management and control are important to lead the system to a desired steady state.

Although a single initial state may lad to two different steady states and the outcome is not predictable, the probability of approaching either steady state can be calculated using Monte Carlo simulations. Define P_W as the probability of the system with the steady state at the stable center **W**. It is obvious that P_W depends on two factors, one is the noise intensity and another is the initial state. P_W was calculated versus varying initial g_0 with initial $w_0 = 0.1$ and for three different noise intensities. The results are depicted in Fig. 4. When there is no stochastic disturbance, i.e., $D = 0$, the point on the separatrix is $g_0 \cong 0.0622$ ($w_0 = 0.1$). The system will approach the steady state **W** if $g_0 < 0.0622$, while it will end up at point **G** if $g_0 > 0.0622$. With a stochastic disturbance present, the separatrix disappears, and there is always a probability to go either steady state starting from a single initial state. With a larger noise intensity, the P_W curve deviates farther from that of the deterministic one.

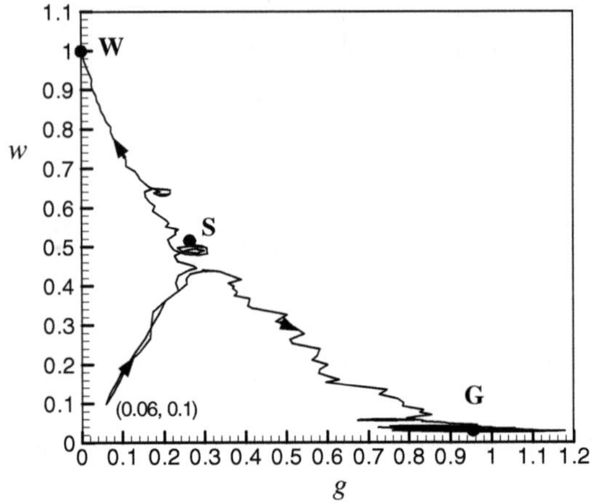

Fig. 3 Two possible trajectories of stochastic system (5) with white noise variation starting from the same point

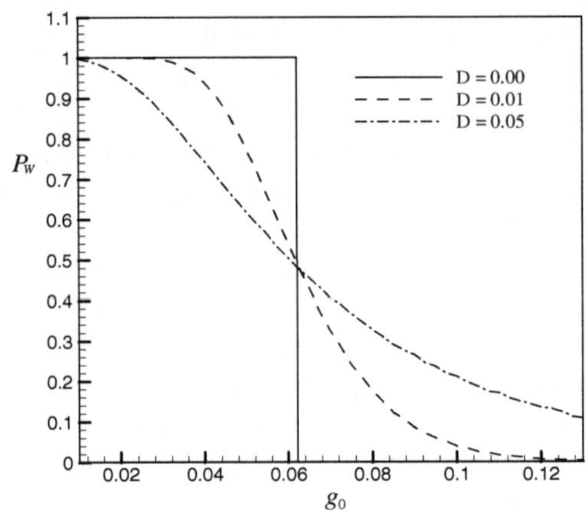

Fig. 4 Probability P_W versus the initial value g_0 for the case of initial w_0 = 0.1 and three different intensities of white noise

Although the outcome for a single trajectory of system (5) is not predictable, the probability distribution of all possible outcomes can be determined. Fig 5 shows the stationary probability density $p(g)$ of the grass density for the system starting from point (0.2, 0.1) with different noise intensities. The starting point (0.2, 0.1) is in the attraction zone of stable state **G** and far from the separatrix. The areas under the probability density curves are 1, 0.999, and 0.985 for

D = 0.01, 0.02, and 0.04, respectively. For the case of weak noise level of D = 0.01, almost all trajectories approach the area around **G**. With an increasing noise intensity, more sample trajectories approach the stable point **W**, resulting in the total probability in the **G** area less than one. The figure also shows that the stronger the noise is, the more diffusive the system state is around point **G**.

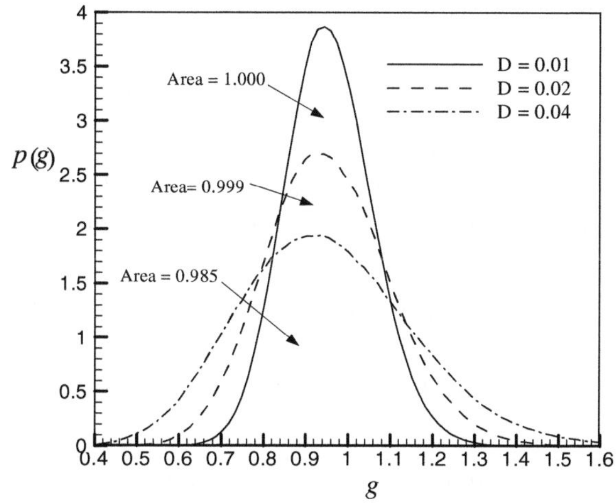

Fig. 5 Probability densities of grass density, computed with starting point (0.2, 0.1), and different intensities of white noise

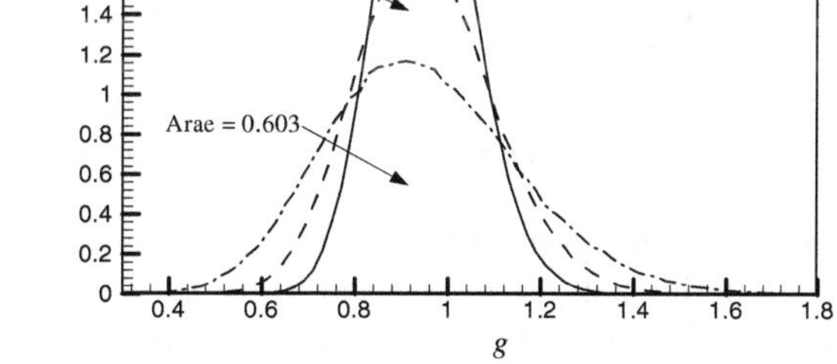

Fig. 6 Probability densities of grass density, computed with starting point (0.07, 0.1), and different intensities of white noise

Fig 6 shows the same stationary probability density $p(g)$ for three different noise intensities. However, the system starts from point (0.07, 0.1) which is closer to the separatrix, although still in the attraction zone of point **G**. Compared with the case of Fig. 5 in which the starting point is farther from the separatrix, the area under each curve is much less than one since more samples end up at the stable point **W**. The stronger the noise is, the less the total probability in the **G** area is.

It is well known that for a linear or nonlinear system, the stationary probability density of the system response, if exists, depends on system parameters and noise intensities. It is generally recognized that effect of the initial state disappear after a long period of time. However, the nonlinear system (5) exhibits an interesting phenomenon that the system stationary probability distribution also depends on the system initial state, as shown in Figs. 5 and 6. The initial state determines the probability of the system approaching one of the two possible steady states; thus, determines the overall system stationary probability distribution.

3.2 Randomized Sinusoidal Process Model

Another feasible model for $\xi(t)$ is randomized sinusoidal process described as follows

$$\xi(t) = s_r \cos\theta(t), \qquad d\theta = \nu dt + \sigma \mathrm{d}B(t) \tag{7}$$

in which s_r is a positive constants indicating the intensity of the noise, ν and σ are also positive constants representing the mean frequency and the level of randomness of the noise, $B(t)$ is a unit Brownian process (also known as the Wiener process). Equation (7) indicates that $\xi(t)$ has a random phase. This randomized sinusoidal process was proposed independently by Dimentberg [1] and Wedig [8], and it has been used in the investigations of a variety of engineering problems.

The stochastic system (5) with a random variation modeled as a randomized sinusoidal process (7) is now investigated for the same system parameters as before. It is found that the system behavior is similar qualitatively to that of the case of white noise model. Two possible outcomes exist for the system with the same initial state, indicating that a definite separatrix does not exist. The stationary state is not the single point **G**, but an area around **G** However, due to the periodic factor in the random noise, the system also exhibit periodic nature. The larger the randomness parameter σ is, the more close the system behavior is to that of the white noise case.

Stationary probability density $p(g)$ of the grass density is calculated for a noise intensity $s_r = 0.2$, an initial state $g_0 = 0.07$ and $w_0 = 0.1$, and three different randomnesses, $\sigma = 0.05$, 0.2 and 1.0. The results are shown in Fig. 7. With a stronger randomness, the probability density is very much different from that with a weak randomness. Moreover, the total probability, i.e., the area under the curve is smaller in the strong randomness case since the total probability of the system arriving at the final state **W** is higher.

Fig. 8 shows the probability density $p(g)$ for $s_r = 0.2$, $\sigma = 0.2$, $w_0 = 0.1$, and three different initial states of the grass density of $g_0 = 0.058$, 0.062 and 0.066, respectively. Although they have similar shapes, the areas under the curves are different. With the initial state closer to the point **G**, i.e. the case of $g_0 = 0.066$, the area is larger, indicating that the system is more likely to approach the states around **G**.

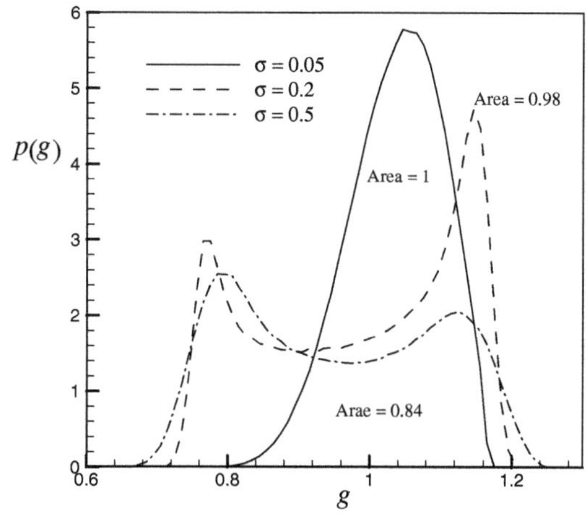

Fig. 7 Probability densities of grass density, computed with starting point (0.07, 0.1), and different σ values of randomized sinusoidal noise

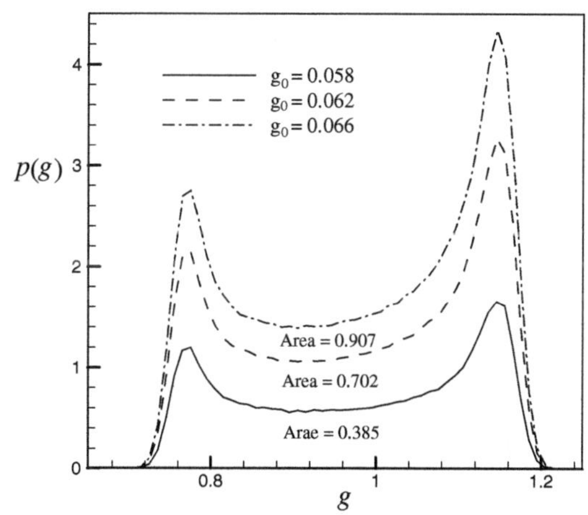

Fig. 8 Probability densities of grass density, computed with noise randomness $\sigma = 0.2$, initial $w_0 = 0.1$, and different initial g_0 values

4 Conclusions

Nonlinearity and uncertain variability are always present in natural ecosystems, which must be accounted for in their mathematical models and in the formulation of effective management policies. The present investigation is an attempt in this direction, using the example of a system with two competing species, specifically, the semi-arid savanna graving system. In particular, the intraspecies- and interspecies- competitions are introduced and the stocking rate is assumed to be a constant plus a random variation. Qualitative changes in the system nonlinear behaviors take place as a result of the presence of the random variation. A single stable state of the deterministic system is expanded into a stable region, the same initial state may lead to multiple final states, and the system stationary probability distribution depends not only on the system parameters and noise intensities, but also on the initial state.

Two different random processes, Gaussian white noise and randomized sinusoidal process, are utilized to model the random variation. Although they are quite different in nature, the qualitative behaviors of the system are similar.

In the present investigation, certain types of nonlinearity are introduced, and only one system parameter is treated as being randomly varying with time. Similar approaches may apply to other types of system nonlinearity and random variation, or for other ecosystems. The system properties, random variations, as well as the system initial state must be estimated from the observation data. Nonetheless, theoretical and numerical investigations are important to find out what types of data are most critically needed and what are not.

References

[1] Dimentberg, M.F.: Statistical Dynamics of Nonlinear and Time-Varying Systems. Wiley, New York (1988)

[2] Lotka, A.J.: Elements of Physical Biology. William and Wilkins, Baltimore (1925)

[3] Ludwig, D., Jones, D., Holling, C.S.: Qualitative analysis of insect outbreak systems: the spruce budworm and the forest. Journal of Animal Ecology 47, 315–332 (1978)

[4] Ludwig, D., Walker, B., Holling, C.S.: Sustainability, stability, and resilience. Conservation Ecology 1, Article 7 (1997)

[5] Volterra, V.: Variaziomi a flucttuazioni del numero d'individui in specie d'animali conviventi. Mem. Acad. Lincei. 2, 31–113 (1926)

[6] Volterra, V.: Lecons sur la theorie mathematique de la lutte pour la vie. Gauthiers-Vilars, Paris (1931)

[7] Walker, B., Ludwig, D., Holling, C.S., Peterman, R.M.: Stability of semi-arid savanna grazing system. Journal of Ecology 69, 473–498 (1981)

[8] Wedig, W.V.: Analysis and simulation of nonlinear stochastic systems. In: Schiehlen, W. (ed.) Nonlinear Dynamics in Engineering Systems, pp. 337–344. Springer, Berlin (1989)

Energy Transition Rate at Peptide-Bond Using Stochastic Averaging Method

M.L. Deng and W.Q. Zhu

Department of Mechanics, State Key laboratory of Fluid Power Transmission and Control, Zhejiang University, Hangzhou 310027, China

Abstract. The stochastic energy transition of peptide-bond (PB) in enzyme-substrate-complex (ESC) is investigated. By introducing a coupled Pippard system and thermal fluctuation as the noise, the two important mechanisms of enhancing PB breaking rate, i.e., the Fermi resonance and the reduction of potential barrier height are studied. The necessary frequency ratio 1:2 for Fermi resonance is verified by using the deterministic averaging method. With the noise terms present, the average energy transition rate of PB is predicted by using the stochastic averaging method. The comparison of the analytical results and Monte Carlo simulation results shows that the stochastic averaging method is promising for predicting the PB breaking rate. A preliminary work to investigate the possible influence of the oxygen ion on the PB is presented. The Morse potential model is introduced to describe the potential landscape of the oxygen ion oscillator, which is coupled with the Pippard system. Monte Carlo simulation results for PB energy transition rate are obtained for the coupled system, and they agree well with the theoretical results in the range of low potential barrier.

Keywords: Peptide bond breaking, Fermi resonance, Energy transition, Stochastic averaging method.

1 Introduction

Recently, research topics concerning protein and its reaction dynamics have attracted considerable attention of physicist [1]. Particularly, many researchers focused on the problem of enzyme-catalyzed PB breaking in protein molecular chain [2-4]. There are two different reaction processes in PB breaking, i.e., the spontaneous breaking in aqueous solution and the catalysis breaking under the action of enzyme. The primary difference between these two processes is that the latter has higher breaking rate than the former by several orders of magnitude. The protein hydrolysis enzyme such as chymotrypsin (ChT) is usually named as molecular scissor in which the active site affects the substrate protein. At the active site the local reactive dynamics connected with potential barrier crossing may be coupled with other oscillating degrees of freedom. As for the research work for the dynamical process in PB breaking, two dynamics mechanisms are usually used to explain the increasing probability of PB breaking in ESC, i.e., the Fermi resonance and the reduction of potential barrier height [1,2].

W.Q. Zhu et al. (Eds.): Nonlinear Stochastic Dynamics & Control, IUTAM BOOK SERIES 29, pp. 13–23.
springerlink.com

Fermi resonance is named according to the classical work of Enrico Fermi on the Raman effect in CO_2 molecule [5]. Fermi resonance was also found in other inorganic matters, most significantly, in the protein macromolecules [6,7]. Fermi established a simple model of 2-DOF system by introducing a coupling term to the potential [5]. The frequent energy exchange between the two coupled oscillators was regarded as the mechanism for promoting energy transition rate over a potential barrier [8,9]. On the other hand, the height of potential barrier is reduced due to the influence of some electronegative atom groups, e.g., the oxygen ion O^- in Ser195 of ChT [2]. The reduction of the potential barrier height increases the probability of PB breaking by several orders of magnitude, and the catalysis efficiency is higher than Fermi resonance.

In general, chemical reactions are stochastic processes because the chemical transformation is effected by inevitable uncertain factors, such as thermal fluctuation [1]. In this paper, we study the stochastic dynamics of enzyme catalyzed PB breaking. Both mechanisms mentioned above are considered. The energy transition rates in both non-resonance and Fermi resonance cases are obtained by using the stochastic averaging method and solving the averaged Pontryagin equation [10]. The analytical results are compared with those from the Monte Carlo simulation. Preliminary work to explore the influence of oxygen ion O^- is also presented. Monte Carlo simulations are carried out to obtain numerical results for some cases.

2 PB Breaking in ESC and Its Stochastic Dynamical Model

PB is a single valence bond jointing aminoacids to construct protein, and its breaking will lead to protein splitting. The structure of PB and its breaking reaction is illustrated in Fig. 1. The left-hand-side is the state before the reaction, while the right-hand-side is that after the reaction. R^1 and R^2 are the residues of the two neighbor aminoacids; the wavy line ~ denotes the PB. The PB breaking reaction in Fig. 1 is a hydrolysis process that can spontaneously take place in aqueous solution at very low rate. In aqueous solution, the hydroxyl ions OH^- casually attack protein molecule, and incur PB breaking with a rate k about $10^{-8}\,s^{-1}$. In vivo, such a reaction process is in general assisted by hydrolytic enzyme (e.g. ChT). Under the support of enzyme, the PB breaking rate increases to about $10^3\,s^{-1}$.

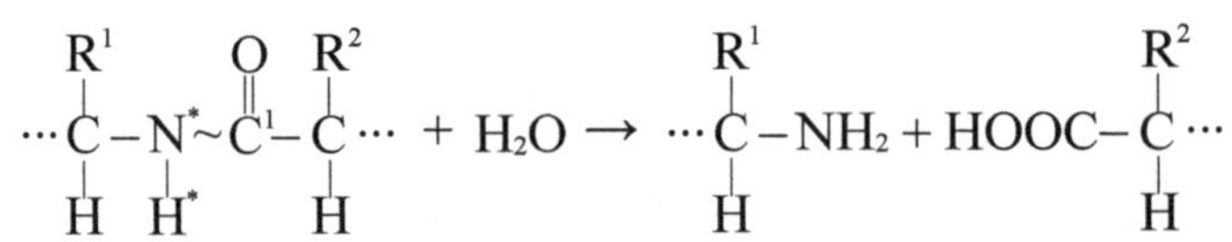

Fig. 1 PB breaking reaction process

ChT is a typical molecular scissor which plays an extremely important role during the hydrolysis process of PB and ester bonds. As shown in Fig. 2, the active site of ChT is located between the two sub-globules. The catalytic active site is composed with aminoacids residues Ser197, His57 and Asp102. Those aminoacid residues in active site are activated by proton transfer which brings electronegative to an oxygen atom in Ser195 to form oxygen ion O^-. Affected by the oxygen ion O^-, the PB in ESC breaks with rather high probability.

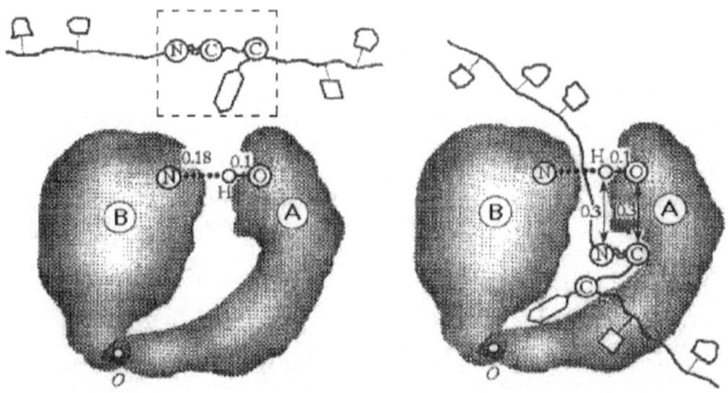

Fig. 2 Chymotrypsin (ChT) molecule consisting of two sub-globules A and B and the substrate (protein chain) from Reference 1

From the dynamics point of view, protein is a complex nonlinear system with thousands degrees of freedom [11,12]. Furthermore, the theory and methods of nonlinear stochastic dynamics should be used since any biomacromolecular reaction is subjected to stochastic fluctuation (e.g. thermal noise) [1]. Presently, two dynamical characteristics are used to explain the increasing probability of PB breaking in ESC, i.e., the Fermi resonance and the reduction of potential barrier height [1,2].

The theory of Fermi resonance was applied in Pippard's book [13] on the theory of oscillations. Volkenstein [14] first discussed the Fermi resonance in the PB. The popular stochastic model for studying Fermi resonance in the PB is a test particle moving in a 2-dimensional potential $U(x_1, x_2)$ under action of noise and damping [1]. The model is governed by the following Langevin equation

$$\begin{aligned} m\ddot{X}_1 + \gamma\dot{X}_1 + \partial U(X_1, X_2)/\partial X_1 &= \sqrt{2D}\xi_1(t) \\ m\ddot{X}_2 + \gamma\dot{X}_2 + \partial U(X_1, X_2)/\partial X_2 &= \sqrt{2D}\xi_2(t) \end{aligned} \tag{1}$$

where X_1 is the displacement of first oscillator representing the $N^* \sim C^1$ bond in PB as shown in Fig. 1, while X_2 the displacement of second oscillator representing $N^* - H^*$ bond; m is the mass of test particle; γ is a constant damping coefficient; $2D$ is the noise intensity; $\xi_1(t)$, $\xi_2(t)$ are independent unit Gaussian white noises originating from the thermal fluctuation. According to the fluctuation dissipation theorem, the Einstein relation $D = \gamma k_B T$ is applied here, where k_B is the Boltzmann constant and T is the temperature. The popular Pippard potential in (1) for investigating Fermi resonance is [13]

$$U(x_1, x_2) = \omega_1^2 x_1^2 / 2 + \omega_2^2 (x_2 - c x_1^2)^2 / 2 \tag{2}$$

where ω_1 and ω_2 are the frequencies of the two oscillators, respectively, and c is a coefficient reflecting the coupling strength between the two oscillators. Since c is generally very small, the two oscillators can be treated as quasi-linear system.

Consider the conservative Pippard model, namely, Eq. (1) without dampings and noises. Denote e_1, e_2 and ϕ_1, ϕ_2 as the energies and phase angles of the two oscillators in system (1), respectively. Introducing the frequency ratio $\omega_1 : \omega_2 = n : m$ where n and m are prime integers, the phase angle difference $\varphi = (m/n)\phi_1 - \phi_2$, and then using the deterministic averaging method yield the following equations for e_1, e_2 and φ

$$\frac{de_1}{dt} = -\frac{de_2}{dt} = \begin{cases} -2ce_1\sqrt{2e_2}\sin\varphi, & n : m = 1 : 2 \\ 0, & n : m \neq 1 : 2 \end{cases} \tag{3}$$

$$\frac{d\varphi}{dt} = \frac{\sqrt{2}c(e_1 - 2e_2)\cos\varphi}{\sqrt{e_2}}. \tag{4}$$

It is seen from Eq. (3) that $de_1/dt = -de_2/dt \neq 0$ only when $\omega_1 : \omega_2 = 1 : 2$. In other words, when frequency ratio $\omega_1 : \omega_2$ of the conservative Pippard system is equal to 1:2, energy exchange between the two oscillators occurs. This case is called Fermi resonance.

With dampings and noises, it can be seen by performing Monte Carlo simulation that the energy process of the first oscillator fluctuates more frequently in the case of Fermi resonance than in the case of non-resonant case [15]. It is point out that the frequent energy exchange between the two oscillators reduces the energy transition time over potential barrier in the first oscillator.

3 Energy Transition Rate in the Case of Fermi Resonance

For the quasi-linear Pippard system (1) with weak coupling and weak dampings, Fermi resonance can be investigated by using the stochastic averaging method. The mean transition time τ in both cases of non-resonance and Fermi resonance have been obtained by solving the averaged Pontryagin equation [16]. Denote $\tau = \tau(E_{10}, E_{20}, \varphi_0)$ as the mean time of energy process $E_1(t)$ of the first oscillator over the potential barrier ΔU for the first time, given the initial energies $0 \le E_{10} < \Delta U$, $0 \le E_{20}$. When $\omega_1 : \omega_2$ is far from 1:2, i.e., in the case of non-resonance, one can obtain the following exact energy transition rate

$$k = \tau^{-1} = \left\{ \frac{-1}{\gamma} \left[\ln(\frac{\Delta U}{k_B T}) + \text{real}\left(\Gamma(0, -\frac{\Delta U}{k_B T}) \right) + \gamma_0 \right] \right\}^{-1} \tag{5}$$

where $\gamma_0 \approx 0.577216$ is the Euler-Mascheroni constant; $\Gamma(\cdot,\cdot)$ is the incomplete Gamma function.

In the resonance case, i.e., $\omega_1 : \omega_2 \approx 1:2$, we replace the frequency ratio by $\omega_2 / \omega_1 = 2 + \sigma$, where σ is a small detuning parameter. The Pontryagin equation governing the mean transition time can be obtained as

$$a_1 \frac{\partial \tau}{\partial E_{10}} + a_2 \frac{\partial \tau}{\partial E_{20}} + a_3 \frac{\partial \tau}{\partial \varphi_0} + b_1 \frac{\partial^2 \tau}{\partial E_{10}^2} + b_2 \frac{\partial^2 \tau}{\partial E_{20}^2} + b_3 \frac{\partial^2 \tau}{\partial \varphi_0^2} = -1 \tag{6}$$

where E_{10}, E_{20} are the initial energies of the two oscillator, and φ_0 is the initial phase angle difference. The drift and diffusion coefficients in Eq. (6) read

$$\begin{aligned}
a_1 &= \gamma k_B T - \gamma E_{10} - 2c(1+\sigma) E_{10} \sqrt{2E_{20}} \sin \varphi_0, \\
a_2 &= \gamma k_B T - \gamma E_{20} + 2c(1+\sigma) E_{10} \sqrt{2E_{20}} \sin \varphi_0, \\
a_3 &= -\sigma \omega_1 + \sqrt{2} c(1+\sigma)(E_{10} - 2E_{20}) \cos \varphi_0 / \sqrt{E_{20}}, \\
b_1 &= \gamma k_B T E_{10}, \quad b_2 = \gamma k_B T E_{20}, \quad b_3 = \gamma k_B T / E_{10} + \gamma k_B T / 4E_{20}.
\end{aligned} \tag{7}$$

The boundary conditions associated with Eq. (6) are

$$\begin{gathered}
\tau(E_{10} = 0, E_{20}, \varphi_0) = \text{finite}, \quad \tau(E_{10} = \Delta U, E_{20}, \varphi_0) = 0, \\
\tau(E_{10}, E_{20} = 0, \varphi_0) = \text{finite}, \quad \tau(E_{10}, E_{20} \to \infty, \varphi_0) = \text{finite}, \\
\tau(E_{10}, E_{20}, \varphi_0 = 2\pi) = \tau(E_{10}, E_{20}, \varphi_0 = 0).
\end{gathered} \tag{8}$$

For the case of Fermi resonance, τ is obtained from numerically solving Eq. (6) together with boundary conditions in Eq. (8) using finite difference method.

Other than solving accurate PB energy transition rate for a particular case, we aim at the feasibility of the present analytical method. Therefore, the values of parameter in system (1) are chosen representatively for real protein molecules. Specifically, mass m is chosen as one unit mass $m = 1[\mathrm{m}]$ where $\mathrm{m} = 12\mathrm{u}$ ($\mathrm{u} = 1.6605\times 10^{-27}\,\mathrm{kg}$ is the unified atomic mass unit) which is equal to the mass of one carbon atom. The length unit is chose as the angstrom A. The energy unit is chosen as $k_{\mathrm{B}}T$ ($T = 300\mathrm{K}$ for normal temperature) for convenience. These chosen units make the time unit to be $\mathrm{t}_0 = 2.193\times 10^{-13}\mathrm{s}$. For $\gamma = \omega_1/Q \approx 5\times 10^{11}\mathrm{s}^{-1}$ and $Q = 100$ [15], the frequency of the first oscillator is $\omega_1 = 10.97$ and damping coefficient $\gamma = 0.1097$. Numerical results for the energy transition rate of PB for frequency ratio $\omega_2 : \omega_1$ varying from 10^{-2} to 10^{2} are obtained, and compared with those from Monte Carlo simulation, as shown in Fig. 3.

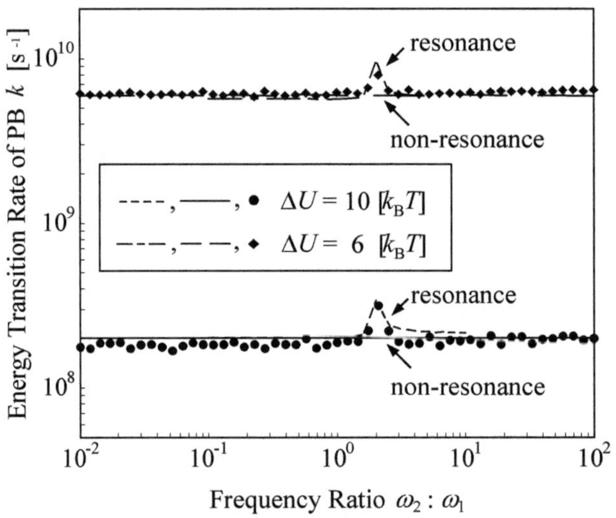

Fig. 3 The energy transition rate of peptide-bond as function of frequency ratio

It is seen that the energy transition rate reaches its maximum at Fermi resonance frequency $\omega_2 : \omega_1 = 2:1$. The agreement between the analytical results and the simulation results indicates the applicable ranges of the exact solution in Eq. (5) for the non-resonant case and the numerical solution of Eq. (6) for the resonant case.

Fig. 4 shows the energy transition rates for varying potential barrier ΔU for the Fermi resonance frequency $\omega_2 : \omega_1 = 2:1$. When ΔU exceeds $12k_B T$, solution of equations become ill-conditioned, and the finite difference method is not applicable. On the other hand, the computer time for Monte Carlo simulation increases exponentially with an increasing ΔU, and simulation results were obtained up to $\Delta U = 24k_B T$. It is seen that both results agree quite well.

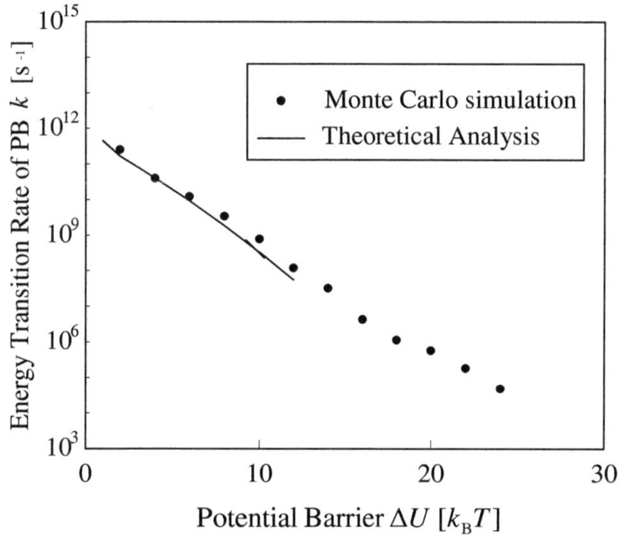

Fig. 4 Energy transition rate of PB for the Fermi resonanve case

4 PB Breaking Rate Affected by Oxygen Ion

As mentioned before, the probability of PB breaking in the presence of oxygen ion O^- of Ser195 of catalytic group is rather greater than the probability of spontaneous breaking in aqueous solution. The quantum mechanics and experimental data have shown that the existence of oxygen ion O^- affects the potential landscape of PB which is changed with the distance between them [2]; thus, the potential barrier for PB breaking changes accordingly. The change of the bond energy and potential barrier is shown in Fig. 5. On the other hand, the oxygen ion O^- is oscillating about its equilibrium, and its potential is also dependent on its distance to PB. X-ray data show that the equilibrium distance between O^- and PB is about 3Å, and that the potential barrier ΔU is extremely sensitive to the distance near 3Å, as shown in Fig. 5. It is also found that the minimum distance between

O^- and PB is about 1.34Å, which implies that the repulsive force is rather high when they approach each other. Therefore, it is important to consider the vibration of O^- in the orthogonal direction to PB and its coupling with the PB.

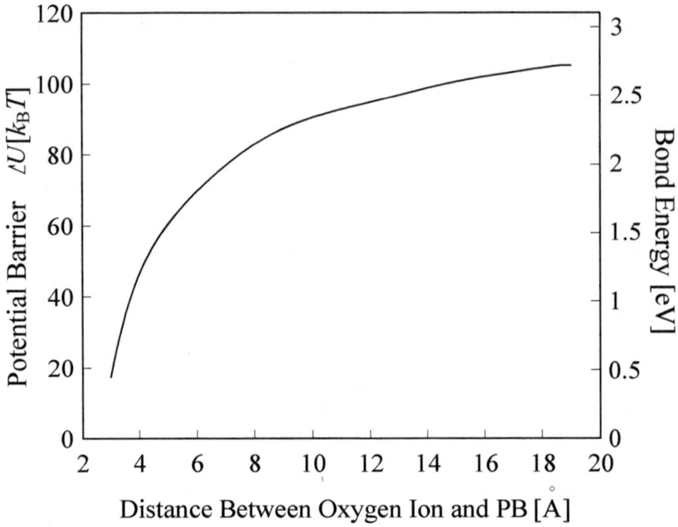

Fig. 5 Bond energy versus the distance between oxygen ion O^- and PB (data from Reference 2)

Denoting L as the distance between the PB and the oxygen ion O^-, the potential barrier is a function of L, i.e.,

$$\Delta U = f(L) \tag{9}$$

where the function f is defined in Fig. 5.

Without taking into account the effect of PB oscillators, a linear model was suggested in [2] for the oxygen ion O^- oscillation. It reads

$$m_0 \ddot{X}_3 + \gamma \dot{X}_3 + \frac{dM(X_3)}{dX_3} = \sqrt{2D}\xi_3(t), \quad M(X_3) = \omega_0^2 X_3^2 / 2 \tag{10}$$

With the linear oscillator model (10), the potential is parabolic. To introduce the effect of the PB on the motion of oxygen ion O^-, we propose to use the Morse potential to replace the parabolic potential in Eq. (10). The Morse potential reads

$$M(X_3) = \beta\left[\exp(-\alpha X_3) - 1\right]^2. \tag{11}$$

which has been used to model the interaction between atoms or molecules generally. By varying parameters α and β in (11) with respect to the distance L, the Morse potential can be used to model the potential landscape of oxygen ion O^-. Fig. 6 shows the parabolic potential with $\omega_0 = 30$, as well as the Morse potentials with several sets of α and β values. Larger α and β values indicates harder stiffness.

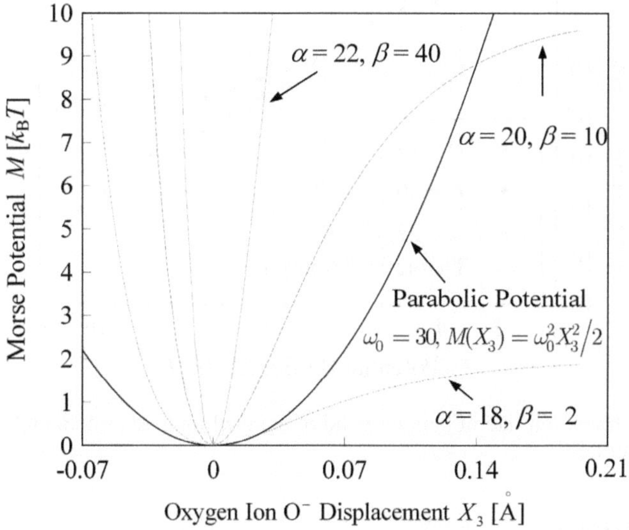

Fig. 6 Morse potential (11) for different parameters and parabolic potential

Combining PB oscillator (2), oxygen ion O^- oscillator (10) with Morse potential (11), and the variation of the potential barrier (9), one can establish a coupled stochastic dynamics model with both the effects of Fermi resonance and oxygen ion O^- taken into account.

Monte Carlo simulations were performed for the coupled stochastic model. Currently, the dependence of parameters α and β on distance L has not been accessed; thus, we consider three sets of α and β values. The other parameters are the same as those in Fig. 4, i.e. the case of Fermi resonance frequency $\omega_2 : \omega_1 = 2 : 1$. The results are shown in Fig. 7. The same theoretical results as in Fig. 4 obtained without considering the effect of oxygen ion O^- are also shown ion Fig. 7. It is seen that the simulation results from the coupled model agree well with those from the theoretical results for certain ranges of potential barrier, especially for the set of large α and β values. In this case, the stiffness of O^- oscillator is hard, its displacement X_3 is small, $L + X_3 \approx L$, and the effect of oxygen ion O^- is negligible.

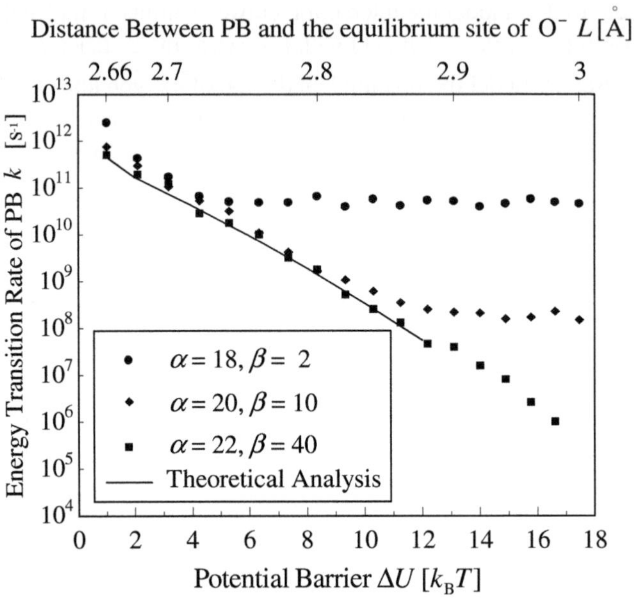

Fig. 7 Energy transition rate of peptide-bond of coupled Pippard system under the effect of oxygen ion oscillator

5 Conclusions

In the present paper, the stochastic dynamics of enzyme-catalyzed PB breaking in protein molecular chain has been studied by using the stochastic averaging method and the Monte Carlo simulation. The two main mechanisms of enhancing the PB breaking rate, i.e., the Fermi resonance and the reduction of potential barrier due to oxygen ion O^-, are investigated. The stochastic Pippard dynamical model of PB breaking was introduced for the Fermi resonance. The energy transition rate of PB has been predicted by applying the stochastic averaging method and solving Pontryagin equation, the analytical results and those from Monte Carlo simulation agree well. For investigating the second mechanism of enhancing PB breaking rate, preliminary work has been done. Specifically, the parabolic potential in oxygen ion O^- oscillator has been replaced by the Morse potential, the PB potential barrier is varying with the distance between PB and oxygen ion O^-, and the Pippard model is coupled with the oscillation of oxygen ion O^-. Some simulation results have shown the feasibility of the proposed coupled model.

Acknowledgments. The work reported in this paper is supported by the National Natural Science Foundation of China under a key Grant No. 10932009 and 10802074 and the Research Fund for the Doctoral Program of Higher Education of China the under Grant No. 20060335125, and the Zhejiang Provincial Natural Science Foundation of China under Grant No. Y7080070.

References

1. Ebeling, W., Schimansky-Geier, L., Romanovsky, Y.M.: Stochastic Dynamics of Reacting Biomolecules. World Scientific, Singapore (2003)
2. Kargovsky, A.V., Khodjer, O.P., Romanovsky, Y.M.: Functional dynamics of hydrolytic enzymes. In: SPIE, vol. 5068, p. 16 (2003)
3. Chichigina, O.A., Ebeling, W., Makarov, V.A., Netrebko, A.V., Romanovsky, Y.M., Schimansky-Geier, L.: Stochastic dynamics of enzymes – molecular scissors. In: Proc. SPIE, vol. 5110, p. 28 (2003)
4. Romanovsky, Y.M., Chikishev, A.Y., Kroo, S.V., Netrebko, A.V.: Computer simulation of the decay of nonlinear subglobular oscillation in aqueous solutions of protein molecules. In: Proc. SPIE, vol. 4624, p. 139 (2002)
5. Fermi, E.: Über den Ramaneffekt des Kohlendioxyds. Zeitschrift für Physik A 71, 250 (1931)
6. Volkenstein, M.V.: Structure of molecules. Moscow. Izd. Akad. Nauk (1947) (in Russian)
7. Shidlovskaya, E., Schimansky-Geier, L., Romanovsky, Y.M.: Nonlinear vibrations in 2-dimensional protein cluster model with linear bonds. Z. Phys. Chem. 214, 65 (2000)
8. Ebeling, W.: Thermodynamics of selforganization and evolution. Biomed. Biochi. Acta 44, 831 (1985)
9. Ebeling, W., Romanovsky, Y.M.: Energy transfer and chaotic oscillations in enzyme catalysis. Z. Phys. Chem. Leipzig 266, 836 (1985)
10. Stratonovich, R.L.: Topics in the Theory of Random Noise, vol. 1. Gordon and Breach, New York (1963)
11. Netrebko, A., Netrebko, N., Romanovsky, Y., Khurgin, Y., Shidlovskaya, E.: Complex modulation regimes and vibration stochastization in cluster dynamics models of macromolecules. Izv 2, 26 (1994) (in Russian)
12. Ebeling, W., Romanovsky, Y., Khurgin, Y., Netrebko, A., Netrebko, N., Shidlovskaya, E.: Complex regimes in the simple models of molecular dynamics of enzymes. In: Proc. SPIE, vol. 2370, p. 434 (1994)
13. Pippard, A.B.: The Physics of Vibration, The simple vibrator in quantum mechanics. Cambridge University Press, Cambridge (1983)
14. Volkenstein, M. V., Golovanov, I.B., Sobolev, V.M.: Molecular orbitals in enzymology. Moscow, Nauka (1982) (in Russian)
15. Ebeling, W., Kargovsky, A., Netrebko, A., Romanovsky, Y.M.: Fermi resonance – new applications of an old effect. Fluctuation and Noise Letters 4, L183 (2004)
16. Deng, M.L., Zhu, W.Q.: Fermi resonance and its effect on the mean transition time and rate. Physical Review E 77, 061114 (2008)

A New Method for the Probabilistic Solutions of Large-Scale Nonlinear Stochastic Dynamic Systems

G.K. Er and V.P. Iu

Department of Civil and Environmental Engineering, University of Macau, Macau SAR, China

Abstract. This paper proposes a novel method for determining the probabilistic stationary solution of multi-dimensional nonlinear stochastic dynamic systems. In general, the PDF solution is governed by Fokker-Planck equations in multi-dimensions. By dividing the space of the state variables into two subspaces and integrating the Fokker-Planck equation over one of the subspaces, a reduced set of Fokker-Planck equations can be obtained in the state variables of the other subspace. This is achieved by manipulating the integrals and approximating the conditional PDFs resulted from integration. Hence, the reduced set of Fokker-Planck equation will have a smaller number of state variables at choices and can be solved by the exponential polynomial closure method. Examples of the nonlinear stochastic dynamic systems with polynomial nonlinearity are given to show the effectiveness of this novel subspace method. The paper attempts to provide a tool for analyzing the probabilistic solutions of some highly multi-dimensional nonlinear stochastic dynamics systems in various areas of science and engineering.

Keywords: Fokker-Planck equation, nonlinear stochastic dynamic system, Subspace, Probability density function.

1 Introduction

The problems of nonlinear random vibrations of multi-degree-of-freedom (MDOF) systems are encountered in many areas of science and engineering. However, obtaining the probabilistic solution of MDOF nonlinear stochastic dynamic (NSD) systems has been a challenge for over one century. Only in some limited cases, the exact probabilistic solutions of two-degree-of-freedom system or few-degree-of-freedom systems are obtainable [Scheurkogel and Elishakoff 1988, Lin and Cai 1995]. It is well known that even the solutions of nonlinear single-degree-of-freedom (SDOF) systems attracted much attention in the last decades. The key problem in nonlinear random vibrations is about obtaining the probability density function (PDF) of system responses because all other statistical analyses are based on it. Even if the PDF solution of NSD system is governed by Fokker-Planck (FP) equation, it is still difficult to obtain the exact solution if the system is nonlinear or there are parametric excitations. Therefore, some methods were employed for the approximate solutions. The most frequently employed

W.Q. Zhu et al. (Eds.): Nonlinear Stochastic Dynamics & Control, IUTAM BOOK SERIES 29, pp. 25–34.
springerlink.com

approximation method is the equivalent linearization (EQL) (Caughey 1959, Lin 1967, Spanos 1981). The advantage of EQL method is that it can be used for analyzing large-scale NSD systems. It is suitable for the system with weak nonlinearity because in this case the probability distribution of the system responses is close Gaussian. To improve the accuracy of the approximate solution in the case of high nonlinearity of system, various non-Gaussian closure methods were proposed, but all the methods are limited to either SDOF systems or some other conditions. The stochastic average method can be used for analyzing MDOF systems. It is suitable for the systems with weak nonlinearity and weak excitations (Stratonovich 1963, Roberts and Spanos 1986). Monte Carlo simulation (MCS) is versatile, but the amount of computation with it is usually unacceptable for estimating the PDF solution of system responses, especially for small probability problems. Exponential polynomial closure (EPC) method was proposed which is suitable for analyzing few-degree-of-freedom system without being limited by parametric excitations and the level of system nonlinearity (Er 1998, Er and Iu 1999). From the above discussion, it is seen that the problem of obtaining the PDF solutions of large-scale NSD systems has been a challenge since the formulation of the FP equation. In this paper, a new method is proposed for obtaining the PDF solutions of some large-scale NSD systems. With the idea of this method, the problem of solving the FP equation in high-dimensional space becomes the problem of solving some FP equations in low-dimensional spaces. Thereafter, the FP equations in the low-dimensional spaces can be solved with the EPC method. Numerical results are presented to show the effectiveness of the proposed method.

2 Problem Formulation

In the following discussion, the summation convention applies unless stated otherwise. The random state variable or vector is denoted with capital letter and the corresponding deterministic state variable or vector is denoted with the same letter in low case.

Many problems in science and engineering can be described with the following multi-dimensional nonlinear stochastic dynamic system:

$$\ddot{Y}_i + h_{i0}(\mathbf{Y}, \dot{\mathbf{Y}}) = h_{ij}(\mathbf{Y}, \dot{\mathbf{Y}}) W_j(t) \qquad i = 1,2,\cdots,n_Y;\ j = 1,2,\cdots,m \tag{1}$$

where $Y_i \in R, (i = 1,2,\cdots,n_Y)$, are components of the vector process $\mathbf{Y} \in R^{n_Y}$, $h_{i0} : R^{n_Y} \times R^{n_Y} \to R$, $h_{ij} : R^{n_Y} \times R^{n_Y} \to R$, and $W_j(t)$ are excitations which are zero-mean white noise with cross-correlation $E[W_j(t)W_k(t+\tau) = S_{jk}\delta(\tau)$ in which $\delta(\tau)$ is Dirac function and S_{jk} are constants, representing the cross spectral density of W_j and W_k. $h_{i0}(\mathbf{Y}, \dot{\mathbf{Y}})$ and $h_{ij}(\mathbf{Y}, \dot{\mathbf{Y}})$ are nonlinear and their functional forms are assumed to be deterministic.

Setting $Y_i = X_{2i-1}$, $\dot{Y}_i = X_{2i}$, $d_{2i-1} = X_{2i}$, $d_{2i} = -h_{io}$, $g_{2i-1,j} = 0$, $g_{2i,j} = h_{ij}$, $(i = 1,2,\cdots,n_Y; j = 1,2,\cdots,m)$, and $n_X = 2n_Y$, then equation (1) can be expressed as follows.

$$\frac{d}{dt}X_i = f_i(\mathbf{X}) + g_{ij}(\mathbf{X})W_j(t) \qquad (i = 1,2,\cdots,n_X) \tag{2}$$

in Ito's form, where $f_i(\mathbf{X}) = d_i(\mathbf{X}) + \frac{1}{2}\frac{\partial g_{ij}(\mathbf{X})}{\partial X_k} g_{kj}(\mathbf{X})$, state vector $\mathbf{X} \in R^{n_x}$, X_i, ($i = 1,2,\ldots n_X$), are components of the state vector process $\mathbf{X}$; $f_i(\mathbf{X}): R^{n_x} \to R$; and $g_{ij}(\mathbf{X}): R^{n_x} \to R$. The state vector $\mathbf{X}$ is Markovian and the PDF $p(\mathbf{x},t)$ of the Markov vector is governed by FP equation. Without loss of generality, consider the case when the white noise are Gaussian. In this case, the stationary PDF $p(\mathbf{x})$ of the Markov vector is governed by the following reduced FP equation [2]:

$$\frac{\partial}{\partial x_j}[f_j(\mathbf{x})\mathrm{p}(\mathbf{x})] - \frac{1}{2}\frac{\partial^2}{\partial x_i \partial x_j}[G_{ij}(\mathbf{x})p(\mathbf{x})] = 0 \tag{3}$$

where $\mathbf{x}$ is the deterministic state vector, $\mathbf{x} \in R^{n_x}$, and $G_{ij}(\mathbf{x}) = S_{ls} g_{il}(\mathbf{x}) g_{js}(\mathbf{x})$.

It is assumed that the PDF solution $p(\mathbf{x})$ of equation (2) fulfill the following conditions:

$$\lim_{x_i \to \pm\infty} f_j(\mathbf{x})p(\mathbf{x}) = 0 \text{ and } \lim_{x_i \to \pm\infty} \frac{\partial[G_{ij}(\mathbf{x})p(\mathbf{x})]}{\partial x_i} = 0 \qquad i,j = 1,2,\cdots,n_X. \tag{4}$$

3 Subspace Method

Separate the state vector $\mathbf{X}$ into two parts $\mathbf{X}_1 \in R^{n_{x_1}}$ and $\mathbf{X}_2 \in R^{n_{x_2}}$, i.e., $\mathbf{X} = \{\mathbf{X}_1, \mathbf{X}_2\} \in R^{n_x} = R^{n_{x_1}} \times R^{n_{x_2}}$.

Denote the PDF of $\mathbf{X}_1$ as $p_1(\mathbf{x}_1)$. In order to obtain $p_1(\mathbf{x}_1)$, integrating equation (2) over $R^{n_{x_2}}$ gives

$$\int_{R^{n_{x2}}} \frac{\partial}{\partial x_j}[f_j(\mathbf{x})\mathrm{p}(\mathbf{x})]d\mathbf{x}_2 - \frac{1}{2}\int_{R^{n_{x2}}} \frac{\partial^2}{\partial x_i \partial x_j}[G_{ij}(\mathbf{x})p(\mathbf{x})]d\mathbf{x}_2 = 0 \tag{5}$$

Because of equation (4), we have

$$\int_{R^{n_{x2}}} \frac{\partial}{\partial x_j}[f_j(\mathbf{x})p(\mathbf{x})]d\mathbf{x}_2 = 0 \qquad x_j \in R^{n_{x_2}} \tag{6}$$

and

$$\int_{R^{n_{x2}}} \frac{\partial^2}{\partial x_i \partial x_j}[G_{ij}(\mathbf{x})p(\mathbf{x})]d\mathbf{x}_2 = 0 \qquad x_i \text{ or } x_j \in R^{n_{x_2}} \tag{7}$$

Equation (5) can then be expressed as

$$\int_{R^{n_{x2}}} \frac{\partial}{\partial x_j}[f_j(\mathbf{x})p(\mathbf{x})]d\mathbf{x}_2 - \frac{1}{2}\int_{R^{n_{x2}}} \frac{\partial^2}{\partial x_i \partial x_j}[G_{ij}(\mathbf{x})p(\mathbf{x})]d\mathbf{x}_2 = 0 \qquad x_i, x_j \in R^{n_{x_1}} \tag{8}$$

which can be further expressed as

$$\frac{\partial}{\partial x_j}\left[\int_{R^{n_{x2}}} f_j(\mathbf{x})p(\mathbf{x})d\mathbf{x}_2\right] - \frac{1}{2}\frac{\partial^2}{\partial x_i \partial x_j}\left[\int_{R^{n_{x2}}} G_{ij}(\mathbf{x})p(\mathbf{x})d\mathbf{x}_2\right] = 0 \qquad x_i, x_j \in R^{n_{x_1}} \tag{9}$$

Separate $f_j(\mathbf{x})$ and $G_{ij}(\mathbf{x})$ into two parts, respectively, as

$$f_j(\mathbf{x}) = f_j^I(\mathbf{x}_1) + f_j^{II}(\mathbf{x}), \quad G_{ij}(\mathbf{x}) = G_{ij}^I(\mathbf{x}_1) + G_{ij}^{II}(\mathbf{x}) \tag{10}$$

Substituting equation (10) into equation (9) gives

$$\begin{aligned}&\frac{\partial}{\partial x_j}\left[f_j^I(\mathbf{x}_1)p_1(\mathbf{x}_1) + \int_{R^{n_{x2}}} f_j^{II}(\mathbf{x})p(\mathbf{x})d\mathbf{x}_2\right] - \\ &\frac{1}{2}\frac{\partial^2}{\partial x_i \partial x_j}\left[G_{ij}^I(\mathbf{x}_1)p_1(\mathbf{x}_1) + \int_{R^{n_{x2}}} G_{ij}^{II}(\mathbf{x})p(\mathbf{x})d\mathbf{x}_2\right] = 0 \qquad x_i, x_j \in R^{n_{x_1}}\end{aligned} \tag{11}$$

For practical systems, normally $f_j^{II}(\mathbf{x})$ and $G_{ij}^{II}(\mathbf{x})$ are functions of only few state variables for given i and j. Denote $f_j^{II}(\mathbf{x}) = f_j^{II}(\mathbf{x}_1, \mathbf{z}_k)$ in which $\mathbf{z}_k \in R^{n_{z_k}} \subset R^{n_{x_2}}$, and $G_{ij}^{II}(\mathbf{x}) = G_{ij}^{II}(\mathbf{x}_1, \mathbf{z}_r)$ in which $\mathbf{z}_r \in R^{n_{z_r}} \subset R^{n_{x_2}}$. n_{z_k} is the number of the state variables in $\mathbf{z}_k$ and n_{z_r} is the number of the state variables in $\mathbf{z}_r$. Therefore, equation (11) can be expressed as

$$\frac{\partial}{\partial x_j}\left[f_j^I(\mathbf{x}_1)p_1(\mathbf{x}_1) + \int_{R^{n_{z_k}}} f_j^{II}(\mathbf{x}_1,\mathbf{z}_k)p_k(\mathbf{x}_1,\mathbf{z}_k)d\mathbf{z}_{\mathrm{k}} \right] - \frac{1}{2}\frac{\partial^2}{\partial x_i \partial x_j}\left[G_{ij}^I(\mathbf{x}_1)p_1(\mathbf{x}_1) + \int_{R^{n_{z_r}}} G_{ij}^{II}(\mathbf{x}_1,\mathbf{z}_r)p_r(\mathbf{x}_1,\mathbf{z}_r)d\mathbf{z}_{\mathrm{r}} \right] = 0 \quad x_i, x_j \in R^{n_{X_1}} \tag{12}$$

in which $p_k(\mathbf{x}_1,\mathbf{z}_k)$ denotes the joint PDF of $\{\mathbf{X}_1,\mathbf{Z}_k\}$ and $p_r(\mathbf{x}_1,\mathbf{z}_r)$ denotes the joint PDF of $\{\mathbf{X}_1,\mathbf{Z}_r\}$. The summation convention not applies on the indexes k and r in equation (12) and in the following discussions.

From equation (12), it is seen that the coupling of $\mathbf{X}_1$ and $\mathbf{X}_2$ comes from $f_j^{II}(\mathbf{x}_1,\mathbf{z}_k)p_k(\mathbf{x}_1,\mathbf{z}_k)$ and $G_{ij}^{II}(\mathbf{x}_1,\mathbf{z}_r)p_r(\mathbf{x}_1,\mathbf{z}_r)$. Express $p_k(\mathbf{x}_1,\mathbf{z}_k)$ as

$$p_k(\mathbf{x}_1,\mathbf{z}_k) = p_1(\mathbf{x}_1)q_k(\mathbf{z}_k;\mathbf{x}_1) \tag{13}$$

where $q_k(\mathbf{z}_k;\mathbf{x}_1)$ is the conditional PDF of $\mathbf{Z}_k$ for given $\mathbf{X}_1 = \mathbf{x}_1$, and express $p_r(\mathbf{x}_1,\mathbf{z}_r)$ as

$$p_r(\mathbf{x}_1,\mathbf{z}_r) = p_1(\mathbf{x}_1)q_r(\mathbf{z}_r;\mathbf{x}_1) \tag{14}$$

where $q_r(\mathbf{z}_r;\mathbf{x}_1)$ is the conditional PDF of Z_r for given $\mathbf{X}_1 = \mathbf{x}_1$.

Substituting equations (13) and (14) into equation (12) gives

$$\frac{\partial}{\partial x_j}\left\{\left[f_j^I(\mathbf{x}_1) + \int_{R^{n_{z_k}}} f_j^{II}(\mathbf{x}_1,\mathbf{z}_k)q_k(\mathbf{z}_k;\mathbf{x}_1)d\mathbf{z}_{\mathrm{k}} \right]p_1(\mathbf{x}_1)\right\} - \frac{1}{2}\frac{\partial^2}{\partial x_i \partial x_j}\left\{\left[G_{ij}^I(\mathbf{x}_1) + \int_{R^{n_{z_r}}} G_{ij}^{II}(\mathbf{x}_1,\mathbf{z}_r)q_r(\mathbf{z}_r;\mathbf{x}_1)d\mathbf{z}_{\mathrm{r}} \right]p_1(\mathbf{x}_1)\right\} = 0 \quad x_i, x_j \in R^{n_{X_1}} \tag{15}$$

Approximately replace the conditional PDFs $q_k(\mathbf{z}_k;\mathbf{x}_1)$ and $q_r(\mathbf{z}_r;\mathbf{x}_1)$ by those from equivalent linearization, then equation (15) is written as

$$\frac{\partial}{\partial x_j}\left\{\left[f_j^I(\mathbf{x}_1) + \int_{R^{n_{z_k}}} f_j^{II}(\mathbf{x}_1,\mathbf{z}_k)\overline{q}_k(\mathbf{z}_k;\mathbf{x}_1)d\mathbf{z}_{\mathrm{k}} \right]\tilde{p}_1(\mathbf{x}_1)\right\} - \frac{1}{2}\frac{\partial^2}{\partial x_i \partial x_j}\left\{\left[G_{ij}^I(\mathbf{x}_1) + \int_{R^{n_{z_r}}} G_{ij}^{II}(\mathbf{x}_1,\mathbf{z}_r)\overline{q}_r(\mathbf{z}_r;\mathbf{x}_1)d\mathbf{z}_{\mathrm{r}} \right]\tilde{p}_1(\mathbf{x}_1)\right\} = 0 \quad x_i, x_j \in R^{n_{X_1}} \tag{16}$$

where $\overline{q}_k(\mathbf{z}_k;\mathbf{x}_1)$ is the conditional PDF of $\mathbf{Z}_k$ from EQL for given $\mathbf{X}_1 = \mathbf{x}_1$, $\overline{q}_r(\mathbf{z}_r;\mathbf{x}_1)$ is the conditional PDF of $\mathbf{Z}_r$ from EQL for given $\mathbf{X}_1 = \mathbf{x}_1$, and $\tilde{p}_1(\mathbf{x}_1)$ is the approximation of $p_1(\mathbf{x}_1)$. Denote

$$\tilde{f}_j(\mathbf{x}_1) = f_j^I(\mathbf{x}_1) + \int_{R^{n_{z_k}}} f_j^{II}(\mathbf{x}_1, \mathbf{z}_k)\bar{q}_k(\mathbf{z}_k;\mathbf{x}_1)d\mathbf{z}_k \tag{17}$$

$$\tilde{G}_{ij}(\mathbf{x}_1) = G_{ij}^I(\mathbf{x}_1) + \int_{R^{n_{z_r}}} G_{ij}^{II}(\mathbf{x}_1, \mathbf{z}_r)\bar{q}_r(\mathbf{z}_r;\mathbf{x}_1)d\mathbf{z}_r \tag{18}$$

Then equation (16) can be expressed as

$$\frac{\partial}{\partial x_j}\left[\tilde{f}_j(\mathbf{x}_1)\tilde{p}_1(\mathbf{x}_1)\right] - \frac{1}{2}\frac{\partial^2}{\partial x_i \partial x_j}\left[\tilde{G}_{ij}(\mathbf{x}_1)\tilde{p}_1(\mathbf{x}_1)\right] = 0 \quad x_i, x_j \in R^{n_{x_1}} \tag{19}$$

which is the approximate FP equation for the joint PDF of state variables in the sub state space $R^{n_{x_1}}$.

If one chooses $\mathbf{X}_1$ to have a few state variables, for instance, two state variables, the resulting FP equations is in low dimensions and the EPC method can be employed to solve equation (19). Therefore, the whole solution procedure is named subspace-EPC method in the following discussions.

4 Numerical Analysis

From the above discussion, it is seen that the subspace method is not limited by the number of state variables in the system. Consider the following 10-DOF system:

$$\{\ddot{\mathbf{Y}}\} + [\mathbf{C}]\{\dot{\mathbf{Y}}\} + [\mathbf{K}]\{\mathbf{Y}\} + \{\mathbf{f}(\mathbf{Y}, \dot{\mathbf{Y}})\} = \{\mathbf{I}\}W(t)$$

where $W(t)$ denote unit white noise with $E[W(t)W(t+\tau) = \delta(t)$; $\{\mathbf{I}\} = \{1 \;\; 1 \;\; \ldots \;\; 1\}^t$,

$$[\mathbf{K}] = \begin{bmatrix} 1 & 0.2 & 0.1 & 0.3 & 0.2 & 0.2 & 0.1 & 0.3 & 0.2 & 0.1 \\ 0.2 & 1 & 0.2 & 0.5 & 0.2 & 0.1 & 0.3 & 0.2 & 0.2 & 0.2 \\ 0.1 & 0.2 & 1 & 0.2 & 0.1 & 0.2 & 0.2 & 0.1 & 0.1 & 0.1 \\ 0.3 & 0.3 & 0.2 & 1 & 0.2 & 0.2 & 0.1 & 0.3 & 0.2 & 0.2 \\ 0.2 & 0.2 & 0.1 & 0.2 & 1 & 0.1 & 0.3 & 0.2 & 0.2 & 0.1 \\ 0.2 & 0.1 & 0.2 & 0.2 & 0.1 & 1 & 0.5 & 0.3 & 0.3 & 0.3 \\ 0.1 & 0.3 & 0.2 & 0.1 & 0.3 & 0.5 & 1 & 0.2 & 0.3 & 0.2 \\ 0.3 & 0.2 & 0.1 & 0.3 & 0.2 & 0.3 & 0.2 & 1 & 0.5 & 0.2 \\ 0.2 & 0.2 & 0.1 & 0.2 & 0.2 & 0.3 & 0.3 & 0.5 & 1 & 0.3 \\ 0.1 & 0.2 & 0.3 & 0.1 & 0.2 & 0.3 & 0.1 & 0.2 & 0.3 & 1 \end{bmatrix}$$

$$
[\mathbf{C}]=\begin{bmatrix}
1 & 0.2 & 0.1 & 0.1 & 0.3 & 0.2 & 0.1 & 0.3 & 0.2 & 0.1 \\
0.2 & 1 & 0.2 & 0.2 & 0.1 & 0.2 & 0.1 & 0.2 & 0.3 & 0.2 \\
0.1 & 0.2 & 1 & 0.1 & 0.1 & 0.2 & 0.2 & 0.1 & 0.1 & 0.3 \\
0.1 & 0.2 & 0.1 & 1 & 0.1 & 0.3 & 0.2 & 0.4 & 0.2 & 0.2 \\
0.3 & 0.1 & 0.1 & 0.1 & 1 & 0.1 & 0.2 & 0.3 & 0.4 & 0.3 \\
0.2 & 0.2 & 0.2 & 0.3 & 0.1 & 1 & 0.3 & 0.2 & 0.1 & 0.2 \\
0.1 & 0.1 & 0.2 & 0.2 & 0.2 & 0.3 & 1 & 0.1 & 0.2 & 0.1 \\
0.3 & 0.2 & 0.1 & 0.4 & 0.3 & 0.2 & 0.1 & 1 & 0.3 & 0.2 \\
0.2 & 0.3 & 0.1 & 0.2 & 0.4 & 0.1 & 0.2 & 0.3 & 1 & 0.3 \\
0.1 & 0.2 & 0.1 & 0.2 & 0.1 & 0.2 & 0.3 & 0.1 & 0.2 & 1
\end{bmatrix}
$$

and

$$
\begin{aligned}
\{\mathbf{f}(\mathbf{Y},\dot{\mathbf{Y}})\}= \{&2Y_1^3+Y_3^3+0.5\dot{Y}_1^3+0.5\dot{Y}_2^3 \quad 2Y_2^3+Y_3^3+0.5\dot{Y}_2^3+0.5\dot{Y}_3^3 \quad 2Y_3^3+Y_1^3+0.5\dot{Y}_3^3+0.5\dot{Y}_1^3 \\
&2Y_4^3+Y_2^3+0.5\dot{Y}_4^3+0.5\dot{Y}_7^3 \quad 2Y_5^3+Y_2^3+0.5\dot{Y}_5^3+0.5\dot{Y}_1^3 \quad 2Y_6^3+Y_3^3+0.5\dot{Y}_6^3+0.5Y_4^3 \\
&2Y_7^3+Y_4^3+0.5\dot{Y}_7^3+0.5\dot{Y}_6^3 \quad 2Y_8^3+Y_5^3+0.5\dot{Y}_8^3+0.5\dot{Y}_9^3 \quad 2Y_9^3+Y_6^3+0.5\dot{Y}_9^3+0.5\dot{Y}_{10}^3 \\
&2\mathrm{Y}_{10}^3+Y_7^3+0.5\dot{Y}_{10}^3+0.5\dot{Y}_8^3\}^t
\end{aligned}
$$

The results obtained with the subspace-EPC method are compared with those from MCS and EQL to verify the effectiveness of the proposed method. The sample size is 10^7 in MCS. The results corresponding to $n=4$ are the results obtained from subspace-EPC when the polynomial order equals 4 in the EPC solution procedure. σ_{yi} denotes the standard deviation of Y_i from EQL, $\sigma_{\dot{y}i}$ denotes the standard deviation of $\dot{Y}_i$ from EQL in the figures. The stationary PDFs obtained with the subspace-EPC method, MCS and EQL methods are compared in order to show the effectiveness of the subspace-EPC method in analyzing the large-scale NSD systems with additive excitations and polynomial nonlinearity. Because there are 10 degrees of freedom or 20 state variables, it is not possible to present all the results in this paper. Only the PDFs and logarithmic PDFs of Y_1, $\dot{Y}_1$, Y_5, $\dot{Y}_5$, Y_{10}, and $\dot{Y}_{10}$ are shown and compared in Figs. (1)-(6). With the proposed method, the stationary PDFs $\tilde{p}_1(\mathbf{x}_1)$ are obtained by taking $\mathbf{X}_{1i}=\{Y_i,\dot{Y}_i\},(i=1,5,10)$. It can be seen that the PDFs and the tails of the PDFs of Y_i and $\dot{Y}_i$ obtained with subspace-EPC are close to MCS while the PDFs from EQL method deviate much from simulation. Similar behavior of the PDFs and the tails of the PDFs of other state variables can also be observed though not being presented here.

It is seen from the solution procedure that the results from EQL are employed in deriving the subspace method. Hence the subspace method is only suitable for the systems for which EQL is applicable for obtaining the approximate covariance matrix of the system responses.

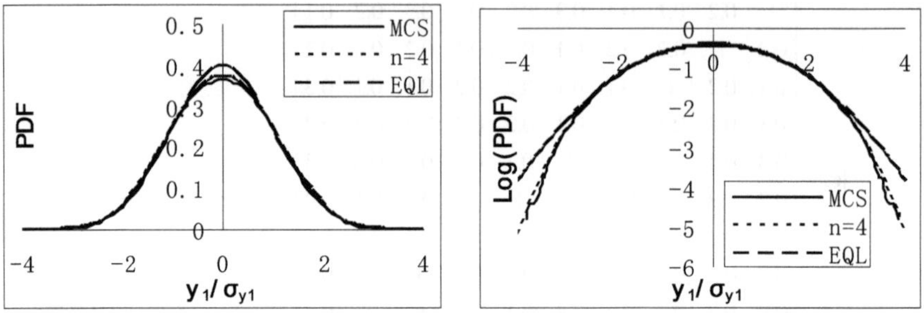

(a) PDF of Y_1

(b) Logarithmic PDF of Y_1

Fig. 1 Comparison of PDFs and logarithmic PDFs of displacement Y_1

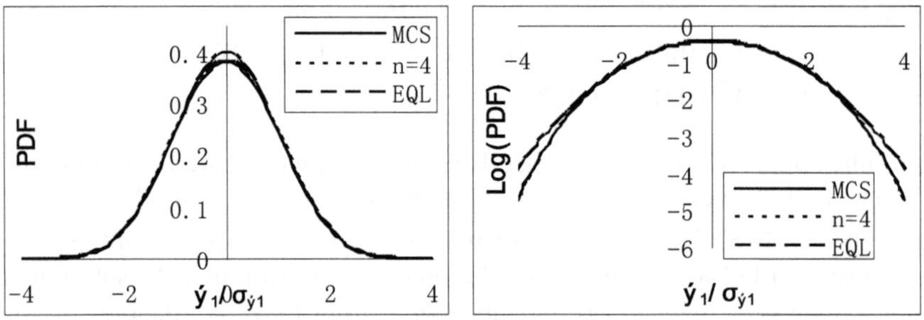

(c) PDF of $\dot{Y}_1$

(d) Logarithmic PDF of $\dot{Y}_1$

Fig. 2 Comparison of PDFs and logarithmic PDFs of velocity $\dot{Y}_1$

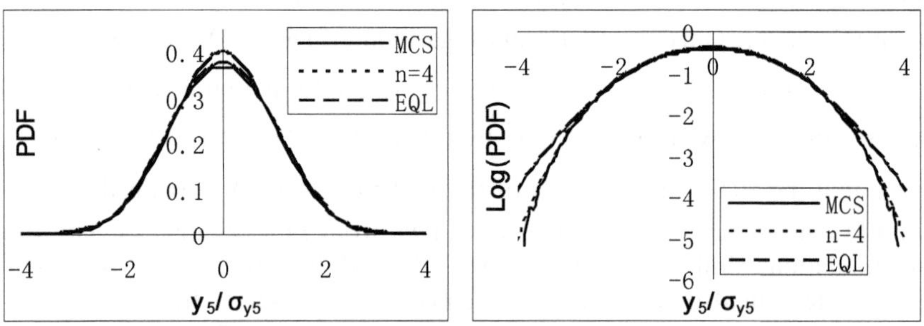

(a) PDF of Y_5

(b) Logarithmic PDF of Y_5

Fig. 3 Comparison of PDFs and logarithmic PDFs of displacement Y_5

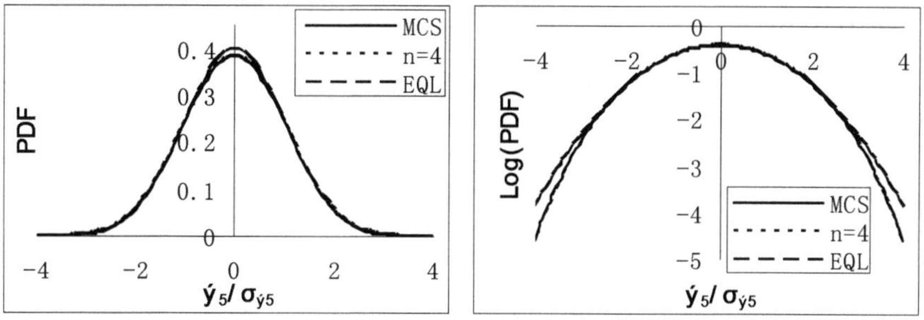

(a) PDF of $\dot{Y}_5$ (b) Logarithmic PDF of $\dot{Y}_5$

Fig. 4 Comparison of PDFs and logarithmic PDFs of displacement $\dot{Y}_5$

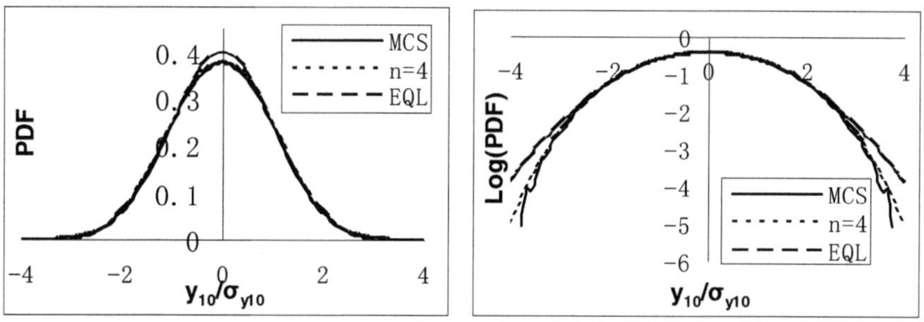

(a) PDF of Y_{10} (b) Logarithmic PDF of Y_{10}

Fig. 5 Comparison of PDFs and logarithmic PDFs of displacement Y_{10}

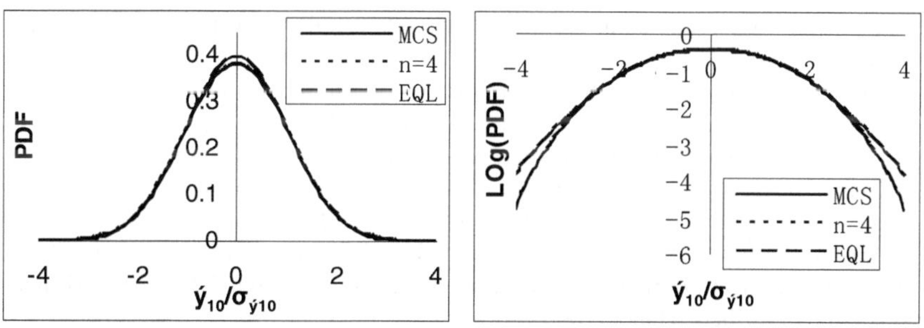

(c) PDF of $\dot{Y}_{10}$ (d) Logarithmic PDF of $\dot{Y}_{10}$

Fig. 6 Comparison of PDFs and logarithmic PDFs of velocity $\dot{Y}_{10}$

5 Conclusions

A novel subspace method is presented to solve the reduced FP equation in highly multi-dimensions. The subspace method reduces the problem of solving the FP equations in high dimensional state space to a set of FP equations in the state space dimensions at choices. The resulting FP equations in lower dimensions can be solved accurately by the EPC method. The proposed method is not limited by high dimensions, high nonlinearity of the systems, and the presence of multiplicative excitations. The responses of the NDS system with 10 degrees of freedom and high nonlinearity are given to show the effectiveness and accuracy of the proposed method in the case of polynomial nonlinearity of system. The results show good agreement with those of MCS. It attempts to provide an effective tool for obtaining the probabilistic solutions of some practical NSD systems in science and engineering. The subspace method is only suitable for the systems for which EQL is applicable for obtaining the approximate covariance matrix of the system responses.

References

1. Scheurkogel, A., Elishakoff, I.: Non-linear random vibration of a two-degree-of-freedom system. In: Ziegler, F., Schuëller, G.I. (eds.) Non-Linear Stochastic Engineering Systems, pp. 285–299. Springer, Berlin (1988)
2. Lin, Y.K., Cai, G.Q.: Probabilistic Structural Dynamics. International edn. McGraw-Hill, New York (1995)
3. Caughey, T.K.: Response of a non-linear string to random loading. ASME J. Appl. Mech. 26, 341–344 (1959)
4. Lin, Y.K.: Probabilistic Theory of Structural Dynamics. McGraw-Hill, New York (1967)
5. Spanos, P.D.: Stochastic linearization in structural dynamics. Appl. Mech. Review 34, 1–8 (1981)
6. Stratonovich, R.L.: Topics in the theory of random noise, vol. 1. Gordon and Breach, New York (1963)
7. Roberts, J.B., Spanos, P.D.: Stochastic averaging: an approximate method of solving random vibration problems. Int. J. Non-Linear Mech. 21, 111–134 (1986)
8. Er, G.-K.: An improved non-Gaussian closure method for randomly excited nonlinear stochastic systems. Nonlinear Dynamics 17(3), 285–297 (1998)
9. Er, G.-K., Iu, V.P.: A consistent and effective method for nonlinear random oscillations of MDOF systems. In: Proceedings of the IUTAM Symposium on Recent Developments in Nonlinear Oscillations of Mechanical Systems, March 2-5, pp. 85–94. Kluwer Academic Publishers, Hanoi (2000)

Nonstationary Probability Densities of Nonlinear Multi-Degree-of-Freedom Systems under Gaussian White Noise Excitations

X.L. Jin and Z.L. Huang

Department of Mechanics, Zhejiang University, Hangzhou, 310027, P.R. China

Abstract. The nonstationary probability densities of system responses are obtained for nonlinear multi-degree-of-freedom systems subject to stochastic parametric and external excitations. First, the stochastic averaging method is used to obtain the averaged Itô equation for amplitude envelopes of the system response. Then, the corresponding Fokker-Planck-Kolmogorov equation governing the nonstationary probability density of the amplitude envelopes is deduced. By applying the Galerkin method, the nonstationary probability density can be expressed as a series expansion in terms of a set of orthogonal base functions with time-dependent coefficients. Finally, the nonstationary probability densities for the amplitude response, as well as those for the state-space response, are solved approximately. To illustrate the applicability, the proposed method is applied to a two-degree-of-freedom van der Pol oscillator subject to external excitations of Gaussian white noises.

Keywords: Nonstationary probability density, Nonlinear stochastic system, Stochastic averaging method, Galerkin method.

1 Introduction

The randomness in structural dynamics is often encountered in various engineering fields. The source of randomness may arise from environmental loads, e.g. ground motion, atmospheric turbulence, sea waves, etc., as well as structural properties, such as materials properties, geometry parameters and so on. Moreover, real structures are generally nonlinear and of multiple degrees of freedom (MDOF). It is difficult to predict the nonstationary response of stochastic nonlinear MDOF system. The common methods so far are the equivalent linearization method and the Monte Carlo simulation [1-5]. In many MDOF systems with intrinsic nonlinear property, the equivalent linearization and lower-order statistical moments can not completely characterize the system response. The Monte Carlo simulation, although being a universal method, usually cannot capture system behavior change with varying system parameters.

In the present paper, a procedure is proposed to predict the nonstationary probability densities of system responses for MDOF nonlinear systems subject to stochastic parametric and external excitations. Conbining the stochastic averaging method, a similar procedures proposed in [6], and the Galerkin method, the

W.Q. Zhu et al. (Eds.): Nonlinear Stochastic Dynamics & Control, IUTAM BOOK SERIES 29, pp. 35–44.
springerlink.com

nonstationary probability densities of the amplitude processes and the state variables are obtained. An example is given to illustrate the feasibility of the proposed procedure.

2 Approximate Nonstationary Probability Density

Consider a nonlinear MDOF system subject to stochastic parametric and external excitations, governed by

$$\ddot{X}_i + \varepsilon c_{ij}\left(\mathbf{X}, \dot{\mathbf{X}}\right)\dot{X}_j + g_i(X_i) = \varepsilon^{1/2} f_{ik}\left(\mathbf{X}, \dot{\mathbf{X}}\right)W_k(t), \quad i, j = 1, \cdots, n; k = 1, \cdots, m \quad (1)$$

where $\mathbf{X} = [X_1, \cdots, X_n]^{\mathrm{T}}$ and $\dot{\mathbf{X}} = \left[\dot{X}_1, \cdots, \dot{X}_n\right]^{\mathrm{T}}$ are the generalized displacements and velocities, respectively, ε is a small positive parameter, $c_{ij}\left(\mathbf{X}, \dot{\mathbf{X}}\right)$ are the damping coefficients, $g_i(X_i)$ are the stiffnesses, which are assumed to be odd functions of the X_i, i.e., $g_i(-X_i) = -g_i(X_i)$, $\varepsilon^{1/2} f_{ik}\left(\mathbf{X}, \dot{\mathbf{X}}\right)$ are magnitudes of the external and/or parametric excitations, and $W_k(t)$ are independent Gaussian white noises with intensities $2D_{kk}$. It is assumed that for each degree of freedom (also called subsystem) in (1), at least one external excitation is present. In the pesent investigation, only the case of non-internal resonance is considered.

Eq. (1) suggests that the stiffness terms are uncoupled, and the dampings and excitations are weak. Then, the following transformations can be introduced [7, 8]

$$X_i(t) = A_i \cos\Theta_i(t), \qquad \dot{X}_i(t) = -A_i\Lambda_i(A_i, \Theta_i)\sin\Theta_i(t) \qquad i = 1, \cdots, n \quad (2)$$

where

$$\begin{aligned}
&\Theta_i(t) = \Phi_i(t) + \Gamma_i(t) \\
&\Lambda_i(A_i, \Theta_i) = \frac{d\Phi_i}{dt} = \sqrt{\frac{2[U_i(A_i) - U_i(A_i\cos\Theta_i)]}{A_i^2 \sin^2\Theta_i}} = b_{i0}(A_i) + \sum_{r=1}^{\infty} b_{ir}(A)\cos r\Theta_i \quad (3) \\
&U_i(X_i) = \int_0^{X_i} g_i(u)du
\end{aligned}$$

In (3), $A_i(t), \Theta_i(t), \Phi_i(t)$, and $\Gamma_i(t)$ are stochastic processes. Substituting Eq.(2) into Eq.(1), one can obtain the stochastic differential equations for the amplitudes A_i and phases Γ_i. Based on the Stratonovich-Khasminskii theorem [9], the slowly varying processes A_i and Γ_i converge weakly into a $2n$-dimensional diffusion Markov process. After stochastic averaging and deterministic averaging, the averaged Itô equations for A_i are independent of Γ_i as follows

$$dA_i = m_i(\mathbf{A})dt + \sigma_{ik}(\mathbf{A})dB_k(t) \qquad i=1,\cdots,n; k=1,\cdots,m \tag{4}$$

where the drift and diffusion coefficients are

$$\begin{aligned} m_i(\mathbf{A}) &= \varepsilon\left\langle F_{1i} + D_{kk}\frac{\partial G_{1ik}}{\partial A_j}G_{1jk} + D_{kk}\frac{\partial G_{1ik}}{\partial \varGamma_j}G_{2jk}\right\rangle_{\mathbf{\Theta}} \\ b_{ij}(\mathbf{A}) &= \sigma_{ik}(\mathbf{A})\sigma_{jk}(\mathbf{A}) = \varepsilon\left\langle 2D_{kk}G_{1ik}G_{1jk}\right\rangle_{\mathbf{\Theta}} \end{aligned} \tag{5}$$

in which $\langle\cdot\rangle_{\mathbf{\Theta}} = \int_0^{2\pi}\langle\cdot\rangle d\mathbf{\Theta}\big/(2\pi)^n$ denotes the deterministic averaging with respect to $\mathbf{\Theta}$, and

$$\begin{aligned} F_{1i}(\mathbf{A},\mathbf{\Gamma}) &= -A_i\varLambda_i(A_i,\varTheta_i)\sin\varTheta_i c'_{ij}(\mathbf{A},\mathbf{\Theta})A_j\varLambda_j(A_j,\varTheta_j)\sin\varTheta_j\big/g_i(A_i) \\ G_{1ik}(\mathbf{A},\mathbf{\Gamma}) &= -A_i\varLambda_i(A_i,\varTheta_i)\sin\varTheta_i f'_{ik}(\mathbf{A},\mathbf{\Theta})\big/g_i(A_i) \\ G_{2ik}(\mathbf{A},\mathbf{\Gamma}) &= -\varLambda_i(A_i,\varTheta_i)\cos\varTheta_i f'_{ik}(\mathbf{A},\mathbf{\Theta})\big/g_i(A_i) \end{aligned} \tag{6}$$

In (4)-(6), the vector processes $\mathbf{A}$, $\mathbf{\Gamma}$ and $\mathbf{\Theta}$ are defined as $\mathbf{A}=[A_1,\cdots,A_n]^{\mathrm{T}}$, $\mathbf{\Gamma}=[\varGamma_1,\cdots,\varGamma_n]^{\mathrm{T}}$, and $\mathbf{\Theta}=[\varTheta_1,\cdots,\varTheta_n]^{\mathrm{T}}$, respectively, and functions $c'_{ij}(\mathbf{A},\mathbf{\Theta})$ and $f'_{ik}(\mathbf{A},\mathbf{\Theta})$ are obtained from functions $c_{ij}(\mathbf{X},\dot{\mathbf{X}})$ and $f_{ik}(\mathbf{X},\dot{\mathbf{X}})$ respectively, according to the transformations (2). The explicit expressions for m_i and b_{ij} can be obtained by expanding F_{1i}, G_{1ik}, and G_{2ik} into Fourier series with respect to $\mathbf{\Theta}$.

The corresponding FPK equation associated with Eq.(4) can be obtained as follows

$$\frac{\partial p(\mathbf{A},t)}{\partial t} = -\frac{\partial}{\partial A_i}\left[m_i(\mathbf{A})p(\mathbf{A},t)\right] + \frac{1}{2}\frac{\partial^2}{\partial A_i\partial A_j}\left[b_{ij}(\mathbf{A})p(\mathbf{A},t)\right] \tag{7}$$

It is assumed that the system (1) is initially at rest, i.e., the initial condition for Eq.(7) is

$$p(\mathbf{A},0) = \hat{\delta}(\mathbf{A}) \tag{8}$$

where $\hat{\delta}(\mathbf{A})$ denotes the one-sided Dirac delta function.

It is difficult to obtain the exact solution of Eq.(7) under initial condition (8). Inspired by the procedures proposed in [6], we express the nonstationary probability density $p(\mathbf{A},t)$ as follows

$$p(\mathbf{A},t) = \sum_{r_1,\cdots,r_n=0}^{\infty}\left(\exp\left(-\sum_{i=1}^{n}\lambda_{ir_i}t\right) + s_{r_1,\cdots,r_n}(t)\right)\prod_{i=1}^{n}R_{ir_i}(A_i) \tag{9}$$

where $s_{r_1,\cdots,r_n}(t)$ are functions to be determined, $\lambda_{ir_i}, R_{ir_i}(A_i)$ are the eigenvalue and eigenfunction, respectively, of the FPK equation corresponding to the uncoupled and linearized ith subsystem [6]:

$$\begin{aligned}
&\lambda_{ir_i} = \left|\varepsilon c_{ii}(\mathbf{0},\mathbf{0})\right| r_i, \qquad R_{ir_i}(A_i) = \frac{1}{r_i!}\frac{A_i}{\sigma_{is}^2}\exp\left(-\frac{A_i^2}{2\sigma_{is}^2}\right)L_{r_i}\left(\frac{A_i^2}{2\sigma_{is}^2}\right) \\
&\sigma_{is}^2 = f_{il_i}^2 D_{l_i l_i} \Big/ \left(\left|c_{ii}(\mathbf{0},\mathbf{0})\right| \left.\frac{dg_i}{dX_i}\right|_{X_i=0}\right), \qquad i=1,\cdots,n;\ r_i = 0,1,\cdots
\end{aligned} \tag{10}$$

in which $L_{r_i}(\bullet)$ is the Laguerre polynomial of order r_i.

Using Eq.(9) and the properties of Laguerre polynomials, the initial conditions for $s_{r_1,\cdots,r_n}(t)$ can be derived from (8) as follows

$$s_{r_1,\cdots,r_n}(0) = 0. \qquad r_1,\cdots,r_n = 0,1,2,\cdots \tag{11}$$

Substitution of Eq.(9) into Eq.(7) yields the residual error

$$\begin{aligned}
R = &\sum_{r_1,\cdots,r_n=0}^{\infty}\left(-\left(\sum_{i=1}^{n}\lambda_{ir_i}\right)\exp\left(-\sum_{i=1}^{n}\lambda_{ir_i}t\right) + \dot{s}_{r_1,\cdots,r_n}(t)\right)\prod_{i=1}^{n}R_{ir_i}(A_i) - \\
&\sum_{r_1,\cdots,r_n=0}^{\infty}\left(\exp\left(-\sum_{i=1}^{n}\lambda_{ir_i}t\right) + s_{r_1,\cdots,r_n}(t)\right)\left(-\frac{\partial}{\partial A_i}\left[m_i(\mathbf{A})\prod_{k=1}^{n}R_{kr_k}(A_k)\right]\right. \\
&\left.+\frac{1}{2}\frac{\partial^2}{\partial A_i\partial A_j}\left[b_{ij}(\mathbf{A})\prod_{k=1}^{n}R_{kr_k}(A_k)\right]\right)
\end{aligned} \tag{12}$$

According to the Galerkin scheme, the unknown functions $s_{r_1,\cdots,r_n}(t)$ $(r_1,\cdots,r_n = 0,1,2,\cdots)$ can be calculated by making the projection of the residual error R to vanish on a proper set of independent functions. For the present case, we select $\prod_{i=1}^{n}R_{ik_i}(A_i)/R_{i0}(A_i)$ as weighting function and obtain

$$\int_0^\infty \prod_{i=1}^{n}\frac{R_{ik_i}(A_i)}{R_{i0}(A_i)}Rd\mathbf{A} = 0 \qquad k_1,\cdots,k_n = 0,1,\cdots \tag{13}$$

Substituting Eq.(12) into Eq.(13) and rearranging the resulting equations with the properties of Laguerre polynomials, a set of linear first-order ordinary differential equations governing the unknown functions $s_{r_1,\cdots,r_n}(t)$ can be derived as follows

$$\dot{s}_{k_1,\cdots,k_n}=\left(\sum_{i=1}^{n}\lambda_{ik_i}\right)\exp\left(-\sum_{i=1}^{n}\lambda_{ik_i}t\right)+\sum_{r_1,\cdots,r_n=0}^{\infty}\left(\exp\left(-\sum_{i=1}^{n}\lambda_{ir_i}t\right)+s_{r_1,\cdots,r_n}(t)\right)$$

$$\int_0^{\infty}\prod_{i=1}^{n}\frac{R_{ik_i}(A_i)}{R_{i0}(A_i)}\left(-\frac{\partial}{\partial A_i}\left[m_i(\mathbf{A})\prod_{k=1}^{n}R_{kr_k}(A_k)\right]+\frac{1}{2}\frac{\partial^2}{\partial A_i\partial A_j}\left[b_{ij}(\mathbf{A})\prod_{k=1}^{n}R_{kr_k}(A_k)\right]\right)d\mathbf{A} \quad (14)$$

$$k_1,\cdots k_n=0,1,\cdots$$

The series in Eq.(9) will be truncated in numerical calculation, and the number of terms for the ith degree is denoted by N_i, i.e., $r_i = 0, 1, \ldots, N_i$. Note that N_i may be different for different i. Then Eq.(14) can be solved numerically with initial condition (11) by using the Runge-Kutta algorithm. An approximate analytical expression of $p(\mathbf{A},t)$ can be obtained by substituting $s_{r_1,\cdots,r_n}(t)$ into Eq.(9). For the stationary case, Eq.(14) is reduced to a set of linear algebraic equations. By determining the eigenvalues and eigenvectors, the stationary probability density $p_s(\mathbf{A})$ can be obtained.

The nonstationary marginal probability density of the amplitude A_i can be obtained by integrating Eq.(9) with respect to $\hat{\mathbf{A}}_i=[A_1,\cdots,A_{i-1},A_{i+1},\cdots,A_n]^{\mathrm{T}}$ as

$$p(A_i,t)=\sum_{r_i=0}^{\infty}\left[e^{-\lambda_{ir_i}t}+s_{0,0,\cdots,r_i,0,\cdots,0}(t)\right]R_{ir_i}(A_i) \quad (15)$$

Then, the joint nonstationary probability density of the generalized displacement X_i and velocity $\dot{X}_i$ can be calculated from

$$p(X_i,\dot{X}_i,t)=p(A_i,t)b_{i0}(A_i)/[2\pi g_i(A_i)]\big|_{A_i=U_i^{-1}(\dot{X}_i^2/2+U_i(X_i))} \quad (16)$$

where $U_i^{-1}(\bullet)$ is the inverse function of $U_i(\bullet)$. The corresponding stationary probability densities can also be obtained similarly.

3 An Illustrative Example

Consider two-degree-of-freedom van der Pol oscillator subject to additive Gaussian white noise excitations. The equation of motion of the system is of the form

$$\begin{aligned}&\ddot{X}_1+(\beta_{10}+\beta_{11}X_1^2+\beta_{12}X_2^2)\dot{X}_1+\omega_1^2X_1=W_1(t)\\&\ddot{X}_2+(\beta_{20}+\beta_{21}X_1^2+\beta_{22}X_2^2)\dot{X}_2+\omega_2^2X_2=W_2(t)\end{aligned} \quad (17)$$

where β_{ij},ω_1, and ω_2 are constants, $W_1(t)$ and $W_2(t)$ are independent Gaussian white noises with intensities $2D_{11},2D_{22}$, respectively.

By using the stochastic averaging method described in Section 2, the averaged Itô equation for $A_i(t)$ is of the form of Eq.(4) with the following averaged drift and diffusion coefficients

$$\begin{aligned} m_1 &= -\beta_{10}A_1/2-\beta_{11}A_1^3/8-\beta_{12}A_1A_2^2/4+D_{11}/\left(2\omega_1^2A_1\right) \\ m_2 &= -\beta_{20}A_2/2-\beta_{22}A_2^3/8-\beta_{21}A_1^2A_2/4+D_{22}/\left(2\omega_2^2A_2\right) \\ b_{11} &= D_{11}/\omega_1^2, b_{22} = D_{22}/\omega_2^2, b_{12}=b_{21}=0 \end{aligned} \tag{18}$$

The associated FPK equation has the form (7) and initial condition (8). According to the procedures proposed in Section 2, the nonstationary probability density of the amplitudes can be approximately expressed as

$$p(A_1,A_2,t)=\sum_{i=0}^{N_1}\sum_{j=0}^{N_2}\left[e^{-(\lambda_{1i}+\lambda_{2j})t}+s_{ij}(t)\right]R_{1i}(A_1)R_{2j}(A_2) \tag{19}$$

where

$$\begin{aligned} &\lambda_{ki}=|\beta_{k0}|i, \qquad R_{ki}(A_k)=A_k/\left(i!\sigma_{ks}^2\right)\exp\left[-A_k^2/\left(2\sigma_{ks}^2\right)\right]L_i\left[A_k^2/\left(2\sigma_{ks}^2\right)\right] \\ &\sigma_{ks}^2=D_{kk}/\left(|\beta_{k0}|\omega_k^2\right), \qquad k=1,2 \end{aligned} \tag{20}$$

Functions $s_{ij}(t)$ are governed by the following set of linear first-order ordinary differential equations,

$$\begin{aligned} &\dot{s}_{kl}=(\lambda_{1k}+\lambda_{2l})e^{-(\lambda_{1k}+\lambda_{2l})t}+\sum_{i=0}^{N_1}\sum_{j=0}^{N_2}\left[e^{-(\lambda_{1i}+\lambda_{2j})t}+s_{ij}\right]\times\{\delta_{jl} \\ &\{(\beta_{10}-|\beta_{10}|)\left[(i+1)\delta_{i+1,k}-i\delta_{ik}\right]+\beta_{11}\sigma_{1s}^2\left[(i+1)I_{i+1,k}-(i-1)I_{ik}\right]/2-\lambda_{1i}\delta_{ik}\} \\ &+\delta_{ik}\{(\beta_{20}-|\beta_{20}|)\left[(j+1)\delta_{j+1,l}-j\delta_{jl}\right]+\beta_{22}\sigma_{2s}^2\left[(j+1)I_{j+1,l}-(j-1)I_{jl}\right]/2 \\ &-\lambda_{2j}\delta_{jl}\}+\beta_{12}\sigma_{2s}^2\left[(i+1)\delta_{i+1,k}-i\delta_{ik}\right]I_{jl}+\beta_{21}\sigma_{1s}^2\left[(j+1)\delta_{j+1,l}-j\delta_{jl}\right]I_{ik}\} \\ &k=0,\cdots,N_1, l=0,\cdots,N_2 \end{aligned} \tag{21}$$

in which δ_{jl} is the Kronecker delta symbol, and

$$I_{jl}=(2l+1)\delta_{jl}-l\delta_{j,l-1}-(l+1)\delta_{j,l+1} \tag{22}$$

Eq.(21) can be solved numerically by using the fourth-order Runge-Kutta method.

The nonstationary marginal probability densities of the amplitudes A_1 and A_2 can be derived from (15), respectively,

$$p(A_1,t)=\sum_{i=0}^{N_1}\left[e^{-\lambda_{1i}t}+s_{i0}(t)\right]R_{1i}(A_1), \quad p(A_2,t)=\sum_{j=0}^{N_2}\left[e^{-\lambda_{2j}t}+s_{0j}(t)\right]R_{2j}(A_2) \tag{23}$$

According to (16), the joint nonstationary probability density of the generalized displacement X_k and velocity $\dot{X}_k$ are obtained as follows

$$p(X_k, \dot{X}_k, t) = p(A_k, t)/(2\pi\omega_k A_k)\big|_{A_k=\sqrt{X_k^2+\dot{X}_k^2/\omega_k^2}}, \qquad k = 1,2 \tag{24}$$

Numerical calculations were carried out for system (17) with parameters β_{10} = -0.04, β_{11} = 0.01, β_{12} = 0.01, β_{20} = -0.085, β_{21} = 0.02, β_{22} = 0.01, ω_1 = 1, $\omega_2 = \sqrt{2}$, D_{11} = 0.1, and D_{22} = 0.3. The numbers of truncated terms are $N_1 = 20$ and $N_2 = 40$. Fig.1 shows the nonstationary and stationary probability densities of the amplitudes. The joint nonstationary probability densities of X_2 and $\dot{X}_2$ (V_2) are shown in Fig. 2. The joint stationary probability densities $p_s(X_1, \dot{X}_1)$, $p_s(X_2, \dot{X}_2)$ are shown in Fig. 3. In these figures, results from Monte Carlo simulations are also depicted.

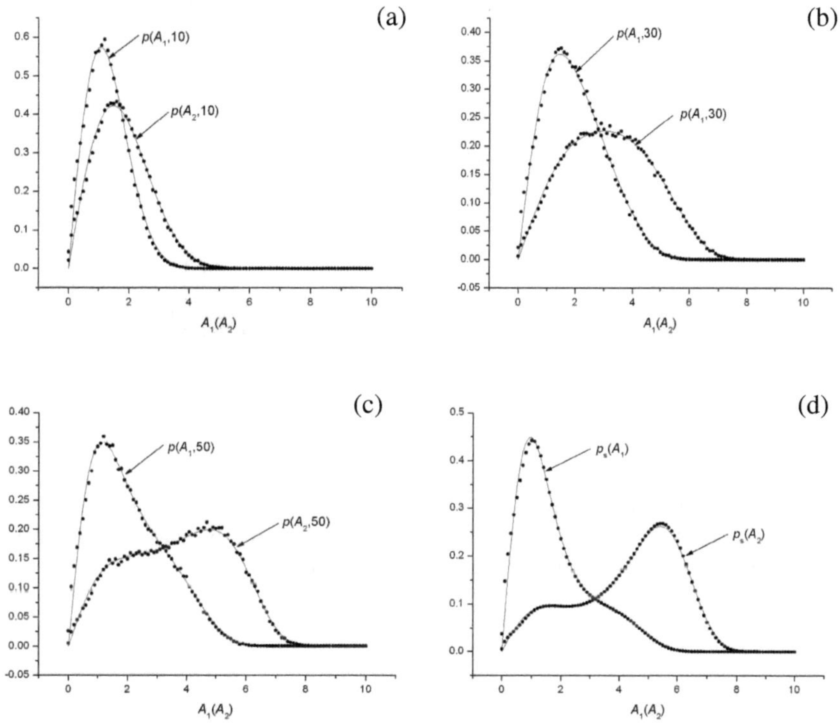

Fig. 1 The nonstationary probability densities of the amplitude at different time instants and the stationary probability densities of the amplitudes. (a) $t = 10$ (b) $t = 30$ (c) $t = 50$ (d)stationary case. The solid lines represent the results obtained by the proposed method while the symbols represent the results from Monte Carlo simulation.

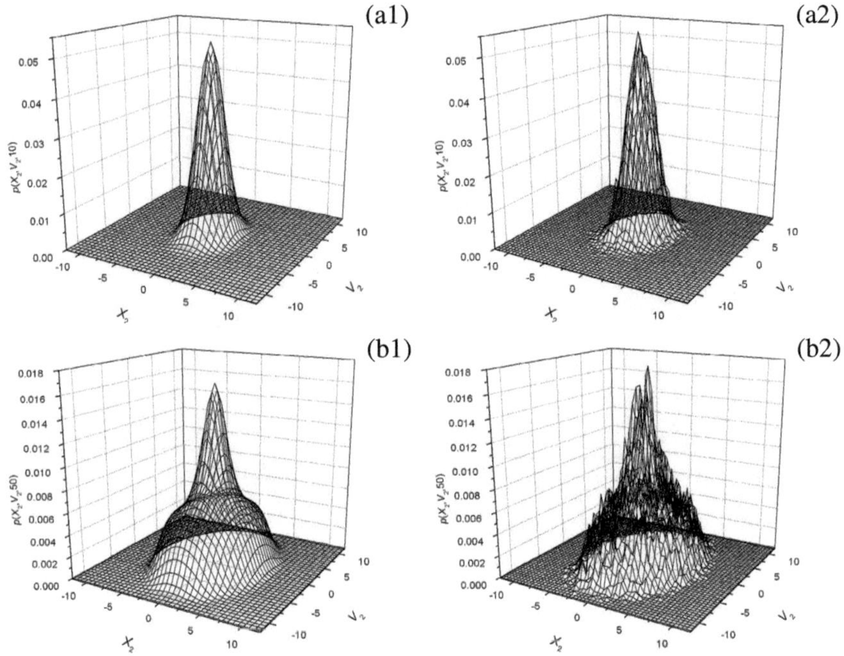

Fig. 2 The joint nonstationary probability densities $p(X_2, \dot{X}_2, t)$ at different time instants. (a1)(a2) $t = 10$ (b1)(b2) $t = 50$. (a1)(b1) depict results obtained by the proposed method while (a2)(b2) show the results from Monte Carlo simulation.

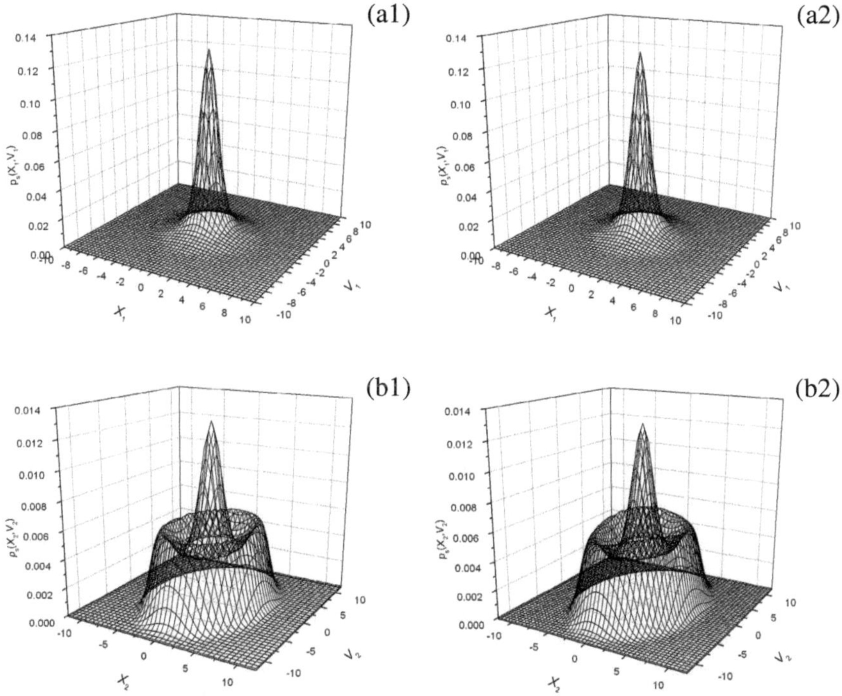

Fig. 3 The joint stationary probability densities of the generalized displacement and velocity $p_s(X_1, \dot{X}_1), p_s(X_2, \dot{X}_2)$. (a1)(b1) depict results obtained by the proposed method while (a2)(b2) show the results from Monte Carlo simulation.

It can be seen from Figs.1-3 that the results obtained from the proposed procedure agree well with those from the Monte Carlo simulation. Fig. 3 shows that the two oscillators have different behaviors at the stationary state although they have similar forms of equations with negative linear dampings (β_{10} = -0.04, β_{20} = -0.085). The first oscillator tends to a diffused region about the trivial solution while the second oscillator approaches to diffused regionas both about the trivial solution and a limit circle. Neverthless, the behavious of the two oscillators depend on the parameter values, and the proposed procedure provides practical means to predict the transient behaviors of the system response.

4 Conclusions

In the present paper, nonstationary probability densities of the system response have been estimated for nonlinear MDOF systems subject to stochastic parametric and external excitations. By using the stochastic averaging method, the system is transformed to one for Markovian amplitude processes, and FPK equation governing the nonstationary probability density of the amplitude processes is obtained. Then, using a similar procedures proposed in [6], the nonstationary probability density is expressed as a series expansion in terms of a set of properly selected base functions with time-dependent coefficients. Using Galerkin method, a set of first-order equations governing these time-dependent coefficients are derived and solved numerically. The nonstationary probability densities for the system state-space response can also be obtained from those of the amplitude processes.

The propose procedures have been applied to a two-degree-of-freedom van der Pol oscillator subject to external excitations of Gaussian white noises. Comparisons of the analytical results with those obtained from Monte Carlo simulation show that the proposed procedure is deasible and accurate. In pronciple, the proposed procedures can be extended to nonlinear MDOF systems subject to nonstationary stochastic excitations.

Acknowledgments. The research work was supported by the ZJNSF under Grant No. Z6090125, the specialized research fund for the Doctoral Program of High Education of China under Grant no. 20070335053.

References

1. Micaletti, R.C., Çakmak, A.Ş., Nielsen, S.R.K., Köylüoğlu, H.U.: A solution method for linear and geometrically nonlinear MDOF systems with random properties subject to random excitation. Probabilist. Eng. Mech. 13, 85–95 (1998)
2. Zhang, L., Zu, J.W., Zheng, Z.: The stochastic Newmark algorithm for random analysis of multi-degree-of-freedom nonlinear systems. Computers & Structures 70, 557–568 (1999)
3. Ohtori, Y., Spencer, B.F.: Semi-Implicit Integration Algorithm for Stochastic Analysis of Multi-Degree-of-Freedom Structures. J. Eng. Mech. 128, 635–643 (2002)
4. Smyth, A.W., Masri, S.F.: Nonstationary response of nonlinear systems using equivalent linearization with a compact analytical form of the excitation process. Probabilist. Eng. Mech. 17, 97–108 (2002)
5. Saha, N., Roy, D.: The Girsanov Linearization Method for Stochastically Driven Nonlinear Oscillators. ASME J. Appl. Mech. 74, 885–897 (2007)
6. Spanos, P.D., Sofi, A., Di Paola, M.: Nonstationary response envelope probability densities of nonlinear oscillators. ASME J. Appl. Mech. 74, 315–324 (2007)
7. Zhu, W.Q., Huang, Z.L., Suzuki, Y.: Response and stability of strongly non-linear oscillators under wide-band random excitation. Int. J. Non-Linear Mech. 36, 1235–1250 (2001)
8. Zhu, W.Q.: Nonlinear stochastic dynamics and control in Hamiltonian formulation. Appl. Mech. Rev. 59, 230–248 (2006)
9. Khasminskii, R.Z.: On the averaging principle for Itô stochastic differential equations. Kibernetika 4, 260–279 (1968) (in Russian)

Feature Extraction within the Fei-Tsui Arch Dam under Environmental Variations

C.H. Loh, J.H. Weng, C.H. Chen, and Y.W. Chang

Department of Civil Engineering, National Taiwan University,
Taipei 10617, Taiwan

Abstract. The objective of this research is to develop methods for analyzing the seismic response data and the long-term static data of the Fei-tsui arch dam, and based on the result of analysis to set an early warning threshold level for dam safety early warning evaluation. First, the input/output subspace identification technique is used to analysis the recorded seismic data from 84 earthquake events in order to identify the modal properties of the dam under different water level. Considering the spatial variability of input excitation, two kinds of system model are applied to subspace identification technique: the single-input and the multiple-input system. The regression curves between the identified system natural frequencies and water level are developed from the statistical analysis of identification results. Second, two different approaches are applied to extract features of the long-term data of the dam. The methods include the singular spectrum analysis with AR model (SSA-AR) and the nonlinear principal component analysis (NPCA) using auto-associate neural network method (AANN). By using these methods, the residual deformation between the estimated and the recorded data was generated, through statistical analysis, the threshold level of the dam static deformation can be determined. Discussion on (1) the difference between two kinds of input model for subspace identification and (2) proposed methods to extract static data are also made in this research.

Keywords: Input/Output Subspace Identification, Singular Spectral Analysis, Autoregressive Model, Auto-Associate Neural Network, Nonlinear principal component analysis.

1 Introduction

Monitoring technology plays an important role in securing integrity of structural system and maintaining the longevity of the structure. It consists of three aspects: (1) instrumentation with sensors, (2) methodologies for obtaining meaningful information concerning the structural health monitoring, (3) early warning from the measured data. Various methods based on the dynamic and static test have been applied to address the structural health monitoring and damage identification. However, the system properties of the structure may be changed by the changing environment conditions and strong earthquakes. As a result, it is necessary to develop some effective and efficient approaches based on the dynamic and static data not only to determine damage occurrences and damage location for structural

W.Q. Zhu et al. (Eds.): Nonlinear Stochastic Dynamics & Control, IUTAM BOOK SERIES 29, pp. 45–54.
springerlink.com

health monitoring in practice but also to set an early warning threshold before the disaster occurred.

Application of system identification techniques to the recorded seismic response of structures had been studied based on discrete-time linear filter approach [1, 2]. The modal properties of the Fei-tsui arch dam located in Taiwan had been identified based on recorded seismic data by using ARX-LS method in 1996 [3]. After Chi-Chi earthquake, a forced vibration test was conducted on the dam to check the resonant frequencies and mode shapes [4]. In this study, the subspace identification algorithm is employed to identify the modal properties of the dam.

Feature extraction from the long-term data of structural health monitoring is another important issue in this research. Hsu and Loh [5] applied the nonlinear principal component analysis by using auto-associative neural network to extract the underlying environmental factors and identify the damage features of the structure. The singular spectrum analysis can be applied to extract tendencies and harmonic components of time series [6]. In this study, these two different approaches are applied to the long-term static deformation data of the dam to extract the tendencies of the deformation. Then, through the statistical analysis of the residual deformation, the threshold level for early warning on dam static deformation can be determined.

2 Measurement Systems of the Fei-Tsui Arch Dam

The Fei-Tsui arch dam is a 122.5 meter high, 510 meter long which is located in the Taipei city, Taiwan. Figure 1 shows the photo, the dynamic and static monitoring systems of the dam. To monitor the dynamic properties of the dam during earthquake, eleven tri-axial accelerographs were deployed in the dam, as shown in Figure 1(a). Five of these instruments are installed along the abutment (SD1 ~ SD5), the others are installed on different level gallery of the dam. Total 84 strong motion recorded data collected during 1999 to 2008 were chosen to analyze in this study. Figure 2(a) shows the relationship of 84 seismic events between water level and peak ground acceleration (PGA) from accelerometer SD1. The most intense one of these earthquakes occurred on March 31, 2002 which was called as 331 Earthquake in Taiwan. The recorded PGA of 331 Earthquake is up 0.028g which is greater than 921 Chi-Chi Earthquake (0.025g). In addition, the static measurement system composed of thirteen sensors is used to monitor of dam deformation, as shown in Figure 1(c). Before analyzing, the deformation data points collected once a day from Jan. 1, 1978 to April 30, 2009 were check one by one at first. All data points at the same time were eliminated if one of these sensors was broken. As a result, total 7719 data points were chosen to analyze for early warning in this study. Figure 2(b) plots the measured deformation of the dam along profile NPL2. Note that the abscissa of Figure 2(b) indicates the number of point not time.

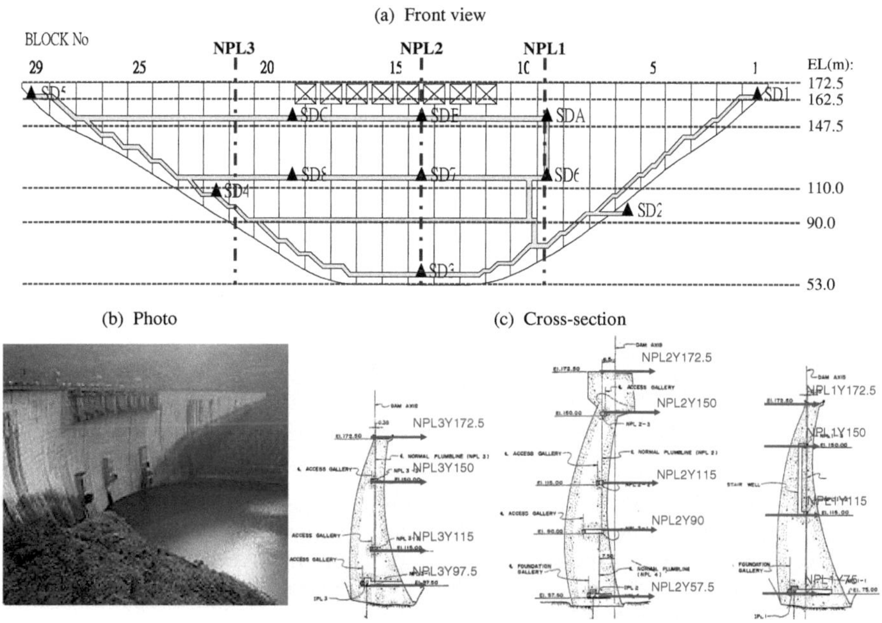

Fig. 1 The Fei-Tsui arch dam ; (a) seismic measurement system , (b) real photo , (c) long-term static measurement system along three plumb lines of the dam

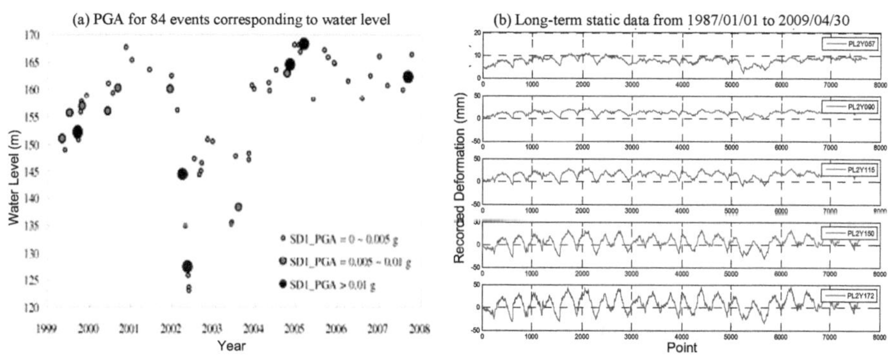

Fig. 2 The measurement data of the Fei-Tsui arch dam ; (a) the relationship between water level and PGA for SD1 for total 84 seismic events during 1999 to 2008 , (b) measured deformation of the dam along profile NPL2 at different level (from Jan. 1, 1978 to April 30, 2009, a total of 7719 data points)

3 System Identification from Seismic Response Data

The subspace identification algorithm [7] was applied to analyze recorded seismic response data to identify dynamic properties of the Fei-Tsui arch dam. This method can effectively identify the modal frequencies and mode shapes of a multi-input/multi-output system in time domain. Note that the input ground motion for the dam is not uniform along the abutment of the dam [4]. As a result, it is suggested that the dam should be considered as a multi-support system for system identification. In this study, the recorded seismic data from five tri-axial accelerographs (SD1~SD5) was considered as five support excitation [8]. Hence, the motion equation for this MIMO system can be written in partitioned form:

$$\begin{bmatrix} \mathbf{m} & \mathbf{m}_g \\ \mathbf{m}_g^T & \mathbf{m}_{gg} \end{bmatrix} \begin{Bmatrix} \ddot{\mathbf{u}}^t \\ \ddot{\mathbf{u}}_g \end{Bmatrix} + \begin{bmatrix} \mathbf{c} & \mathbf{c}_g \\ \mathbf{c}_g^T & \mathbf{c}_{gg} \end{bmatrix} \begin{Bmatrix} \dot{\mathbf{u}}^t \\ \dot{\mathbf{u}}_g \end{Bmatrix} + \begin{bmatrix} \mathbf{k} & \mathbf{k}_g \\ \mathbf{k}_g^T & \mathbf{k}_{gg} \end{bmatrix} \begin{Bmatrix} \mathbf{u}^t \\ \mathbf{u}_g \end{Bmatrix} = \begin{Bmatrix} \mathbf{0} \\ \mathbf{p}_g \end{Bmatrix} \tag{1}$$

where $\mathbf{u}_g$ is a displacement vector which includes the DOFs of the supports (such as SD1~SD5), $\mathbf{u}^t$ is also a displacement vector which includes all DOFs of the dam except the DOFs of the supports, and $\mathbf{p}_g$ is earthquake loading which is applied to the dam. Eq.(1) can then be rewritten by focusing on the dynamic displacements $\mathbf{u}$ on the DOFs of the dam:

$$\mathbf{m}\ddot{\mathbf{u}} + \mathbf{c}\dot{\mathbf{u}} + \mathbf{k}\mathbf{u} = \mathbf{p}_{eff} \tag{2}$$

$$\mathbf{p}_{eff} = -\left(\mathbf{m}\ddot{\mathbf{u}}^s + \mathbf{m}_g\ddot{\mathbf{u}}_g\right) - \left(\mathbf{c}\dot{\mathbf{u}}^s + \mathbf{c}_g\dot{\mathbf{u}}_g\right) \tag{3}$$

where $\mathbf{u}^s = \mathbf{u}^t - \mathbf{u}$ is the quasi-static displacement, and $\mathbf{p}_{eff}$ is the effective earthquake forces. Assume that the dam is a small damped and lump mass system. Eq.(3) can be further simplified by these two assumptions for real application:

$$\mathbf{m}\ddot{\mathbf{u}} + \mathbf{c}\dot{\mathbf{u}} + \mathbf{k}\mathbf{u} = -\mathbf{m}\boldsymbol{\iota}\ddot{\mathbf{u}}_g \tag{4}$$

where $\boldsymbol{\iota} = -\mathbf{k}^{-1}\mathbf{k}_g$ is the influence matrix which describes the influence of support displacements on the structural displacements. On the other hand, assume that the input ground motion is uniform and one of the records from SD1~SD5 is selected to be the representative ground motion. Eq.(4) can be simplified for a dynamic system with uniform support excitations:

$$\mathbf{m}\ddot{\mathbf{u}} + \mathbf{c}\dot{\mathbf{u}} + \mathbf{k}\mathbf{u} = -\mathbf{m}\mathbf{1}\ddot{u}_g \tag{5}$$

where u_g is a scalar of the uniform input and $\mathbf{1}$ is a vector with each element equal to unity.

In this study, these two kinds of dynamic model will be applied to identify system properties of the Fei-Tsui arch dam by using subspace identification. The first step of the subspace identification to identify the system is to transform the motion equation into the continuous-time state space model [7]:

$$\dot{\mathbf{x}} = \mathbf{A}_c\mathbf{x} + \mathbf{B}_c\ddot{\mathbf{u}}_g + \mathbf{w} \tag{6}$$

$$\ddot{\mathbf{u}}^t = \mathbf{C}\mathbf{x} + \mathbf{v} \tag{7}$$

where $\mathbf{x} = \begin{bmatrix} \mathbf{u} \\ \dot{\mathbf{u}} \end{bmatrix}$; $\mathbf{A}_c = \begin{bmatrix} \mathbf{0} & \mathbf{I} \\ -\mathbf{m}^{-1}\mathbf{k} & -\mathbf{m}^{-1}\mathbf{c} \end{bmatrix}$; $\mathbf{B}_c = \begin{bmatrix} \mathbf{0} \\ -\mathbf{\iota} \end{bmatrix}$; $\mathbf{C} = \begin{bmatrix} -\mathbf{m}^{-1}\mathbf{k} & -\mathbf{m}^{-1}\mathbf{c} \end{bmatrix}$.

$\mathbf{w}$ and $\mathbf{v}$ represent the disturbances vector and the measurement noise vector individually which are assumed to be zero mean and white noise. Besides, $\ddot{\mathbf{u}}_g$ and $\ddot{\mathbf{u}}^t$ represent inputs and outputs of this system.

The basic concept of subspace algorithms is exploitation of the state as a finite-dimensional interface between the past part and the future part. First, the input and output data are arranged into the Hankel matrices. Then projection theorem is employed to avoid the influence of noise and extract the observability matrix $\mathbf{\Gamma}_i$:

$$\mathbf{Y}_f /_{\mathbf{U}_f} \mathbf{W}_p = \mathbf{\Gamma}_i \mathbf{X}_f^d \tag{8}$$

Then, we can obtain the system parameters $\mathbf{A}_c$ and $\mathbf{C}$ by using singular value decomposition to extract the observability matrix. The natural frequencies, damping ratios and mode shapes of the dam can be identified at last. The detail procedure of the subspace identification can be found in reference 3.

In this study, the seismic response data from 331 Earthquake was chosen to identify the modal properties for multi-input and single-input model at first. The corresponding water level is 144.4 m during 331 Earthquake. For multi-input case (Input = SD1~SD5), the identified modal frequencies are equal to 2.59, 2.90 and 3.55 Hz. For the single-input cases, the record of SD1 and SD5 were considered as the input excitation individually. The modal frequencies for SD1-input model are equal to 2.59, 2.90 and 3.55 Hz. Besides, the results for SD5-input model are very similar to the results for SD5-input model. The modal frequencies for SD5-input model are equal to 2.47, 2.58 and 3.45 Hz. According to the same procedure, total 84 recorded seismic response data were used to identify the modal frequencies under different water level. The identified modal frequencies under different water level were plotted in the Figure 3. Then, the relationships between water level and modal frequencies were regressed by using the general power-2 function:

$$f(x) = ax^b + c \tag{9}$$

where f is the calculated modal frequency according to the regression model and x is the corresponding water level. It is observed that the modal frequencies of dam decreases when the reservoir level increases. Figure 4 shows the first 3 identified mode shapes for SD1-input model using the data from 331 Earthquake.

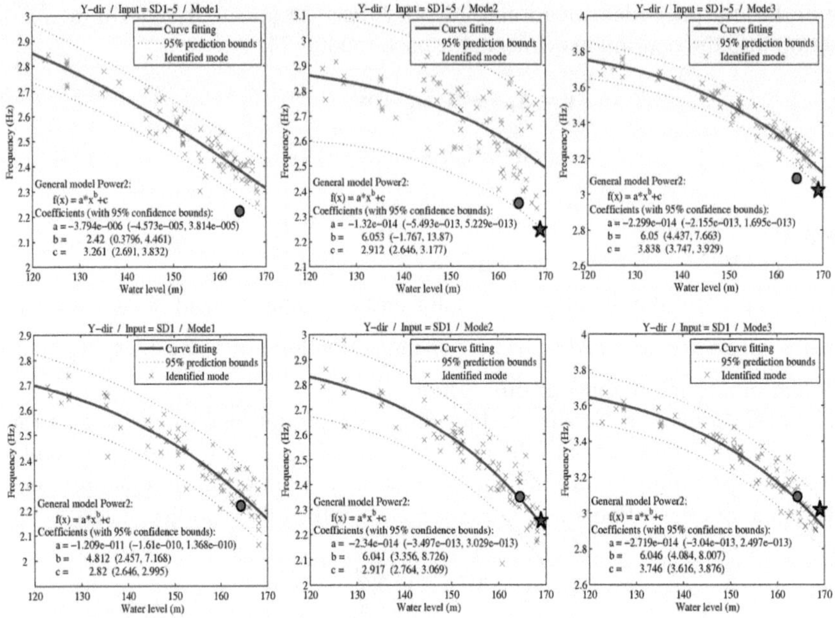

Fig. 3 Regression analysis of the relationship between water level and first 3 modal frequencies of the Fei-Tsui arch dam. (☆ are the results from force vibration test [4])

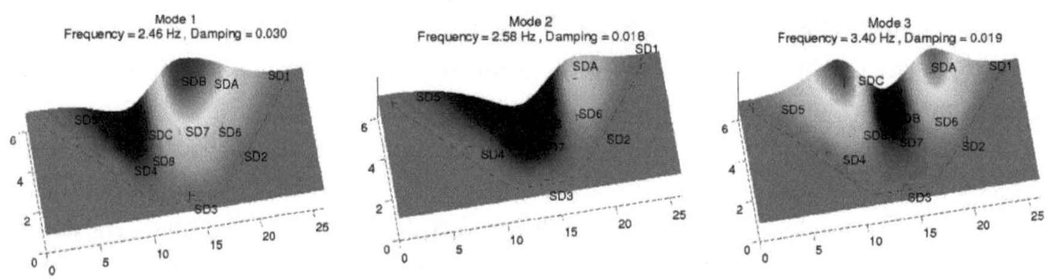

Fig. 4 Identified mode shapes for SD1-input model using the data from 331 earthquake

4 Deformation Monitoring from Long-Term Data

Two different methods are applied to extract the response feature of the long-term static deformation data of the dam. One is the singular spectrum analysis with AR model (SSA-AR) [9] and the other is the nonlinear principal component analysis (NPCA) using auto-associate neural network method (AANN) [10].

SSA-AR model is composed of singular spectrum analysis (SSA) and Autoregressive (AR) model. SSA can be applied to smooth a noisy signal, extraction of tendency or seasonality component, or to detect the singularities. In this study, SSA is used to decompose the original time series into a series of principle component time series. The basic concept of SSA is to produce a Hankel matrix from the recorded time series by a sliding window at first. Note that the length of the sliding window must be smaller than the length of recorded time series. The second step is to decompose the Hankel matrix into element matrices by using singular value decomposition. The last step is to determine a parameter M and reconstruct first M principal components (PCs) corresponding to first M element matrix. The detail procedure of SSA can be found in reference 6. After reconstructing PCs by SSA, each PC was modeled by an AR model. The short term deformation forecasting for each PC can then be calculated based on each fitting AR model. Finally, the summation of PCs after forecasting means predicted response of the structure which can be used for determining the threshold value of early warning. Figure 5 shows the predicted procedure of the SSA-AR Model.

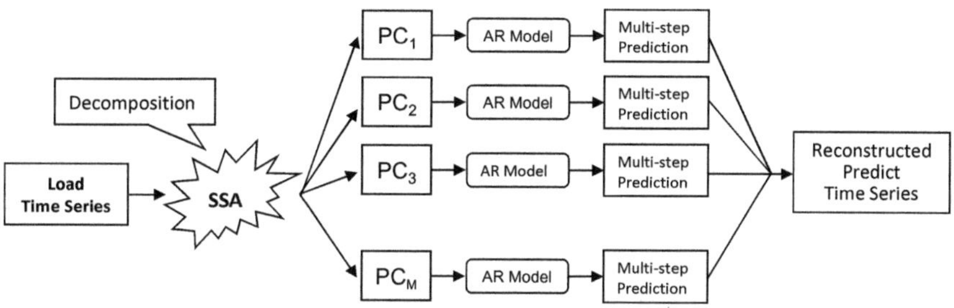

Fig. 5 Diagram of developing short time deformation forecasting for SSA-AR Model

The other method is the NPCA algorithm to perform feature extraction by applying the auto-associative neural network (AANN). The AANN is a particular class of neural networks in which the target output pattern is identical to the input pattern and the NPCA can be used to extract the intrinsic environmental factors causing the variation of measured features. There are five layers in the typical AANN as illustrated in Figure 6, where **s** represents sigmoid transfer function and ***l*** represents the linear transfer function. Supervised learning is applied to train AANN. Once the AANN is trained to reconstruct the original data and it can be used to extract the underlying nonlinear principal components simultaneously. As a result, the latent relationship between the identified features and the unknown intrinsic features causing the variations of the identified features is revealed. The residual error due to the mapping and de-mapping of AANN plays an important rule to determine the uncertainty level with respect to the underlying principle components. In other words, through statistical analysis of the residual error the threshold level for the uncertainty level of abnormal data can be estimated.

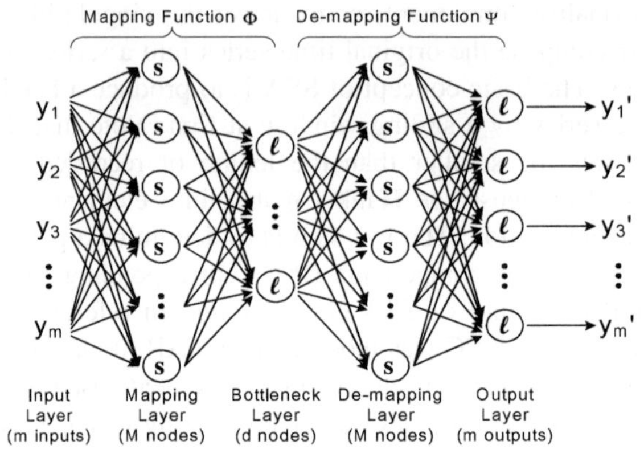

Fig. 6 Auto-Associate Neural Network (AANN)

These two proposed methods were applied to the long-term static data (from January 1, 1987 to April 30, 2009) to establish the model for forecasting the deformation behavior of the dam. Figure 7 shows the comparison between recorded deformation and reconstructed deformation by using SSA-AR model and AANN-NPCA methods from location NPL2Y172.5. The statistical distribution of the residual error between recorded and reconstructed deformation was also calculated. Figure 8 shows the histograms of residual error from two different methods. The standard deviation of the residual error from SSA-AR model and AANN method along the measurement line of NPL2Y is shown in Table 1. From Figure 8 it is found that the distribution of residual error does not exactly follow the normal distribution. As a result, the deviation value at 1% and 99% (σ_{lower} and σ_{lower}) of the cumulative probability density function of the residual error was also calculated in this study, as shown in Table 1. The monitoring displacement can be compared with the estimated displacement from SSA-AR Model and AANN method. Then, the error residual can be generated for early warning.

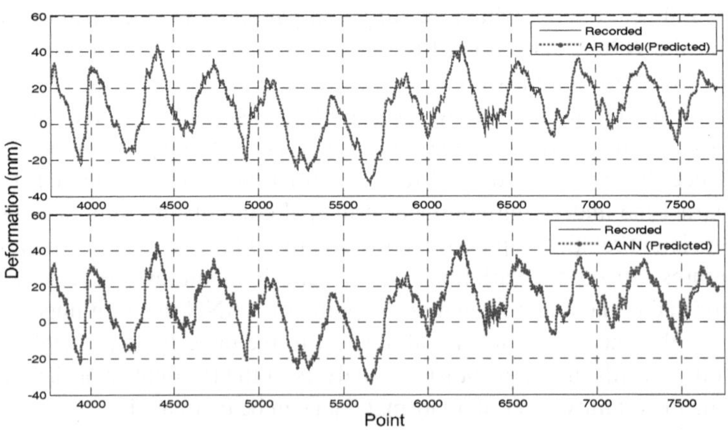

Fig. 7 Comparison with the recorded displacement and the estimated displacement by using (1) SSA-AR model and (2) NPCA-AANN method. (NPL2Y172.5)

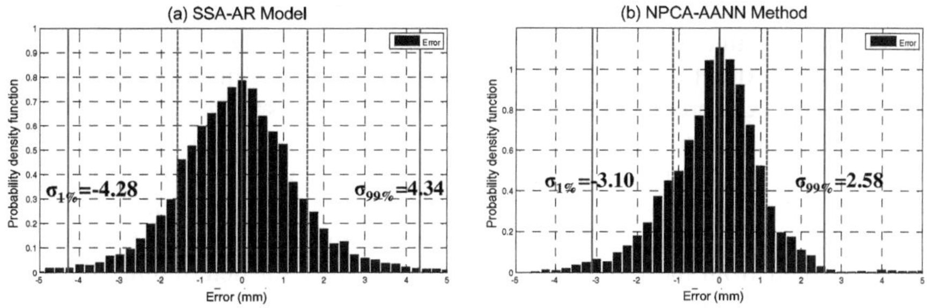

Fig. 8 Probability density function (histogram) of residual errors (NPL2Y172.5)

Table 1 Comparison the estimated standard deviation of deformation residual by using SSA-AR model and NPCA-AANN method

	SSA-AR Model			AANN-method (without "T" and "D")			AANN-method (with "T" and "D")	
	σ (mm)	σ_{upper} (mm)	σ_{lower} (mm)	σ (mm)	σ_{upper} (mm)	σ_{lower} (mm)	σ_{upper} (mm)	σ_{lower} (mm)
NP2Y172.5	**1.59**	4.34	-4.28	**1.15**	2.58	-3.18	4.86	-3.94
NPL2Y150	**1.24**	3.44	-3.6	**1.20**	3.18	-2.76	4.31	-3.45
NPL2Y115	**0.72**	2.07	-2.12	**0.78**	1.98	-1.85	2.03	-1.89
NPL2Y90	**0.43**	1.26	-1.33	**0.48**	1.43	-1.28	1.64	-1.42
NPL2Y57.5	**0.14**	0.37	-0.41	**0.20**	0.53	-0.53	0.86	-0.88

Note: "T" indicates temperature and "D" indicates water level.

5 Conclusions

In this study, the input/output subspace identification is applied to seismic response data to identify the modal properties of the Fei-Tsui arch dam. The dam is considered as a lumped mass system with small damping. In order to compare the accurately of multi-input and single-input system, the result from the forced vibration test [4] was also plotted as the star in Figure 4. It shows that the identified modal frequencies are closer to the result from the forced vibration test than multi-input model under ambient vibration with specific frequency. Based on these results, the regression curves between the identified system natural frequencies and water level can be used to monitor the health condition of the dam. If the identified modal frequencies are out of the confidence interval based on the regression curves, an early warning signal can be sent for close inspection of the structure.

Besides, SSA-AR Model and NPCA-AANN method were used to extract long-term trends of structural health monitoring data in this study. The SSA is used to decompose original time series into principle components and AR model is optimized for each PC. Then, the multi-step predicted values are recombined to make the time series. However, it is necessary to decide the number of principle component to be extracted from the trajectory matrix. In this study, it is suggested that 99% norm value of the trajectory matrix retained. The NPCA-AANN method is also used to extract the underlying features of time series, and then a prediction model for NPCA is proposed to estimate the abnormal response data. Besides, not only the static deformation data were used, but also the temperature and water level data were considered as input by using NPCA-AANN method in this paper. The σ_{lower} and σ_{lower} shown in Table 1 are different with and without considering the effect of environmental conditions. The probability distribution of the residual error generated from the trained data can also be used for determining the structural health monitoring threshold value. If the predicted deformation value from the proposed method larger than the prescribed threshold value, an early warning signal can be sent for close inspection of the structure.

References

1. Loh, C.H., Tou, I.C.: A system identification approach to the detection of changes in both linear and non-linear structural parameters. Earthquake Engineering & Structural Dynamics 24(1), 85–97 (1995)
2. Safak, E.: Adaptive modeling, identification, and control of dynamic structural systems. I. Theory. Journal of Engineering Mechanics 115(11), 2386–2405 (1989)
3. Loh, C.H., Wu, T.S.: Identification of Fei-Tsui arch dam from both ambient and seismic response data. Soil Dynamics and Earthquake Engineering 15, 465–483 (1996)
4. Loh, C.H., Wu, T.C.: System identification of Fei-Tsui arch dam from forced vibration and seismic response data. J. Earthquake Engineering 4(4), 511–537 (2000)
5. Hsu, T.Y., Loh, C.H.: Damage Detection Accommodating Nonlinear Environmental Effects by Nonlinear Principal Component Analysis. Structural Control & Health Monitoring (2009) (accepted for Publication)
6. Golyandina, N.: Analysis of time series structure: SSA and related techniques. Chapman & Hall/CRC, Boca Raton (2001)
7. Van Overschee, P., De Moor, B.: Subspace identification for linear systems: Theory - Implementation – Applications. Kluwer, Dordrecht (1996)
8. Chopra, A.K.: Dynamics of structures – Theory and Applications to Earthquake Engineering, 2nd edn., pp. 384–388. Prentice Hall, New Jersey (2001)
9. Hossein Vahabie, A., Mahdi RezaeiYousefi, M., Araabi, B.N., Lucas, C., Barghinia, S.: Combination of singular spectrum analysis and autore-gressive model for short term load forecasting. In: Proc. 2007 IEEE Lausanne POWERTECH, pp. 1090–1093 (2007)
10. Kramer, M.A.: Nonlinear principal component analysis using auto-associative neural networks. AIChE Journal 37, 233–243 (1991)

About Some Schemes of Study for Systems with Different Forms of Time Aftereffect

V.V. Malanin and I.E. Poloskov

Faculty of Mechanics and Mathematics, Perm State University,
Perm, 614990, GSP, Russia

Abstract. One of types of stochastic retarded systems is under consideration. Our scheme of analysis is applicable for investigation of linear and nonlinear differential difference equations with single and multiple constant delays, linear differential equations with variable delays, linear neutral delay differential equations, and separate linear differential difference equations. In addition, a problem of sensitivity estimation for linear dynamic systems described by stochastic differential difference equations can be explored too. All these schemes are based on extensions of phase spaces.

Keywords: Stochastic system, Differential equations, Discrete and continuous delays, Moment functions, Symbolic and numeric calculations.

1 Introduction

Functional differential equations (FDE) and their special forms such as differential difference equations (DDE), neutral delay differential equations (NDDE), integro-differential equations (IDE), etc. [3,5] have been attracting an increased interest both from theoretical and practical viewpoints since the middle of the last century. Such equations are encountered in those areas where the properties of an object depend on the hereditary effect, and serve as models for different processes, viz., retarded mechanical vibrations in engineering structures, automatic control for technical processes, development of economic and social systems, combustion in liquid jet engines, neutron moderation, effects of radiations, a radio-location, radar and radio-navigation, autonomous vessel course stabilization, oscillations in vacuum-tube generators, struggle for survival in biology, etc. [6,9,13].

Such phenomena arise as a result of deterministic, stochastic, transport, technological, information, inertial and other forms of delays (in long-distance transmission of matter, energy, signals, information), finiteness of speed of charge carriers, and a lag of response delay in man-machine systems. Delays in systems induce new effects, for example, self-excitation of oscillations, increased overcontrol, instability of objects, etc.

As developments of methods for deterministic systems have become important for theory and practice as nowadays significant interest is paid to stochastic FDE (SFDE) of various types (SDDE, SNDDE, SIDE) [10,11].

W.Q. Zhu et al. (Eds.): Nonlinear Stochastic Dynamics & Control, IUTAM BOOK SERIES 29, pp. 55–64.
springerlink.com

An analysis of such systems causes considerable difficulties. Therefore only few areas are known now where developments are produced. Among them there are qualitative researches of existence and stability of SFDE solutions, exact results for linear systems on the base of the method of steps, approximate techniques including Monte Carlo methods, a usage of averaging schemes with respect to small delays, numeric integrators, etc. [1,2,8].

Our scheme for study of such systems is based on an extension of the phase space [7]. We apply this scheme to linear SNDDE with multiple constant delays (Section 2). Examples (Section 3) show the scheme afoot. A tool in our calculations is the computer algebra package *Mathematica* [12], a well-known powerful instrument for different sciences.

2 Neutral Systems with Multiple Constant Delays

Let us consider a linear system of SNDDE

$$\begin{aligned} \boldsymbol{x}'(t) = P_k(t)\,\boldsymbol{x}(t) + \sum_{\nu=1}^{k}\left[Q_{k\nu}(t)\,\boldsymbol{x}(t-\nu\tau) + R_{k\nu}(t)\,\boldsymbol{x}'(t-\nu\tau)\right] + \\ + \boldsymbol{c}_k(t) + H_k(t)\,\boldsymbol{\xi}(t), \quad t > t_k = t_0 + k\tau. \end{aligned} \tag{1}$$

Here $\boldsymbol{x} = \{x_i\} \in \mathbf{R}^n$ is the phase vector, $\boldsymbol{\xi} = \{\xi_i\} \in \mathbf{R}^m$ is a vector of independent Gaussian white noises with $\mathbf{M}[\boldsymbol{\xi}(t)] = 0$, $\mathbf{M}[\boldsymbol{\xi}(t)\, \boldsymbol{\xi}^T(t')] = E\cdot\delta(t-t')$, τ is a constant delay, $k > 0$ is an integer, $\boldsymbol{c}_k = \{c_{ki}\}$, $P_k = \{p_{kij}\}$, $Q_{k\nu} = \{q_{k\nu ij}\}$, $R_{k\nu} = \{r_{k\nu ij}\}$, $H_k = \{h_{kij}\}$ are deterministic vector- and matrix-functions of t, T is a symbol of the transposition, $\mathbf{M}$ stands for the mathematical expectation, $'$ is a symbol of differentiation with respect to t, E is the identity matrix.

We suppose that on the intervals $(t_0, t_1]$, $(t_1, t_2]$, ..., $(t_{k-1}, t_k]$ the phase vector $\boldsymbol{x}$ satisfies the following systems of stochastic differential equations

$$\begin{aligned} \boldsymbol{x}'(t) = P_\mu(t)\,\boldsymbol{x}(t) + \sum_{\nu=1}^{\mu}\left[Q_{\mu\nu}(t)\,\boldsymbol{x}(t-\nu\tau) + R_{\mu\nu}(t)\,\boldsymbol{x}'(t-\nu\tau)\right] + \\ + \boldsymbol{c}_\mu(t) + H_\mu(t)\,\boldsymbol{\xi}(t), \quad t \in (t_\mu, t_{\mu+1}], \quad \boldsymbol{x}(t_0) = \boldsymbol{y}^0. \end{aligned} \tag{2}$$

Let's elements of the mean values vector $\boldsymbol{m}(t) = \mathbf{M}[\boldsymbol{x}(t)]$ and components of the covariance matrix $D(t) = \mathbf{M}[\{\boldsymbol{x}(t) - \boldsymbol{m}(t)\}\{\boldsymbol{x}(t) - \boldsymbol{m}(t)\}^T]$ for the phase vector $\boldsymbol{x}(t)$ are defined for $t = t_0$:

$$\boldsymbol{m}(t_0) = \boldsymbol{m}^0, \quad D(t_0) = D^0. \tag{3}$$

The main task of this research is to obtain a set of ordinary differential equations (ODE) satisfied by elements of the vector $\boldsymbol{m}(t)$ and components of the matrix $D(t)$ for $t > t_0$.

2.1 Scheme of Study

To derive these equations, we expand the phase space of the system and transform a non-Markovian vector process to a Markovian one. For this purpose we introduce the following notation

$$\begin{aligned}
&s\in[0,\tau],\quad \Delta_k=(t_k,t_{k+1}],\quad k=0,1,2,\ldots,\quad \boldsymbol{x}_k(s)=\boldsymbol{x}(s_k),\\
&z_0=\boldsymbol{x}_0,\quad z_1=\mathrm{col}(\boldsymbol{x}_0,\boldsymbol{x}_1),\quad z_2=\mathrm{col}(\boldsymbol{x}_0,\boldsymbol{x}_1,\boldsymbol{x}_2),\quad \ldots,\quad s_k=t_k+s,\\
&\boldsymbol{y}_k\equiv\boldsymbol{x}_k(0)=\boldsymbol{x}_{k-1}(\tau),\quad \boldsymbol{\xi}_k(s)=\boldsymbol{\xi}(s_k),\quad \boldsymbol{\xi}_k(0)=\boldsymbol{\xi}_{k-1}(\tau),\\
&\mathrm{col}(\boldsymbol{x}_0,\boldsymbol{x}_1,\ldots,\boldsymbol{x}_{N-1},\boldsymbol{x}_N)=\{x_{01},x_{02},\ldots,x_{0n},x_{11},x_{12},\ldots,x_{1n},\ldots,\\
&\qquad x_{N-1,1},x_{N-1,2},\ldots,x_{N-1,n},x_{N1},x_{N2},\ldots,x_{Nn}\}^T.
\end{aligned}\tag{4}$$

Using this notation, we construct a chain of sets of ODE for the mean values vectors and the covariance matrices of the vectors z_0, z_1, z_2, …, z_N, … belonging to the family of embedded phase spaces $\mathbf{R}^n\subset\mathbf{R}^{2n}\subset\mathbf{R}^{3n}\subset\ldots\subset\mathbf{R}^{(N+1)n}\subset\ldots$.

Let's consider a sequence of segments $\{\Delta_i\}$.

0°. Let's start from the segment Δ_0. The random vector $\boldsymbol{x}_0(s)$ defined on Δ_0 satisfies the system

$$\boldsymbol{x}_0'(s)=P_0(s_0)\boldsymbol{x}_0(s)+\boldsymbol{c}_0(s_0)+H_0(s_0)\boldsymbol{\xi}_0(s).$$

Here and below symbol ′ stands for differentiation with respect to s. Therefore

$$\begin{aligned}
&z_0'(s)=\overline{P}_0(s)z_0(s)+\overline{\boldsymbol{f}}_0(s)+\overline{H}_0(s)\boldsymbol{\eta}_0(s),\\
&z_0(s)=\boldsymbol{x}_0(s),\quad \boldsymbol{\eta}_0(s)=\boldsymbol{\xi}_0(s),\quad \overline{P}_0(s)=P_{00}(s)=P_0(s_0),\\
&\overline{\boldsymbol{f}}_0(s)=\boldsymbol{f}_0(s)=\boldsymbol{c}_0(s_0),\quad \overline{H}_0(s)=H_{00}(s)=H_0(s_0).
\end{aligned}\tag{5}$$

1°. Let's consider the intervals Δ_0 and Δ_1. It is possible to present the system of SDE for calculation of the vector $\mathrm{col}(\boldsymbol{x}_0(s),\boldsymbol{x}_1(s))$ as follows

$$\begin{aligned}
&\boldsymbol{x}_0'(s)=P_0(s_0)\boldsymbol{x}_0(s)+\boldsymbol{c}_0(s_0)+H_0(s_0)\boldsymbol{\xi}_0(s),\\
&\boldsymbol{x}_1'(s)=Q_{11}(s_1)\boldsymbol{x}_0(s)+P_1(s_1)\boldsymbol{x}_1(s)+R_{11}(s_1)\boldsymbol{x}_0'(s)+\boldsymbol{c}_1(s_1)+H_1(s_1)\boldsymbol{\xi}_1(s)
\end{aligned}$$

Then we have got

$$z_1'(s)=\overline{P}_1(s)z_1(s)+\overline{\boldsymbol{f}}_1(s)+\overline{H}_1(s)\boldsymbol{\eta}_1(s),\tag{6}$$

$$z_1(s)=\begin{bmatrix}\boldsymbol{x}_0(s)\\ \boldsymbol{x}_1(s)\end{bmatrix},\quad \boldsymbol{\eta}_1(s)=\begin{bmatrix}\boldsymbol{\xi}_0(s)\\ \boldsymbol{\xi}_1(s)\end{bmatrix},\quad \bar{\boldsymbol{f}}_1(s)=\begin{bmatrix}\boldsymbol{f}_0(s)\\ \boldsymbol{f}_1(s)\end{bmatrix},$$
$$\bar{P}_1(s)=\begin{bmatrix}P_{00}(s) & 0\\ P_{10}(s) & P_{11}(s)\end{bmatrix},\quad \bar{H}_1(s)=\begin{bmatrix}H_{00}(s) & 0\\ H_{10}(s) & H_{11}(s)\end{bmatrix},$$
$$\boldsymbol{f}_1(s)=\boldsymbol{c}_1(s_1)+R_{11}(s_1)\boldsymbol{f}_0(s),$$
$$P_{10}(s)=Q_{11}(s_1)+R_{11}(s_1)P_{00}(s),\quad P_{11}(s)=P_1(s_1),$$
$$H_{10}(s)=R_{11}(s_1)H_{00}(s),\quad H_{11}(s)=H_1(s_1).$$

....

$\boldsymbol{k}^{\mathrm{o}}$. Now let's pay attention to the time intervals Δ_0, Δ_1, Δ_2 , …, Δ_k and construct the set of SDE for the vector col($\boldsymbol{x}_0(s)$, $\boldsymbol{x}_1(s)$, .., $\boldsymbol{x}_k(s)$) by the way

$$\boldsymbol{x}_0'(s)=P_0(s_0)\boldsymbol{x}_0(s)+\boldsymbol{c}_0(s_0)+H_0(s_0)\boldsymbol{\xi}_0(s),$$
$$\boldsymbol{x}_1'(s)=Q_{11}(s_1)\boldsymbol{x}_0(s)+P_1(s_1)\boldsymbol{x}_1(s)+R_{11}(s_1)\boldsymbol{x}_0'(s)+\boldsymbol{c}_1(s_1)+H_1(s_1)\boldsymbol{\xi}_1(s),$$

....

$$\boldsymbol{x}_k'(s)=\sum_{\mu=0}^{k-1}Q_{k,\mu+1}(s_k)\boldsymbol{x}_\mu(s)+P_k(s_k)\boldsymbol{x}_k(s)+\sum_{\mu=0}^{k-1}R_{k,\mu+1}(s_k)\boldsymbol{x}_\mu'(s)+$$
$$+\boldsymbol{c}_k(s_k)+H_k(s_k)\boldsymbol{\xi}_k(s).$$

These equations can be converted to the form

$$z_k'(s)=\bar{P}_k(s)z_k(s)+\bar{\boldsymbol{f}}_k(s)+\bar{H}_k(s)\boldsymbol{\eta}_k(s), \tag{7}$$
$$z_k(s)=\begin{bmatrix}z_{k-1}(s)\\ \boldsymbol{x}_k(s)\end{bmatrix},\quad \boldsymbol{\eta}_k(s)=\begin{bmatrix}\boldsymbol{\eta}_{k-1}(s)\\ \boldsymbol{\xi}_k(s)\end{bmatrix},\quad \bar{\boldsymbol{f}}_k(s)=\begin{bmatrix}\bar{\boldsymbol{f}}_{k-1}(s)\\ \boldsymbol{f}_k(s)\end{bmatrix},$$
$$\bar{P}_k(s)=\begin{bmatrix}\bar{P}_{k-1}(s) & 0\\ \tilde{P}_k(s) & P_k(s_k)\end{bmatrix},\quad \bar{H}_k(s)=\begin{bmatrix}\bar{H}_{k-1}(s) & 0\\ \tilde{H}_k(s) & H_k(s_k)\end{bmatrix},$$
$$\boldsymbol{f}_k(s)=\boldsymbol{c}_k(s_k)+\sum_{\mu=0}^{k-1}R_{k,\mu+1}(s_k)\boldsymbol{f}_\mu(s),$$
$$\tilde{P}_k(s)=\begin{bmatrix}Q_{k1}(s_k)+\sum_{\mu=0}^{k-1}R_{k,\mu+1}(s_k)P_{\mu 0}(s)\\ Q_{k2}(s_k)+\sum_{\mu=1}^{k-1}R_{k,\mu+1}(s_k)P_{\mu 1}(s)\\ \cdots\cdots\cdots\cdots\\ Q_{kk}(s_k)+R_{kk}(s_k)P_{k-1,k-1}(s)\end{bmatrix}^T,\quad \tilde{H}_k(s)=\begin{bmatrix}\sum_{\mu=0}^{k-1}R_{k,\mu+1}(s_k)H_{\mu 0}(s)\\ \sum_{\mu=1}^{k-1}R_{k,\mu+1}(s_k)H_{\mu 1}(s)\\ \cdots\cdots\cdots\\ R_{kk}(s_k)H_{k-1,k-1}(s)\end{bmatrix}^T.$$

....

$\boldsymbol{N}^{\mathrm{o}}$. Similarly for the vector col($\boldsymbol{x}_0(s)$, $\boldsymbol{x}_1(s)$, …, $\boldsymbol{x}_k(s)$, …, $\boldsymbol{x}_N(s)$) we obtain equations

$$\begin{aligned}
\boldsymbol{x}_0'(s) &= P_0(s_0)\,\boldsymbol{x}_0(s) + \boldsymbol{c}_0(s_0) + H_0(s_0)\boldsymbol{\xi}_0(s),\\
\boldsymbol{x}_1'(s) &= Q_{11}(s_1)\boldsymbol{x}_0(s) + P_1(s_1)\,\boldsymbol{x}_1(s) +\\
&\qquad R_{11}(s_1)\,\boldsymbol{x}_0'(s) + \boldsymbol{c}_1(s_1) + H_1(s_1)\boldsymbol{\xi}_1(s),
\end{aligned}$$

....

$$\begin{aligned}
\boldsymbol{x}_k'(s) &= \sum_{\mu=0}^{k-1} Q_{k,\mu+1}(s_k)\boldsymbol{x}_\mu(s) + P_k(s_k)\,\boldsymbol{x}_k(s) +\\
&\qquad \sum_{\mu=0}^{k-1} R_{k,\mu+1}(s_k)\boldsymbol{x}_\mu'(s) + \boldsymbol{c}_k(s_k) + H_k(s_k)\boldsymbol{\xi}_k(s),\\
\boldsymbol{x}_{k+1}'(s) &= \sum_{\mu=0}^{k-1} Q_{k,\mu+1}(s_{k+1})\boldsymbol{x}_{\mu+1}(s) + P_k(s_{k+1})\,\boldsymbol{x}_{k+1}(s) +\\
&\qquad \sum_{\mu=0}^{k-1} R_{k,\mu+1}(s_k)\boldsymbol{x}_{\mu+1}'(s) + \boldsymbol{c}_k(s_{k+1}) + H_k(s_{k+1})\boldsymbol{\xi}_{k+1}(s),
\end{aligned}$$

....

$$\begin{aligned}
\boldsymbol{x}_N'(s) &= \sum_{\mu=0}^{k-1} Q_{k,\mu+1}(s_N)\boldsymbol{x}_{N-k+\mu}(s) + P_k(s_N)\,\boldsymbol{x}_N(s) +\\
&\qquad \sum_{\mu=0}^{k-1} R_{k,\mu+1}(s_N)\boldsymbol{x}_{N-k+\mu}'(s) + \boldsymbol{c}_k(s_N) + H_k(s_N)\boldsymbol{\xi}_N(s)
\end{aligned}$$

and their more suitable form

$$\begin{aligned}
&\boldsymbol{z}_N'(s) = \overline{P}_N(s)\boldsymbol{z}_N(s) + \overline{\boldsymbol{f}}_N(s) + \overline{H}_N(s)\boldsymbol{\eta}_N(s), \qquad (8)\\
&\boldsymbol{z}_N(s) = \begin{bmatrix} \boldsymbol{z}_{N-1}(s) \\ \boldsymbol{x}_N(s) \end{bmatrix}, \quad \boldsymbol{\eta}_N(s) = \begin{bmatrix} \boldsymbol{\eta}_{N-1}(s) \\ \boldsymbol{\xi}_N(s) \end{bmatrix}, \quad \overline{\boldsymbol{f}}_N(s) = \begin{bmatrix} \overline{\boldsymbol{f}}_{N-1}(s) \\ \boldsymbol{f}_N(s) \end{bmatrix},\\
&\overline{P}_N(s) = \begin{bmatrix} \overline{P}_{N-1}(s) & 0 \\ \tilde{P}_N(s) & P_k(s_N) \end{bmatrix}, \quad \overline{H}_N(s) = \begin{bmatrix} \overline{H}_{N-1}(s) & 0 \\ \tilde{H}_N(s) & H_k(s_N) \end{bmatrix},\\
&\boldsymbol{f}_N(s) = \boldsymbol{c}_k(s_N) + \sum_{\mu=0}^{k-1} R_{k,\mu+1}(s_N)\boldsymbol{f}_{N-k+\mu}(s),\\
&\tilde{P}_N(s) = \begin{bmatrix} Q_{k1}(s_N) + \sum_{\mu=0}^{k-1} R_{k,\mu+1}(s_N)P_{N-k+\mu,N-k}(s) \\ Q_{k2}(s_N) + \sum_{\mu=1}^{k-1} R_{k,\mu+1}(s_N)P_{N-k+\mu,N-k+1}(s) \\ \cdots\cdots\cdots\cdots\cdots\cdots\cdots \\ Q_{kk}(s_N) + R_{kk}(s_N)P_{N-1,N-1}(s) \end{bmatrix}^T,
\end{aligned}$$

$$\tilde{H}_N(s) = \begin{bmatrix} \sum_{\mu=0}^{k-1} R_{k,\mu+1}(s_N) H_{N-k+\mu,N-k}(s) \\ \sum_{\mu=1}^{k-1} R_{k,\mu+1}(s_N) H_{N-k+\mu,N-k+1}(s) \\ \dots\dots\dots\dots\dots\dots \\ R_{kk}(s_N) H_{N-1,N-1}(s) \end{bmatrix}^T .$$

2.2 Equations for Moments

Now the chain of linear SDE without delay can be used to derive new sequence of equations, i.e., ODE for the first moments of the vectors z_0, z_1, z_2, …, z_N, … and

$$z_0^+ = \text{col}(\boldsymbol{y}^0, z_0(s)), \quad z_1^+ = \text{col}(\boldsymbol{y}^0, z_1(s)), \quad ..., \quad z_N^+ = \text{col}(\boldsymbol{y}^0, z_N(s)), \quad ...$$

Let's denote

$$\begin{aligned} &\boldsymbol{m}_k(s) = \mathbf{M}[z_k] = \text{col}(\boldsymbol{m}_{x_0}, \boldsymbol{m}_{x_1}, ..., \boldsymbol{m}_{x_k}), \\ &\boldsymbol{m}_k^+(s) = \mathbf{M}[z_k^+] = \text{col}(\boldsymbol{m}^0, \boldsymbol{m}_k(s)), \\ &D_k(s) = \mathbf{M}\left[(z_k - \boldsymbol{m}_k)(z_k - \boldsymbol{m}_k)^T\right] = \begin{bmatrix} D_{x_0x_0} & D_{x_0x_1} & \dots & D_{x_0x_k} \\ D_{x_1x_0} & D_{x_1x_1} & \dots & D_{x_1x_k} \\ \dots & \dots & \dots & \dots \\ D_{x_kx_0} & D_{x_kx_1} & \dots & D_{x_kx_k} \end{bmatrix}, \end{aligned} \tag{9}$$

$$D_k^+(s) = \begin{bmatrix} D_{y_0y_0} & D_{y_0x_0} & D_{y_0x_1} & \dots & D_{y_0x_k} \\ D_{x_0y_0} & D_{x_0x_0} & D_{x_0x_1} & \dots & D_{x_0x_k} \\ D_{x_1y_0} & D_{x_1x_0} & D_{x_1x_1} & \dots & D_{x_1x_k} \\ \dots & \dots & \dots & \dots & \dots \\ D_{x_ky_0} & D_{x_kx_0} & D_{x_kx_1} & \dots & D_{x_kx_k} \end{bmatrix}, \quad k = 0,1,2,...,N,...$$

Obviously, the vector $\boldsymbol{m}_k(s)$ and the matrix $D_k(s)$ are blocks of $\boldsymbol{m}_k^+(s)$ and $D_k^+(s)$. Therefore it is sufficiently to calculate the last functions and then to choose their required components.

Using equations (4)-(7) and notations (8), the necessary ODE can be presented as follows:

$$\begin{aligned} &\boldsymbol{m}_0^{+\prime}(s) = \overline{P}_0^+(s)\boldsymbol{m}_0^+(s) + \overline{\boldsymbol{f}}_0^+(s), \\ &D_0^{+\prime}(s) = \overline{P}_0^+(s) D_0^+(s) + \left[\overline{P}_0^+(s) D_0^+(s)\right]^T + \overline{H}_0^+(s)\overline{H}_0^{+T}(s), \end{aligned} \tag{10}$$

$$\boldsymbol{m}_0^+(0) = \begin{bmatrix} \boldsymbol{m}^0 \\ \boldsymbol{m}^0 \end{bmatrix}, \quad D_0^+(0) = \begin{bmatrix} D^0 & D^0 \\ D^0 & D^0 \end{bmatrix},$$

where

$$\overline{P}_0^+ = \begin{bmatrix} 0 & 0 \\ 0 & \overline{P}_0 \end{bmatrix}, \quad \overline{\boldsymbol{f}}_0^+ = \begin{bmatrix} 0 \\ \overline{\boldsymbol{f}}_0 \end{bmatrix}, \quad \overline{H}_0^+ = \begin{bmatrix} 0 & 0 \\ 0 & \overline{H}_0 \end{bmatrix}.$$

Similar equations for the vector z_1^+ take the form

$$\begin{aligned} \boldsymbol{m}_1^{+\prime}(s) &= \overline{P}_1^+(s)\boldsymbol{m}_1^+(s) + \overline{\boldsymbol{f}}_1^+(s), \\ D_1^{+\prime}(s) &= \overline{P}_1^+(s)\,D_1^+(s) + \left[\overline{P}_1^+(s)\,D_1^+(s)\right]^T + \overline{H}_1^+(s)\overline{H}_1^{+T}(s), \end{aligned} \tag{11}$$

$$\boldsymbol{m}_1^+(0) = \begin{bmatrix} \boldsymbol{m}^0 \\ \boldsymbol{m}^0 \\ \boldsymbol{m}_{x_0}(\tau) \end{bmatrix}, \quad D_1^+(0) = \begin{bmatrix} D^0 & D^0 & D_{y_0x_0}(\tau) \\ D^0 & D^0 & D_{y_0x_0}(\tau) \\ D_{x_0y_0}(\tau) & D_{x_0y_0}(\tau) & D_{x_0x_0}(\tau) \end{bmatrix},$$

$$\overline{P}_1^+ = \begin{bmatrix} 0 & 0 \\ 0 & \overline{P}_1 \end{bmatrix}, \quad \overline{\boldsymbol{f}}_1^+ = \begin{bmatrix} 0 \\ \overline{\boldsymbol{f}}_1 \end{bmatrix}, \quad \overline{H}_1^+ = \begin{bmatrix} 0 & 0 \\ 0 & \overline{H}_1 \end{bmatrix}.$$

....

At last, for characteristics of the vector z_N^+ we have

$$\begin{aligned} \boldsymbol{m}_N^{+\prime}(s) &= \overline{P}_N^+(s)\boldsymbol{m}_N^+(s) + \overline{\boldsymbol{f}}_N^+(s), \\ D_N^{+\prime}(s) &= \overline{P}_N^+(s)\,D_N^+(s) + \left[\overline{P}_N^+(s)\,D_N^+(s)\right]^T + \overline{H}_N^+(s)\overline{H}_N^{+T}(s), \end{aligned} \tag{12}$$

$$\boldsymbol{m}_N^+(0) = \begin{bmatrix} \boldsymbol{m}_{N-1}^+(0) \\ \boldsymbol{m}_{x_{N-1}}(\tau) \end{bmatrix}, \quad D_N^+(0) = \begin{bmatrix} D_{N-1}^+(0) & D_{y_0x_{N-1}}(\tau) \\ D_{x_{N-1}y_0}(\tau) & D_{x_{N-1}x_{N-1}}(\tau) \end{bmatrix},$$

$$\overline{P}_N^+ = \begin{bmatrix} 0 & 0 \\ 0 & \overline{P}_N \end{bmatrix}, \quad \overline{\boldsymbol{f}}_N^+ = \begin{bmatrix} 0 \\ \overline{\boldsymbol{f}}_N \end{bmatrix}, \quad \overline{H}_N^+ = \begin{bmatrix} 0 & 0 \\ 0 & \overline{H}_N \end{bmatrix}.$$

3 Examples

The first system under investigation was as follows:

$$\begin{aligned} &x'(t) = p\,x(t) + q\,x(t-\tau) + r\,x'(t-\tau) + h\xi(t), \quad t > 0; \\ &x'(t) = 0, \quad t \le 0; \quad p = q = r = h = \text{const}. \end{aligned} \tag{13}$$

Figs. 1 and 2 show behavior of the mean value and the variance of $x(t)$. Values of parameters were

$$r=-1,\quad h=0.2,\quad \tau=0.5,\quad t_0=-\tau,\quad m^0=5,\quad D^0=0.25,$$
$$1)\ p=-1,\ q=0.4;\quad 2)\ p=-1.25,\ q=0.4;\quad 3)\ p=-1.25,\ q=0.2.$$

The cases 1, 2, 3 are represented by dashed, dot-and-dashed and continuous lines respectively.

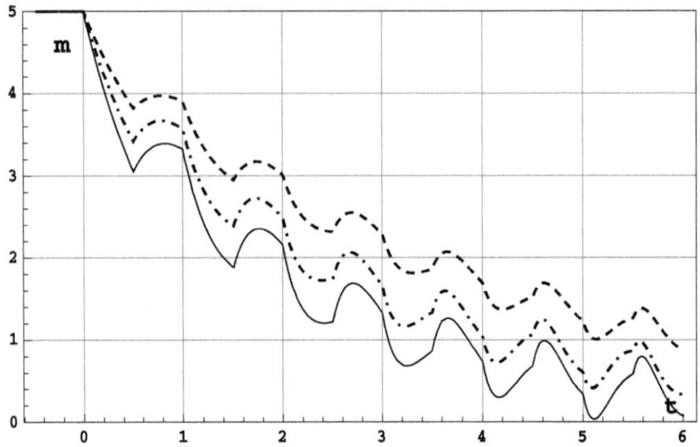

Fig. 1 The mean value of $x(t)$ in the system (12)

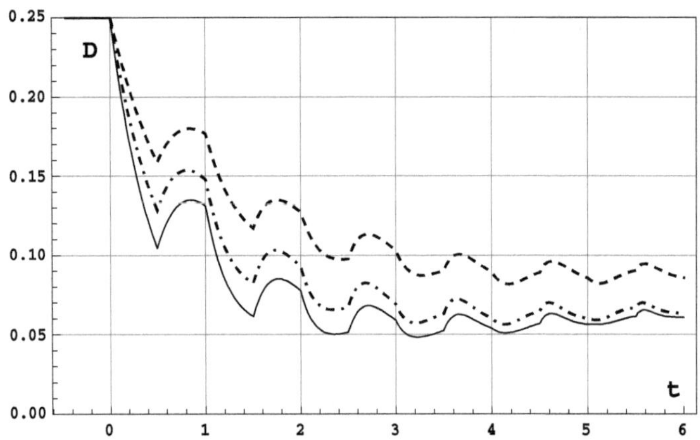

Fig. 2 The variance of $x(t)$ in the system (12)

The second example demonstrates as the scheme was applied to study a system in the form

$$\begin{aligned}&\ddot{x}(t)+\gamma\ddot{x}(t-\tau)+2\alpha_1\,\dot{x}(t)+2\alpha_2\,\dot{x}(t-\tau)+\\&\qquad\qquad\beta_1\,x(t)+\beta_2\,x(t-\tau)=\sigma\,\xi(t),\quad t>0;\\&\ddot{x}(t)=0,\quad t\le 0;\quad \gamma,\alpha_1,\alpha_2,\beta_1,\beta_2,\sigma=\text{const.}\end{aligned} \tag{14}$$

Figs. 3 and 4 display a time evolution of the first moments of the phase vector $\{x_1, x_2\} = \{x, x'\}$ for the following values:

$$\alpha_1 = 0.125, \quad \alpha_2 = 0.167, \quad \beta_1 = 3, \quad \beta_2 = 2,$$
$$\gamma = 0.75, \quad \sigma = 0.1, \quad \tau = 0.5, \quad t_0 = -\tau.$$

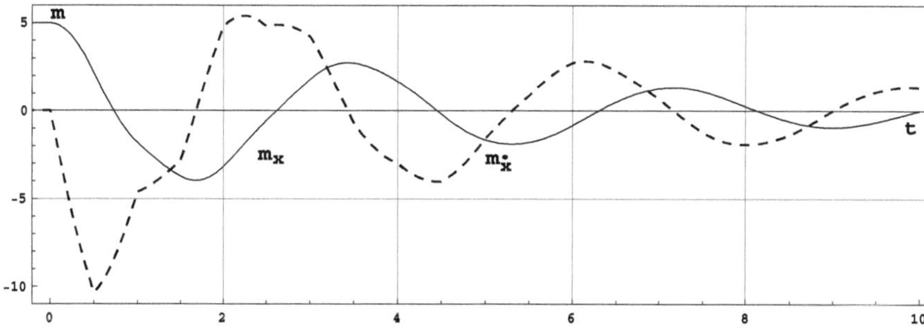

Fig. 3 The mean values of $\{x(t), x'(t)\}$ in the system (13)

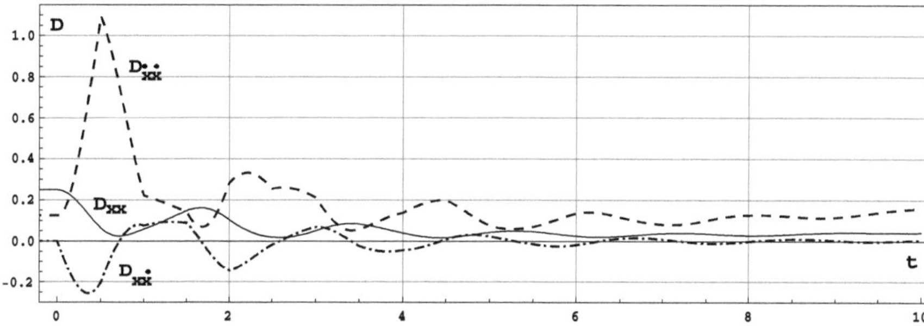

Fig. 4 The covariances of $\{x(t), x'(t)\}$ in the system (13)

4 Conclusions

In the paper the scheme of the phase space extension to study linear SNDDE is described. Moreover this scheme was successfully applied to analyze linear and nonlinear differential difference equations with single and multiple constant delays, linear differential equations with variable delays, a problem of sensitivity estimation for linear dynamic systems described by SDDE, and linear system excited by continuous and discrete fluctuations and constant delay. In contrast to a number of known techniques [4] our scheme doesn't require to change equations with the aim of delay withdrawal and special numeric integration methods, has a transparent algorithm and can be simply combined with Monte Carlo method in case of complicated nonlinear problems.

This work is supported by RFBR grant No.09-01-99006.

References

[1] Allen, E.: Modeling with Ito Stochastic Differential Equations. Springer, Dordrecht (2007)
[2] Baker, C.T.H., Buckwar, E.: Numerical analysis of explicit one-step method for stochastic delay differential equations. London Mathematical Society, Journal of Comput. Math. 3, 315–335 (2000)
[3] Bellman, R.E., Cooke, K.L.: Differential-Difference Equations. Academic, New York (1962)
[4] Elbeyli, O., Sun, J.Q., Ünal, G.: A semi-discretization method for delayed stochastic systems. Communications in Nonlinear Science and Numerical Simulation 10, 85–94 (2005)
[5] Hale, J.: Theory of Functional Differential Equations. Springer, New York (1977)
[6] Kuang, Y.: Delay Differential Equations with Applications in Population Dynamics. Academic Press, Boston (1993)
[7] Poloskov, I.E.: Phase space extension in the analysis of differential-difference systems with random input. Automation and Remote Control 63(9), 1426–1438 (2002)
[8] Rubanik, V.P.: Vibrations of compound quasi-linear systems with delays. Universitetskoe, Minsk (1985) (in Russian)
[9] Stépán, G.: Retarded Dynamical Systems: Stability and Characteristic Functions. Longman Scientific & Technical, Harlow (1989)
[10] Tsar'kov, E.F.: Random Fluctuations of Functional Differential Equations. Zinatne, Riga (1989) (in Russian)
[11] Tsokos, P.C., Padgett, W.J.: Random Integral Equations with Applications to Stochastic Systems. Springer, Berlin (1971)
[12] Wolfram, S.: The Mathematica Book, 5th edn. Wolfram Media, Champaign, Il (2003)
[13] Liao, X., Wang, L.Q., Yu, P.: Stability of Dynamical Systems. Elsevier, Amsterdam (2007)

Reliability of Linear and Nonlinear Dynamic Systems under Multiplicative and Additive Noise

A. Naess[1], D. Iourtchenko[2], and O. Batsevych[1]

[1] Norwegian University of Science and Technology, Trondheim, Norway
[2] Department of Mathematical Sciences, SPBSTU, St. Petersburg, Russia

Abstract. The paper considers a first passage time reliability problem for systems subjected to multiplicative and additive white noises. For numerical calculations of the reliability function and the first passage time the path integration method is properly adapted and used. Some results of numerical calculations are compared to approximate analytical results, obtained by the stochastic averaging method.

Keywords: Probability density function, path integration, reliability, upcrossing rate.

1 Introduction

Among different reliability criteria the *first passage problem* [1] is widely used and studied by a number of authors. It may be defined as the probability that a system's response stays within a prescribed domain, an outcrossing of which leads to immediate failure. It has been shown that the first passage problem is directly related to a solution of the corresponding Pontryagin equation, written with respect to the first excursion time T. An exact analytical solution to this problem, even for a linear system, is yet to be found. During the last decades a few strategies have been proposed to deal with this type of problems. The averaging procedure with further problem reformulation for the system's response amplitude or energy has been used for a linear system [2], systems with nonlinear stiffness [3] or nonlinear damping [4].

Solving the corresponding Pontryagin equation numerically has been proposed in [5], whereas a numerical solution to the backward Kolmogorov-Feller equation, for a system subjected to a Poisson driven train of impulses, has been found in [6]. Different novel analytical as well as numerical strategies were proposed recently by a number of authors [7, 8].

It has been shown recently in [9] that it is possible to adapt the path integration (PI) method [10] for problems of reliability, including the first passage problem. This paper focuses on numerical investigation of reliability of systems, subjected to multiplicative and additive white noise. The PI method is used to construct the reliability function and the first passage time for two systems. The first system considered is a single-degree-of-freedom (SDOF) linear oscillator, considered earlier in [13], see also [1]. Using the averaging procedure it was possible to provide approximate formulas for the first passage time of the system. Using the PI method one is in the position to verify the accuracy of these formulas. As it turns out, the

W.Q. Zhu et al. (Eds.): Nonlinear Stochastic Dynamics & Control, IUTAM BOOK SERIES 29, pp. 65–75.
springerlink.com

approximate first passage times given by stochastic averaging may be quite wrong even for parameter values for which one would expect reasonable agreement. The second system investigated is a nonlinear oscillator with a Duffing type nonlinearity.

2 Path Integration Approach to Reliability

The motion of a stochastic dynamic system may be expressed as an Itô stochastic differential equation (SDE):

$$dZ(t) = h(Z(t))\,dt + b\,dB(t), \tag{1}$$

where the state space vector process $Z(t) = (X(t), Y(t))^T = (X(t), \dot{X}(t))^T$ has been introduced; $h = (h_1, h_2)^T$ with $h_1(Z) = Y$ and $h_2(Z) = -g(X,Y)$; $b = (0, \sqrt{D})^T$, and $B(t)$ denotes a standard Brownian motion process. From Eq. (1) it follows immediately that $Z(t)$ is a Markov process, and it is precisely the Markov property that will be used in the formulation of the PI procedure.

The reliability is defined in terms of the displacement response process $X(t)$ in the following manner, assuming that all events are well defined,

$$R(T\,|\,x_0, 0, t_0) = \{x_l < X(t) < x_c;\, t_0 < t \le T | X(t_0) = x_0,\, Y(t_0) = 0\}, \tag{2}$$

where x_l, x_c are the lower and upper threshold levels defining the safe domain of operation. Hence the reliability $R(T|x_0, 0, t_0)$, as we have defined it here, is the probability that the system response $X(t)$ stays above the threshold x_l and below the threshold x_c throughout the time interval (t_0, T) given that it starts at time t_0 from x_0 with zero velocity ($x_l < x_0 < x_c$). In general, it is impossible to calculate the reliability exactly as it has been specified here since it is defined by its state in continuous time, and for most systems the reliability has to be calculated numerically, which inevitably will introduce a discretization of the time. Assuming that the realizations of the response process $X(t)$ are piecewise differentiable with bounded slope with probability one, the following approximation is introduced

$$R(T\,|\,x_0, 0, t_0) \approx \{x_l < X(t_j) < x_c,\, j = 1, \ldots, n | X(t_0) = x_0,\, Y(t_0) = 0\}, \tag{3}$$

where $t_j = t_0 + j\Delta t$, $j = 1, \ldots, n$, and $\Delta t = (T - t_0)/n$. With the assumptions made, the rhs of this equation can be made to approximate the reliability as closely as desired by appropriately choosing Δt, or, equivalently, n. Within the adopted approximation, it is realized that the reliability can now be expressed in terms of the joint conditional PDF $f_{X(t_1)\ldots X(t_n)|X(t_0),Y(t_0)}(\cdot, \ldots, \cdot | x_0, 0)$ as follows, which is just a rephrasing of Eq. (3),

$$R(T\,|\,x_0, 0, t_0) \approx \int_{x_l}^{x_c} \cdots \int_{x_l}^{x_c} f_{(\ldots)}(x_1, \ldots, x_n | x_0, 0)\, dx_1 \cdots dx_n. \tag{4}$$

Due to the Markov property of the state space vector process $Z(t) = (X(t),Y(t))^T$, we may express the joint PDF of $Z(t_1),\ldots,Z(t_n)$ in terms of the transition probability density function

$$p(z,t|z',t') = f_{Z(t)|Z(t')}(z|z') = f_{Z(t)Z(t')}(z,z')/f_{Z(t')}(z'), \quad (f_{Z(t')}(z') \neq 0) \tag{5}$$

in the following way

$$f_{Z(t_1)\ldots Z(t_n)|Z(t_0)}(z_1,\ldots,z_n|z_0) = \prod_{j=1}^{n} p(z_j,t_j|z_{j-1},t_{j-1}). \tag{6}$$

This leads to the expression ($z_0 = (x_0,0)^T$, $dz_j = dx_j\,dy_j$, $j = 1,\ldots,n$)

$$R(T\,|\,x_0,0,t_0) \approx \int_{-\infty}^{\infty}\int_{x_l}^{x_c}\cdots\int_{-\infty}^{\infty}\int_{x_l}^{x_c}\prod_{j=1}^{n} p(z_j,t_j|z_{j-1},t_{j-1})\,dz_1\cdots dz_n, \tag{7}$$

which is the path integration formulation of the reliability problem. The numerical calculation of the reliability is done iteratively in an entirely analogous way as in standard path integration. To show that, let us introduce a reliability density function (RDF) $q(z,t|z_0,t_0)$ as follows,

$$q(z_2,t_2|z_0,t_0) = \int_{-\infty}^{\infty}\int_{x_l}^{x_c} p(z_2,t_2|z_1,t_1)\,p(z_1,t_1|z_0,t_0)\,dz_1, \tag{8}$$

and ($n > 2$)

$$\begin{aligned} &q(z_k,t_k|z_0,t_0) = \\ &\int_{-\infty}^{\infty}\int_{x_l}^{x_c} p(z_k,t_k|z_{k-1},t_{k-1})\cdot q(z_{k-1},t_{k-1}|z_0,t_0)\,dz_{k-1},\ k = 3,\ldots,n. \end{aligned} \tag{9}$$

The reliability is then finally calculated approximately as ($T = t_n$)

$$R(T\,|\,x_0,0,t_0) \approx \int_{-\infty}^{\infty}\int_{x_l}^{x_c} q(z_n,t_n|z_0,t_0)\,dz_n. \tag{10}$$

The complementary probability distribution of the time to failure T_e, i.e. the first passage time, is given by the reliability function. The mean time to failure $\langle T_e\rangle$ can thus be calculated by the equation

$$\langle T_e\rangle = \int_0^{\infty} R(\tau\,|\,x_0,0,t_0)\,d\tau \tag{11}$$

To evaluate the reliability function it is required to know the transition probability density function $p(z,t|z',t')$, which is unknown for the considered nonlinear systems. However, from Eq. (1) it is seen that for small $t-t'$ it can be determined approximately, which is what is needed for the numerical calculation of the reliability. A detailed discussion of this, and the iterative integrations of Eqs. (8) and (9), is given in [9, 11]. Concerning the integrations, there is, however, one small difference between the present formulation and that described in these references. In Eqs. (8) and (9), the integration in the x-variable only extends over the interval (x_l, x_c). The infinite upper and lower limits on the y-variable are replaced by suitable constants determined by e.g. an initial Monte Carlo simulation.

If the system response $Z(t)$ has a stationary response PDF $f_Z(z)$ as $t \to \infty$, it follows that the conditional response PDF $f_{\{Z(t_n)|Z(0),x_l<X(t_j)<x_c;0\leq j\leq n-1\}}(z)$ also reaches a stationary density, say $q^*(z)$, when $t_n \to \infty$. This means that the reliability process eventually becomes memoryless, and hence the RDF converges $q(z,t_n|z_0,t_0) \to q^*(z)Ke^{-\nu t_n}$ for some constants K and ν as $t_n \to \infty$. Also the numerical method should reach stationarity in the conditional density. This also implies that the numerically estimated reliability function must be exponential, since the same relative amount of probability mass leaves the system at every iteration. So, in the end, the only thing one should need for a good reliability estimate is the behavior in the transient phase, and the exponential decay thereafter. The implication of this for highly reliable systems is that the important parameter to determine is the decay rate, since the transient phase can then usually be neglected. The decay rate can be conveniently estimated using the ACER method [12].

3 Numerical Examples

3.1 A Linear Oscillator with Multiplicative and Additive Noise

Consider the following linear oscillator under both additive and multiplicative random excitations:

$$\ddot{X} + \omega_0[2\zeta + W_2(t)]\dot{X} + \omega_0^2[1 + W_1(t)]X = W_3(t), \tag{12}$$

where $W_j(t))$, $j = 1,2,3$, are wide band stationary processes with zero mean values. Earlier, this model has been studied by Ariaratnam and Tam [13] under the assumption that ζ is of order ε and the $W_j(t))$ are of order $\sqrt{\varepsilon}$, where ε is a small parameter. By applying the stochastic averaging procedure, it was argued that the amplitude process $A(t) = \left(X^2 + \dot{X}^2/\omega_0^2\right)^{1/2}$ is approximately a Markov diffusion process governed by the (Itô) stochastic differential equation (SDE)

$$dA = m(A)dt + \sigma(A)dB(t)\,. \tag{13}$$

The drift coefficient $m(A)$ and the diffusion coefficient $\sigma(A)$ are given by the equations,

$$m(A) = -\alpha A + \frac{\delta}{2A}, \tag{14}$$

$$\sigma(A) = (\gamma A^2 + \delta)^{1/2}, \tag{15}$$

in which

$$\alpha = \zeta\omega_0 - \frac{\pi\omega_0^2}{8}\left[2\Phi_{22}(0) + 3\Phi_{22}(2\omega_0) + 3\Phi_{11}(2\omega_0) - 6\Psi_{12}(2\omega_0)\right], \tag{16}$$

$$\delta = \frac{\pi}{\omega_0^2} \Phi_{33}(\omega_0), \tag{17}$$

$$\gamma = \frac{\pi \omega_0^2}{4} \left[2\Phi_{22}(0) + \Phi_{22}(2\omega_0) + \Phi_{11}(2\omega_0) + 2\Psi_{12}(2\omega_0) \right], \tag{18}$$

and

$$\Phi_{ij}(\omega) = \frac{1}{2\pi} \int_{-\infty}^{\infty} \mathrm{e}[W_i(t)W_j(t+\tau)] \cos(\omega\tau)\, d\tau, \; i,j = 1,2,3, \tag{19}$$

$$\Psi_{ij}(\omega) = \frac{1}{2\pi} \int_{-\infty}^{\infty} \mathrm{e}[W_i(t)W_j(t+\tau)] \sin(\omega\tau)\, d\tau, \; i,j = 1,2,3. \tag{20}$$

Ariaratnam and Tam [13] showed that the expected time $\langle T_f \rangle$ to first failure of the amplitude process $A(t)$ is given by the formulas

$$\langle T_f \rangle = \frac{1}{\eta\gamma} \int_{a_0}^{a_c} \frac{1}{u} \left[\left(1 + \frac{\gamma}{\delta} u^2\right)^{\eta} - 1 \right] du, \; \eta = \frac{\alpha}{\gamma} + \frac{1}{2} \neq 0 \tag{21}$$

$$\langle T_f \rangle = \frac{1}{\gamma} \int_{a_0}^{a_c} \frac{1}{u} \ln\left(1 + \frac{\gamma}{\delta} u^2\right) du, \; \eta = 0 \tag{22}$$

Here a_0 denotes the initial condition and a_c the critical level ($a_0 < a_c$). This approach would usually represent an approximation in the sense that failure for the original problem would typically be when $X(t)$ exceeds a critical region bounded by the thresholds $\pm x_c$. An approximate solution for this is obtained by studying the exceedance of $a_c = x_c$ by the amplitude process $A(t)$.

For the numerical calculations in this paper the $W_j(t)$ are assumed to be independent Gaussian white noise processes, with $\mathrm{e}[W_j(t)W_j(t+\tau)] = \sigma_j^2 \delta(\tau)$. Using numerical PI we have calculated the reliability function associated with the linear oscillator model in Eq. (12) for three case studies with different values of the ω_0 parameter. Since PI can be done for any choice of parameter values, it provides a means of studying the limitations of the amplitude diffusion model adopted in [13], and thereby also the limitations of stochastic averaging in this context.

To provide a means for verification of the PI results, we have calculated the stationary part of the reliability function by the ACER method [14]. This method makes it possible to estimate the exact extreme value distribution, and hence the reliability function, of the response process provided the transient response can be neglected. From Eq. (10) and the following discussion, it is obtained that the tail behaviour of the reliability function is given as,

$$R(t) = R(t_0) \exp\{-\nu(x_c)(t - t_0)\}, \; t \geq t_0, \tag{23}$$

for a suitable t_0. The ACER method provides an estimate and a 95% confidence interval of the parameter $\nu(x_c)$ for each critical level x_c. An approximate mean time to failure is then given by $1/\nu(x_c)$.

3.1.1 Case 1

In this case the system parameters are $\omega_0 = 0.1$, $\zeta = 0.01$, and $\sigma_1 = \sigma_2 = \sigma_3 = 0.1$. The standard deviation of the response was found to be $\sigma_X = 16.271$, while $\sigma_{\dot{X}} = 1.628$. The reliability functions calculated by the numerical PI method for three different critical levels are shown in Fig. 1. Some numerical results are summarized in Table 1. It is seen that the mean time to failure calculated by PI agrees with the approximate time to failure provided by $1/\nu(x_c)$ for all three critical levels, while the corresponding failure times obtained by stochastic averaging are orders of magnitude wrong. This seems to be caused by a large value of the parameter η, cf. Eq. (21), which assumes the value 19.5 for this case.

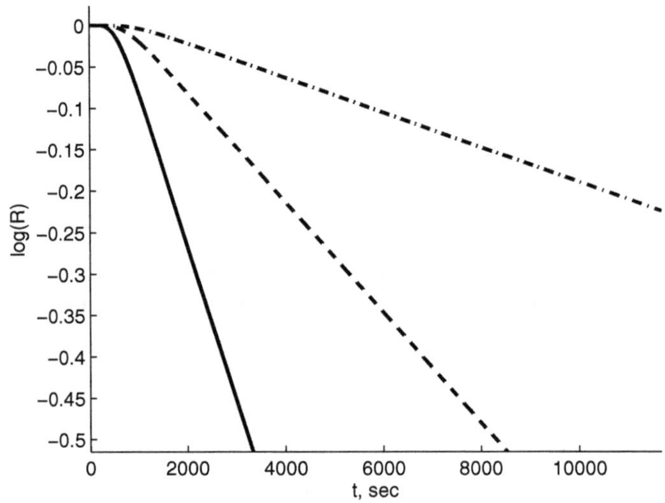

Fig. 1 Two-sided reliability function of system (12). —— : $x_c/\sigma_X = 2.5$, – – –: $x_c/\sigma_X = 3.0$, – · –: $x_c/\sigma_X = 3.5$.

Table 1 Mean time to failure for $\omega_0 = 0.1$, $\zeta = 0.01$

x_c/σ_X	$< T_{SA} >$ (sec.)	$< T_{PI} >$ (sec.)	ν_{PI}	ν_{ACER}	95% CI
2.5	$1.67 \cdot 10^6$	$5.90 \cdot 10^3$	$1.88 \cdot 10^{-4}$	$2.08 \cdot 10^{-4}$	$(1.81 \cdot 10^{-4}, 2.36 \cdot 10^{-4})$
3.0	$4.22 \cdot 10^6$	$1.58 \cdot 10^4$	$6.64 \cdot 10^{-5}$	$7.56 \cdot 10^{-5}$	$(5.64 \cdot 10^{-5}, 9.48 \cdot 10^{-5})$
3.5	$1.20 \cdot 10^7$	$4.87 \cdot 10^4$	$2.10 \cdot 10^{-5}$	$2.21 \cdot 10^{-5}$	$(1.32 \cdot 10^{-5}, 3.09 \cdot 10^{-5})$

3.1.2 Case 2

In this case the system parameters are $\omega_0 = 1.0$, $\zeta = 0.05$, and $\sigma_1 = \sigma_2 = \sigma_3 = 0.1$. The standard deviation of the response was found to be $\sigma_X = 1.832$, while $\sigma_{\dot{X}} = 1.836$. The reliability functions calculated by the numerical PI method for three different critical levels are shown in Fig. 2. Numerical results are summarized in Table 2. Again it is seen that the mean time to failure calculated by PI agrees with the approximate time to failure provided by $1/\nu(x_c)$ for all three critical levels, while the corresponding failure times obtained by stochastic averaging are approximately two orders of magnitude wrong. In this case $\eta = 9.5$, which is still quite large.

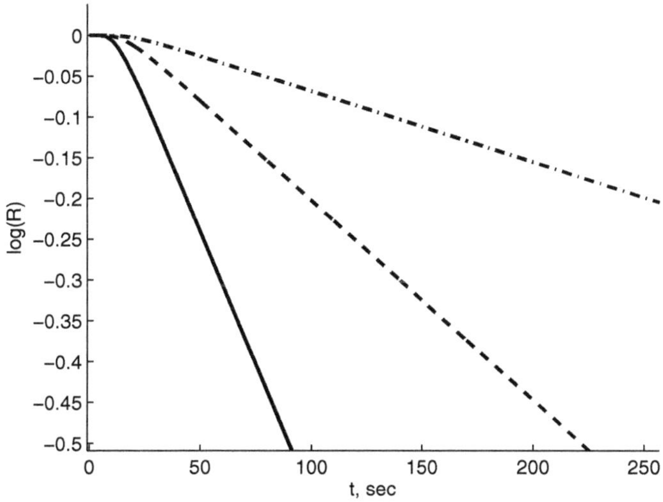

Fig. 2 Two-sided reliability function of system (12). —— : $x_c/\sigma_X = 2.5$, – – –: $x_c/\sigma_X = 3.0$, – · –: $x_c/\sigma_X = 3.5$.

Table 2 Mean time to failure for $\omega_0 = 1.0$, $\zeta = 0.05$

x_c/σ_X	$< T_{SA} >$ (sec.)	$< T_{PI} >$ (sec.)	ν_{PI}	ν_{ACER}	95% CI
2.5	$7.45 \cdot 10^3$	$1.66 \cdot 10^2$	$6.54 \cdot 10^{-3}$	$6.15 \cdot 10^{-3}$	$(5.83 \cdot 10^{-3}, 6.48 \cdot 10^{-3})$
3.0	$1.69 \cdot 10^4$	$4.26 \cdot 10^2$	$2.45 \cdot 10^{-3}$	$2.37 \cdot 10^{-3}$	$(2.19 \cdot 10^{-3}, 2.54 \cdot 10^{-3})$
3.5	$4.03 \cdot 10^4$	$1.17 \cdot 10^3$	$8.68 \cdot 10^{-4}$	$7.98 \cdot 10^{-4}$	$(7.31 \cdot 10^{-4}, 8.65 \cdot 10^{-4})$

3.1.3 Case 3

In this case the system parameters are $\omega_0 = 10.0$, $\zeta = 0.15$, and $\sigma_1 = \sigma_2 = \sigma_3 = 0.1$. The standard deviation of the response was found to be $\sigma_X = 0.0410$, while $\sigma_{\dot{X}} = 0.4261$. The reliability functions calculated by the numerical PI method for three different critical levels are shown in Fig. 3. Numerical results are summarized in Table 3. Also in this case it is seen that the mean time to failure calculated by PI agrees with the approximate time to failure provided by $1/\nu(x_c)$ for all three critical levels, while the corresponding failure times obtained by stochastic averaging are now more on the same level as the correct values. In this case $\eta = 2.5$.

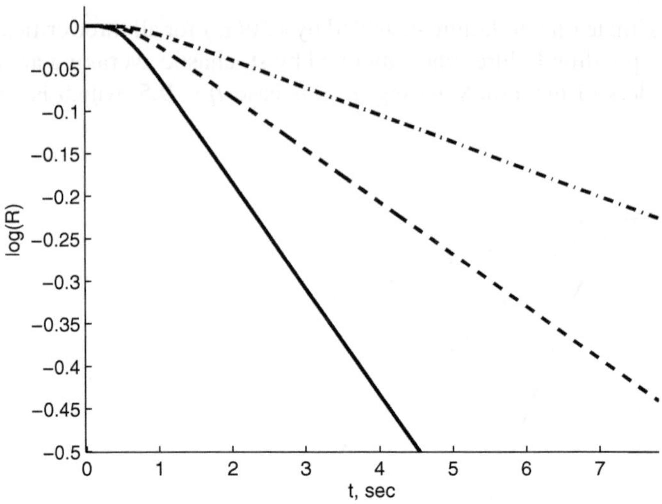

Fig. 3 Two-sided reliability function of system (12). —— : $x_c/\sigma_X = 2.5$, – – –: $x_c/\sigma_X = 3.0$, – · –: $x_c/\sigma_X = 3.5$.

Table 3 Mean time to failure for $\omega_0 = 10.0$, $\zeta = 0.15$

x_c/σ_X	$< T_{SA} >$ (sec.)	$< T_{PI} >$ (sec.)	ν_{PI}	ν_{ACER}	95% CI
2.5	16.8	8.54	$1.25 \cdot 10^{-1}$	$1.16 \cdot 10^{-1}$	$(1.09 \cdot 10^{-1}, 1.24 \cdot 10^{-1})$
3.0	29.4	16.9	$6.16 \cdot 10^{-2}$	$5.85 \cdot 10^{-2}$	$(5.33 \cdot 10^{-2}, 6.36 \cdot 10^{-2})$
3.5	48.9	31.9	$3.21 \cdot 10^{-2}$	$2.89 \cdot 10^{-2}$	$(2.50 \cdot 10^{-2}, 3.28 \cdot 10^{-2})$

3.2 A Duffing Oscillator with Multiplicative and Additive Noise

The following specific version of a nonlinear Duffing oscillator with multiplicative and additive noise is adopted here:

$$\ddot{X} + 2\zeta\omega_0\dot{X} + \omega_0^2[1 + X^2 + W_1(t)]X = W_2(t), \tag{24}$$

where $W_j(t))$, $j = 1,2$, are stationary Gaussian white noise processes with zero mean values, with $\mathrm{e}[W_j(t)W_j(t+\tau)] = \sigma_j^2\,\delta(\tau)$.

Again, using numerical PI we have calculated the reliability function associated with the nonlinear oscillator model in Eq. (24) for five case studies with different values of the σ_1 parameter, that is, the intensity of the parametric noise. For the calculations, $\omega_0 = 1.0$, $\zeta = 0.1$, and $\sigma_2 = 0.1$. Table 4 summarizes the calculated values of the mean time to failure as depending on the parametric noise level σ_1 and the size of the critical level x_c. It is seen that while the mean time to failure is rather insensitive to changes in the parametric noise intensity expressed by σ_1 for $x_c/\sigma_X = 2.5$, this situation changes significantly for $x_c/\sigma_X = 3.5$. At this level, the mean time to failure is reduced by a factor of about 5 when the noise intensity σ_1 is increased by a factor of 5. This effect can also be seen from the calculated reliability functions for the two cases $\sigma_1 = 0.1$ and 0.5, which have been plotted in Figure 5.

Table 4 Mean time to failure $\langle T_e \rangle$ for different values of σ_1

σ_1		0.1	0.2	0.3	0.4	0.5
σ_X		0.155	0.160	0.170	0.186	0.211
x_c/σ_X	2.5	$1.16\cdot10^2$	$1.14\cdot10^2$	$1.11\cdot10^2$	$1.10\cdot10^2$	$1.11\cdot10^2$
	3.0	$4.36\cdot10^2$	$3.80\cdot10^2$	$3.16\cdot10^2$	$2.64\cdot10^2$	$2.30\cdot10^2$
	3.5	$2.34\cdot10^3$	$1.64\cdot10^3$	$1.04\cdot10^3$	$6.81\cdot10^2$	$4.90\cdot10^2$

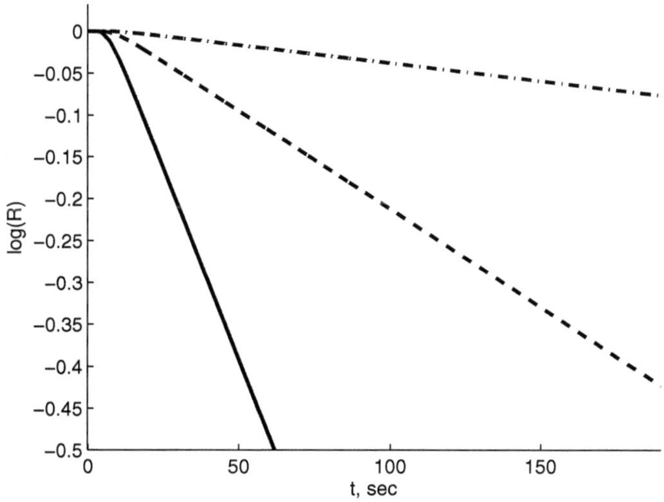

Fig. 4 Two-sided reliability function of system (24) with $\sigma_1 = 0.1$. —— : $x_c/\sigma_X = 2.5$, - - -: $x_c/\sigma_X = 3.0$, – · –: $x_c/\sigma_X = 3.5$.

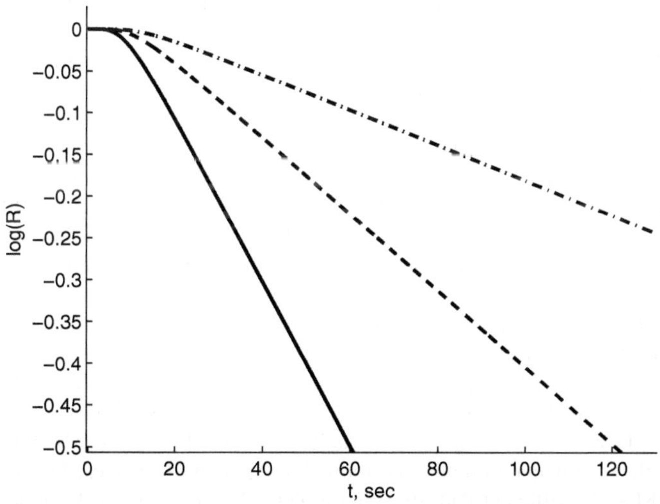

Fig. 5 Two-sided reliability function of system (24) with $\sigma_1 = 0.5$. —— : $x_c/\sigma_X = 2.5$, - - -: $x_c/\sigma_X = 3.0$, – · –: $x_c/\sigma_X = 3.5$.

4 Conclusions

In the paper the authors have considered a first passage type reliability problem for two types of systems: linear and nonlinear oscillators with parametric and additive white noises. The numerical results presented in the paper are obtained by the path integration method, which was reformulated from its standard form to handle reliability problems. The results were verified by Monte-Carlo simulations through the use of the ACER method.

For the linear system it has been shown that the use of stochastic averaging has its limitations especially for calculating the reliability. The results calculated by numerical PI were verified by using Monte-Carlo simulations in combination with the ACER method, which allows the estimation of the exact extreme value distribution for the stationary part of the response process. This provides a means of determining an approximate value of the mean time to failure. In all the case studies investigated there were agreement between the results calculated by PI and estimated by the ACER method.

The applicability of the PI method for calculating the reliability of the nonlinear oscillator with parametric noise was also demonstrated.

Acknowledgement. The financial support from the Research Council of Norway (NFR) through the Centre for Ships and Ocean Structures (CeSOS) at the Norwegian University of Science and Technology (NTNU) is gratefully acknowledged.

References

1. Lin, Y.K., Cai, G.Q.: Probabilistic Structural Dynamics. McGraw-Hill, New York (1995)
2. Roberts, J.B.: First passage time for the envelope of a randomly excited linear oscillator. Journal of Sound and Vibration 46(1), 1–14 (1976)
3. Roberts, J.B.: First passage time for oscillators with non-linear restoring forces. Journal of Sound and Vibration 56(1), 71–86 (1978)
4. Roberts, J.B.: First passage time for oscillators with nonlinear dampimg. ASME Journal of Applied Mechanics 45, 175–180 (1978)
5. Bergman, L.A., Heinrich, J.C.: On the reliability of the linear oscillator and systems of coupled oscillators. Int. J. for Numerical Methods in Engineering 18, 1271–1295 (1982)
6. Koyluoglu, H.U., Nielsen, S.R.K., Iwankiewicz, R.: Reliability of non-linear oscillators subjected to Poisson driven impulses. Journal of Sound and Vibration 176(1), 19–33 (1994)
7. Chen, J.B., Li, J.: Dynamic response and reliability analysis of non-linear stochastic structures. Probabilistic Engineering Mechanics 20, 33–44 (2005)
8. Zhu, W.Q., Deng, M.L., Huang, Z.L.: First-passage failure of quasi-integrable Hamiltonian systems. ASME Journal of Applied Mechanics 69, 274–282 (2002)
9. Iourtchenko, D.V., Mo, E., Naess, A.: Reliability of strongly nonlinear SDOF dynamic systems by the path integration method. Journal of Applied Mechanics, 75 (2008)
10. Naess, A., Moe, V.: Efficient Path Integration methods for nonlinear dynamic systems. Probabilistic Engineering Mechanics 15(2), 221–231 (2000)

11. Naess, A., Kolnes, F.E., Mo, E.: Stochastic spur gear dynamics by numerical path integration. J. of Sound and Vibration 302, 936–950 (2007)
12. Naess, A., Saha, N.: Prediction of first passage failure by the ACER method. In: Proceedings 37th Solid Mechanics Conference, Warsaw (2010)
13. Ariaratnam, S.T., Tam, D.S.F.: Random Vibration and Stability of a Linear Parametrically Excited Oscillator. Z. Angew. Math. Mech. 59(2), 79–84 (1979)
14. Naess, A., Gaidai, O.: Estimation of extreme values from sampled time series. Structural Safety 31, 325–334 (2009)

Numerical Solution of Fokker-Planck Equation for Nonlinear Stochastic Dynamical Systems

S. Narayanan[1] and Pankaj Kumar[2]

[1] Department of Mechanical Engineering, Indian Institute of Technology, Madras, Chennai, India
[2] Gas Turbine Design Department, Bharat Heavy Electricals Limited, Hyderabad, India

Abstract. A new approach for efficient numerical implementation of the path integral (PI) method based on non-Gaussian transition probability density function (PDF) and the Gauss-Legendre integration scheme is developed. The PI method is used to solve the Fokker-Planck (FP) equation and to study the nature of the stochastic and chaotic response of the nonlinear systems. The steady state PDF, jump phenomenon, noise induced state changes of PDF are studied by the method. A computationally efficient higher order, finite difference (FD) technique is derived for the solution of higher dimensional FP equation. A two degree of freedom (DOF) nonlinear system having Coulomb damping with variable friction coefficient is considered representative of bladed disk assembly of turbo-machinery blades. Effects of normal force and viscous damping on the mean square response of a blade are investigated.

Keywords: Random Vibration, Fokker-Planck equation, Path integration method, Nonlinear stochastic dynamics, Finite Difference Method.

1 Introduction

Nonlinear dynamical systems subjected to random excitation occur in many fields of science and engineering. The determination of the PDF of their response is a difficult problem and they are available only for a very restricted class of dynamical systems. Several approximate methods such as the equivalent linearization, perturbation method, Gaussian closure, stochastic averaging, equivalent nonlinearization have been developed for obtaining the second order statistics of the nonlinear system based on the Gaussian assumption of the PDF. The Monte Carlo Simulation (MCS) is a brute force method involving the generation of a large number of sample functions of the excitations and numerically integrating the nonlinear equations to obtain the response and the associated probability structure. This method is computationally intensive especially for higher dimensional systems.

The response of a dynamical system to white noise excitation constitutes a Markov vector process whose transition PDF is governed by the FP equation. Analytical solutions of the FP equation are available only for a few nonlinear oscillators. Numerical techniques like the path integration (PI) [1,2], finite element

W.Q. Zhu et al. (Eds.): Nonlinear Stochastic Dynamics & Control, IUTAM BOOK SERIES 29, pp. 77–86.
springerlink.com

(FE) [3,4], finite difference (FD) [5], and multi scale FE [6] have been developed to get approximate solutions for the FP equation. However, their application to higher dimensional systems has been limited. The PI method is based on the Gaussian assumption of the response process. While second order statistics are predicted reasonably accurately, determination of higher order moments and reliability estimates by this method is fraught with error especially for systems with strong nonlinearities. The application of this method to nonlinear systems subjected to narrow band excitation and combined white noise and harmonic excitations is also very limited. When a nonlinear system is subjected to combined harmonic and white noise excitation, the response is a Markov process with one of the drift coefficients being periodic and may exhibit multi-valued behaviour, with random jumps between two pseudo-stable states [7]. Due to the explicit time dependence of the drift coefficient the PDF is not stationary for such a case.

This paper presents a new modified PI method for the solution of the FP equation and the evaluation of the response statistics of nonlinear systems subjected to wide band, parametric and combined white noise and harmonic excitations. The modified PI method is based on a non-Gaussian transition PDF and the Gauss-Legendre integration scheme. The method captures the changes in the PDF due to noise, jump behaviour of the response and the rotating periodicity of the steady state PDF. The modified PI method is applied to a number of examples. This paper also presents a sixth order FD stencil scheme in conjunction with GEAR time integration for the solution of a four dimensional FP equation. The FD method is applied to a two degree-of-freedom (DOF) nonlinear system having Coulomb as well as viscous damping with variable friction coefficients representative of a bladed disk assembly subjected to random excitation.

2 Path Integration Method

Extremely small time steps and space intervals should be used in the PI method based on the Gaussian assumption of the joint PDF to obtain the PDF with reasonable accuracy leading to large computational effort as the response of nonlinear systems to Gaussian excitation is non-Gaussian. A modified PI method based on an adjustable non-Gaussian transition PDF expressed as a product of a Gaussian PDF and a polynomial function is derived in this paper [8]. The polynomial function is obtained by truncating an infinite expansion [9]. The Gauss-Legendre integration scheme [1] is used. Thus the method takes into account the non-Gaussian nature of the response and makes use of the knowledge of higher order moments allowing a coarse space grid size and longer time steps in the computation, leading to significant reduction in the computational effort. This approach has also the advantage that it provides for the deviation of the response distribution from Gaussian right in the beginning of the PI procedure. The evaluation of the PDF $p(\mathbf{X},t)$ from t_{i-1} to t_i can be expressed as and is obtained from

$$p\left(\mathbf{X}^i,t_i\right)=\int_{R_S} p(\mathbf{X}^i,t_i\,|\,\mathbf{X}^{i-1},t_{i-1})\,p(\mathbf{X}^{i-1},t_{i-1})\,d\mathbf{X}^{i-1} \tag{1}$$

where R_s represents the reduced state range, within an n -dimensional rectangle. The integral of Eq. (1) can be discretized into the following composite Gauss-Legendre quadrature form [1]

$$p(x_{mn}^i,t_i)=\sum_{k=1}^{K}\frac{z_k}{2}\sum_{l=1}^{L_K} q_{kl}\,p(x_{kl}^{i-1},t_{i-1})\,p(x_{mn}^i,t_i\,|\,x_{kl}^{i-1},t_{i-1}) \tag{2}$$

where K is the number of sub-intervals, L_k is the number of quadrature points in the sub-interval k , z_k is the length of the sub interval K, each x_{kl} is the position of a Gauss quadrature point and q_{kl} is its corresponding weight. Eq. (2) can be used to calculate the PDF at any point x^i at step i, provided that the PDF values are known at step i-1.

The best known approach for approximating a non-Gaussian PDF is to express it as a product of a Gaussian PDF and a polynomial function, which is a truncated version of an infinite expansion [9]. Hence, the transition PDF from x_{kl} at t_{i-1} to x_{mn} at t_i is assumed in the non-Gaussian form as

$$p(x_{mn}^i,t_i\,|\,x_{kl}^{i-1},t_{i-1})=\frac{1}{\sqrt{2\pi}\sigma(t_i)}\exp\left\{-\frac{[x_{mn}^i-m_1(t_i)]^2}{2\sigma^2(t_i)}\right\}\left[\sum_{n=0}^{\infty}a_n\,He_n\left\{\frac{[x_{mn}-m_1(t_i)]}{\sigma(t_i)}\right\}\right] \tag{3}$$

The adjustable polynomial function must be consistent with the orthogonality and also the normalization conditions with respect to the other terms in the series. The polynomial is written as a sum of Hermite polynomials because their orthogonality with respect to the normalized Gaussian PDF makes it relatively easy to obtain the expression for the coefficients a_n. The most attractive property of the Hermite approximation is that it transforms the partial differential equation into approximate integral kernels which when evaluated by quadrature allows their calculation through a matrix vector multiplication. Eq. (3) can be written in terms of the Gram-Charlier series of type A[3]

$$p(x_{mn}^i,t_i\,|\,x_{kl}^{i-1},t_{i-1})=\frac{1}{\sqrt{2\pi}\sigma(t_i)}\exp\left\{-\frac{[x_{mn}^i-m_1(t_i)]^2}{2\sigma^2(t_i)}\right\}\left[1+\sum_{n=3}^{N}a_n\,He_n\left\{\frac{[x_{mn}-m_1(t_i)]}{\sigma(t_i)}\right\}\right] \tag{4}$$

The next three coefficents in the series are simply related to the cumulants of **X** as $a_n=\kappa_n(X)/(n!\sigma_X^n)$ for n = 3, 4 and 5, where $\kappa_n(X)$ is the nth cumulant of X. For $n\geq 6$ the relationships are more complicated. Each transition PDF $p(x_{mn}^i,t_i\,|\,x_{kl}^{i-1},t_{i-1})$ in Eq. (2) is now assumed to be non-Gaussian of the form given in Eq. (4), which depends only on the conditional mean, variance and few higher moments of $X_{mn}(t_i)$.

The modified PI method is applied to the following examples showing the efficacy of the method. The Dimentberg oscillator subjected to both external and parametric random excitations is considered as the first example.

$$\ddot{X}+2\alpha\left(1+\sigma_1 W_1(t)\right)\dot{X}+\beta\left(X^2+\dot{X}^2\big/\omega^2\right)\dot{X}+\omega^2\left(1+\sigma_2 W_2(t)\right)X=\sigma_3 W_3(t) \tag{5}$$

The corresponding FP equation for this system is given by

$$\frac{\partial p}{\partial t}=\frac{\partial\left\{2\alpha X_2+\beta X_2(X_1^2+\frac{X_2^2}{\omega^2})+\omega^2 X_1-2\alpha^2\sigma_1^2 X_2\right\}p}{\partial X_2}-X_2\frac{\partial p}{\partial X_1}+\frac{1}{2}\frac{\partial^2\{Q\}p}{\partial X_2^2} \tag{6}$$

where $Q=4\alpha^2\sigma_1^2 X_2^2+\omega^4\sigma_2^2 X_1^2+\sigma_3^2$. The term $2\alpha^2\sigma_1^2 X_2$ appearing in the expression is a Wong-Zakai correction term. The probability densities are calculated from Eq. (2) using the approximate non-Gaussian transition PDF given by Eq. (4) within a reduced state space [-10, 10] x [-10, 10] which is divided into 50 uniform subintervals in each direction, with two quadrature points in each sub-interval and an initial Gaussian density with mean value zero. A uniform time step $\Delta t = 0.60$ sec. is found to be sufficient for good accuracy. For the parameters $\alpha=-0.1, \beta=0.1, \omega=1.0, \sigma_1^2=0.1, \sigma_2^2=0.1$,and $\sigma_3^2=0.3$,the stationary marginal PDFs of the displacement are shown in Fig. 1 on linear as well as on the logarithmic scales. The non-Gaussian based PI method gives impressively-low error of 0.025% at extremely small probability levels of order 10^{-7}. There is very good agreement between the PI and the MCS results. The method shows good convergence with reduced number of grids leading to a reduction of CPU time by a factor of almost 2 compared to the traditional PI method.

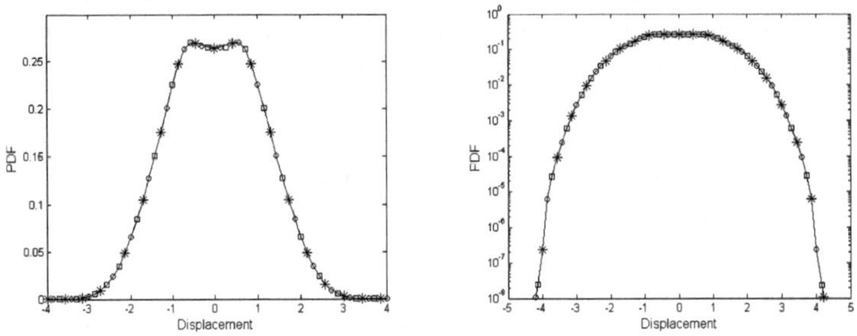

Fig. 1 Linear and logarithmic plot of the Marginal PDF of Displacement; ——— MCS results; ○○○○○ ,PI with non-Gaussian transition PDF; ▫▫▫▫▫ , PI with Gaussian transition PDF; ****,FEM.

The Duffing oscillator subjected to combined white noise and harmonic excitation is considered as the next example to study the jump and bifurcation phenomena.

$$\ddot{X}+2\beta\dot{X}+\alpha X+\lambda X^3=W(t)+f_0\cos\omega t \tag{10}$$

where $W(t)$ is a stationary, zero mean Gaussian white noise satisfying $E[W(t)W(t+\tau)]=2D\delta(\tau)$. For the parameters $\beta=0.05, \lambda=0.3, \alpha=1.0, f_0=0.2$, and $\omega=1.2$, a time step Δt = *T/4*, and 100×100 Gaussian quadrature points in 50×50 subinterval within the reduced domain [-10, 10]×[-10, 10] are selected. The snapshots of joint PDF of X vs. $\dot{X}$ captured by using the modified PI technique at $t = 60.5$ sec. are shown in Fig. 2. The presence of the nonlinearity causes multi-valued regions where more than one mean square value of the response is possible. When the intensity of noise changes from 0.002 to 0.01, the joint PDF is still bimodal implying jumps may occur. At smaller intensity, the two peaks are well separated, jumps occur rarely and the most probable motion is around the lower branch of amplitude response curve. At higher intensity, on the other hand, the peaks connect and even merge, jumps occur more frequently and the most probable motion is around the upper branch of amplitude response curve.

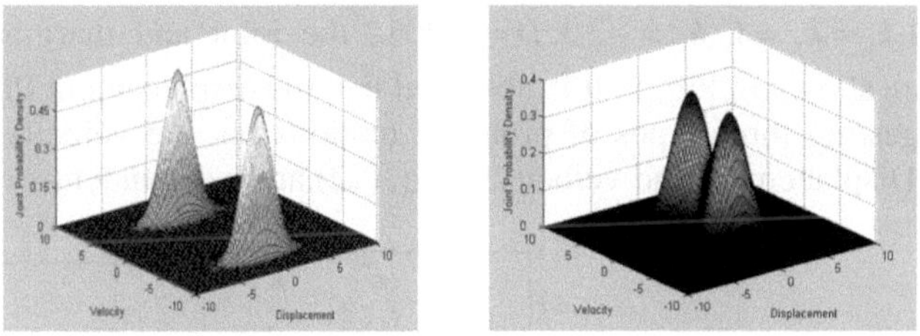

(a) (b)

Fig. 2 Joint PDF of Duffing oscillator (a) D = 0.002, (b) D = 0.01

For $\beta = 0.075, \alpha = -1, \lambda = \omega = 1.0$ and $f_0 = 0.3$ the above Duffing oscillator exhibits chaotic behaviour under harmonic excitation (Fig. 3) and if an addition a stochastic excitation is also present the response may consist of coexisting chaotic and stochastic orbits. It is difficult to distinguish between the two behaviours. The periodic attractor is relatively stronger compared to the chaotic attractor. Periodicity in the PDF can be removed by time averaging. Jung and Hänggi [10] introduced a time averaged density as an invariant measure for both deterministic and noisy chaos. The invariant measure on x_1 can be defined as

$$p_{av}(x_1) = \int_{-\infty}^{\infty} \left\{ \frac{1}{t_n} \int_0^{t_n} p(x_1, x_2, t) dt \right\} dx_2 \tag{11}$$

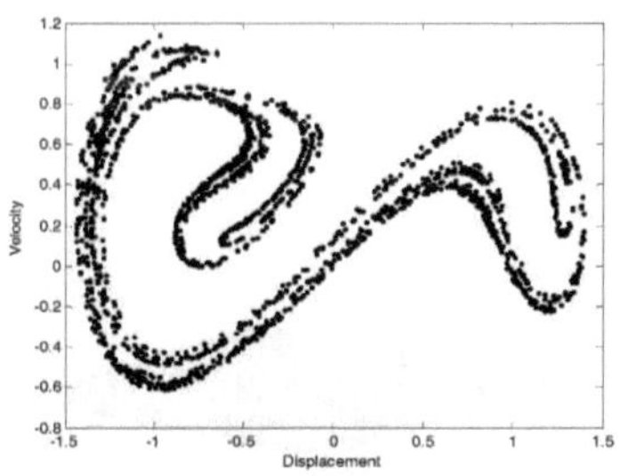

Fig. 3 A strange attractor for the Duffing oscillator

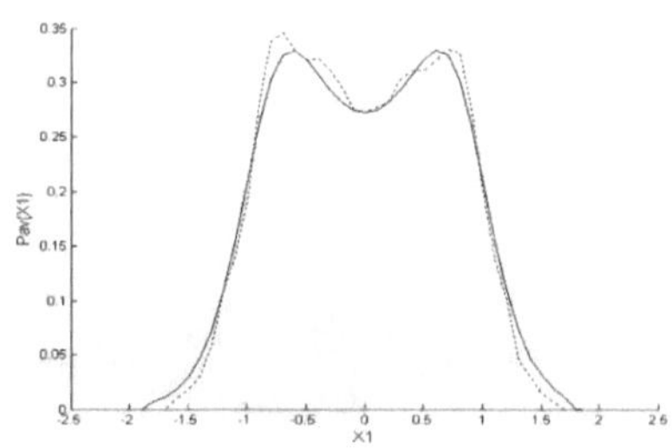

Fig. 4 Invariant Measure on X_1

Fig. 4 shows the time averaged probability density for the noisy chaotic attractors with noise intensity at 0.02 (dashed line) and 0.07 (solid line). The invariant measure indicates that the probability of large excursion is increased as noise intensity increases.

Next, we consider the equation of motion of the system with restoring force having more than one zero subjected to wide band random excitation.

$$\ddot{X} + c\dot{X} + k_1 X + k_3 X^3 + k_5 X^5 = W(t) \tag{12}$$

The computation is carried out in a reduced state space of [-2, 2]×[-2, 2] with a total of 40×40 Gauss-quadrature points in 20×20 subintervals. For the parameters $c = 0.1, k_1 = k_5 = 1,\ k_3 = -2.3, D = 0.005$, the joint and marginal PDF of the response are shown in Fig. 5. It is observed that the displacement PDF shows three peaks suggesting three types of probable motions but the velocity PDF has only one peak. Displacement and velocity in Eq. (12) are independent.

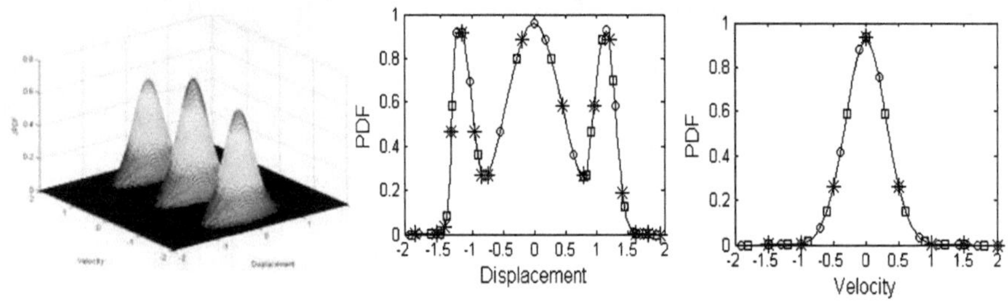

(a) (b) (c)

Fig. 5 (a) Joint PDF, (b) Marginal PDF of displacement, (c) Velocity; Key as Fig. 1

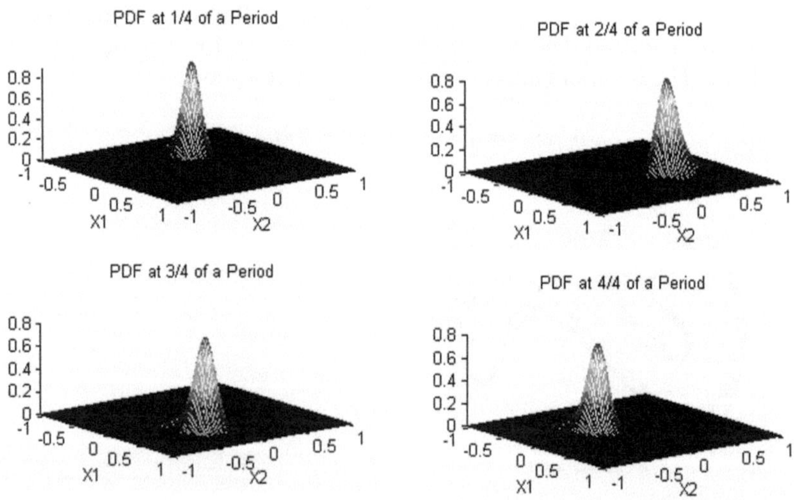

Fig. 6 Surface plot of PDF of the Duffing-Rayleigh oscillator at t = 1/4, 2/4, 3/4, 4/4 of a period

The Duffing-Rayleigh oscillator subjected to harmonic and white noise excitation is investigated next.

$$\ddot{X}+\gamma\left(1+\frac{\beta}{\omega_1^2}\dot{X}^2\right)\dot{X}+\omega_1^2\left(1+\xi X^2\right)X=f_o\cos\omega t+W(t) \tag{13}$$

For the dynamical system subjected to harmonic and stochastic excitations, one of the drift coefficients of the corresponding FP equation is periodic [7]. A steady state solution may exist at $t\rightarrow\infty$. The approximate methods cannot capture the periodicity in the instantaneous PDF before averaging. The modified PI method based on the Gauss-Legendre integration formula [1] is a powerful method which can capture the evolution of the instantaneous PDF. For the parameters $\gamma=0.5$, $\beta = \xi = 0.3$, $\omega_1 = f_0 = 1.0$, $\omega = 2.0$, and $D = 0.05$, the instantaneous PDFs at 1/4, 2/4, 3/4, and 4/4 of the period after reaching periodicity are shown in Fig. 6. From the single peak it may be concluded that for this system there is no multi-valued amplitude as the frequency varies.

3 Finite Difference Method

In this section, higher order discretization schemes using FD methods are proposed to solve four dimensional nonlinear problems. The scheme involves the manipulation of the Taylor series expansion of the PDF about uniformly distributed nodal points in a $n = 4$ dimensional mesh as shown in Fig. 7.

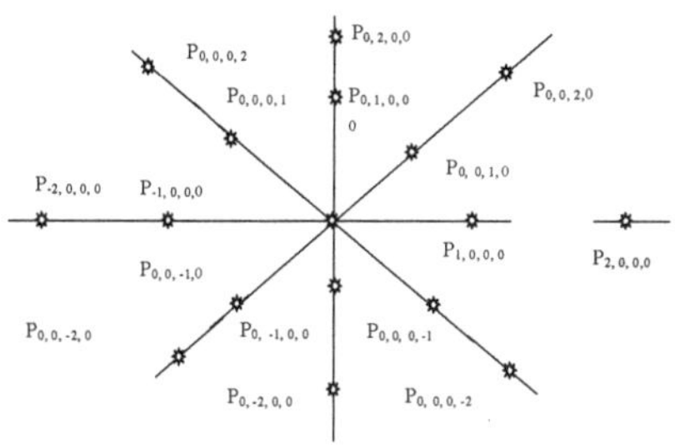

Fig. 7 Finite difference scheme for 4D spatial discretization

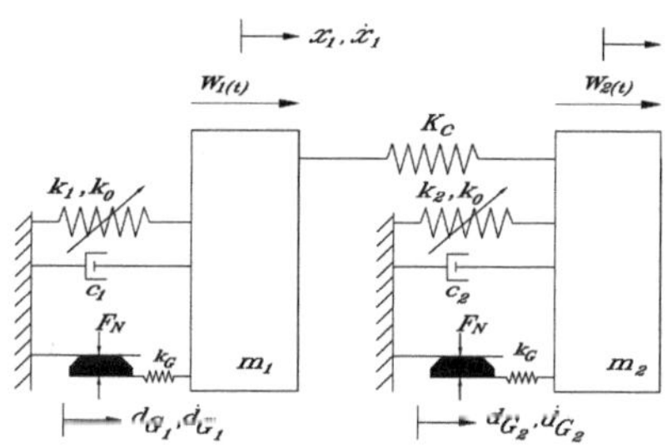

Fig. 8 Two degree of freedom model

Let $p_{i,j,k,l}$ denote the PDF at a discrete location relative to a candidate point as shown in Fig. 7. By taking more and more neighboring nodal points and solving for the derivative terms at the candidate points a 6th order FD stencil is derived as

$$\frac{\partial P(X_1,X_2,X_3,X_4)}{\partial X_1}=\frac{45\alpha_1^{X_1}-9\alpha_2^{X_1}+\alpha_3^{X_1}}{60\Delta X_1},\alpha_k^{X_1}=p_{k,0,0,0}-p_{-k,0,0,0} \tag{14}$$

$$\frac{\partial P(X_1,X_2,X_3,X_4)}{\partial X_2}=\frac{45\alpha_1^{X_2}-9\alpha_2^{X_2}+\alpha_3^{X_2}}{60\Delta X_2},\alpha_k^{X_2}=p_{0,k,0,0}-p_{0,-k,0,0} \tag{15}$$

$$\frac{\partial P(X_1,X_2,X_3,X_4)}{\partial X_3}=\frac{45\alpha_1^{X_3}-9\alpha_2^{X_3}+\alpha_3^{X_3}}{60\Delta X_3},\quad \alpha_k^{X_3}=p_{0,0,k,0}-p_{0,0,-k,0} \tag{16}$$

$$\frac{\partial P(X_1,X_2,X_3,X_4)}{\partial X_4}=\frac{45\alpha_1^{X_4}-9\alpha_2^{X_4}+\alpha_3^{X_4}}{60\Delta X_4},\quad \alpha_k^{X_4}=p_{0,0,0,k}-p_{0,0,0,-k} \tag{17}$$

$$\frac{\partial^2 P(X_1,X_2,X_3,X_4)}{\partial X_2^2}=\frac{270\Sigma_1-27\Sigma_2+2\Sigma_3-245\Sigma_0}{100(\Delta X_2)^2},\quad \Sigma_k=p_{0,k,0,0}+p_{0,-k,0,0} \tag{18}$$

By substituting into the FP equation a set of n-discrete first order differential equations is given by

$$[\mathbf{M}]\{\dot{\mathbf{p}}\}+[\mathbf{K}]\{\mathbf{p}\}=0,\qquad \mathbf{p}\in R^l \tag{19}$$

Using the GEAR time integration scheme [11] the time derivative of the PDF is approximated by a linear combination of the density function evaluated at the current and several previous points given by [11]

$$\dot{p}(t)=\sum_{i=0}^{N_G}\psi_j p(t_i-i\Delta t) \tag{20}$$

where N_G is the order of GEAR method and ψ_j is the integration weight. The resulting system of linear algebraic equations is solved at each time step employing the non-symmetric iterative solver. The PDF obtained by the GEAR method is further integrated by higher order quadrature rules to compute the response moments.

The FD method is applied to the two DOF system consisting of a nonlinear spring, viscous damper and nonlinear Coulomb damper shown in Fig. 8 representative of a tip shrouded bladed disk assembly involving the solution of a four dimensional FP equation. The equations of motion for the two DOF model subjected to white noise excitation (Fig. 8) can be expressed as

$$m_1\ddot{x}_1+c_1\dot{x}_1+k_1x_1+k_0x_1^3+k_c(x_1-x_2)+\mu F_N Sgn(\dot{d}_{G_1})=W_1(t) \tag{21}$$

$$m_2\ddot{x}_2+c_2\dot{x}_2+k_2x_2+k_0x_2^3-k_c(x_1-x_2)+\mu F_N Sgn(\dot{d}_{G_2})=W_2(t) \tag{22}$$

where m_1 and m_2 are corresponding masses, k_1 and k_2 are linear stiffness, k_0 is cubic stiffness, k_c is coupling stiffness, k_G is the stiffness of friction damper, c_1, c_2 are viscous damping coefficients, μ is the coefficient of friction, F_N is the normal preload, d_{G_1} and d_{G_2} are displacements of friction dampers, $W_1(t)$ and $W_2(t)$ are Gaussian white noises and *sgn(.)* is the signum function. The corresponding FP equation is given by

$$\begin{aligned}\frac{\partial p}{\partial t}&=m_1^{-2}D_1\frac{\partial^2 p}{\partial X_2^2}+m_2^{-2}D_2\frac{\partial^2 p}{\partial X_4^2}-X_2\frac{\partial p}{\partial X_1}-X_4\frac{\partial p}{\partial X_3}+m_1^{-1}(k_1+k_c)X_1\frac{\partial p}{\partial X_2}+m_1^{-1}c_1X_2\frac{\partial p}{\partial X_2}\\&-m_1^{-1}k_cX_3\frac{\partial p}{\partial X_2}-m_2^{-1}k_cX_1\frac{\partial p}{\partial X_4}+m_2^{-1}(k_c+k_2)X_3\frac{\partial p}{\partial X_4}+m_2^{-1}c_2X_4\frac{\partial p}{\partial X_4}\\&+m_1^{-1}[k_0X_1^3+\mu F_N Sgn(\dot{d}_{G_1})]\frac{\partial p}{\partial X_2}+m_2^{-1}[k_0X_3^3+\mu F_N Sgn(\dot{d}_{G_2})]\frac{\partial p}{\partial X_4}+(m_1^{-1}c_1+m_2^{-1}c_2)p\end{aligned} \tag{23}$$

The friction coefficient is assumed of the form

$$\mu = A + B\exp^{-\alpha \dot{d}_G} \tag{24}$$

where A, B, and α are constants. The following parameters are assumed for the numerical study $k_1 = k_2 = 4{,}15{,}000 N/m$, $k_c = 75{,}000 N/m$, $k_G = 42{,}000 N/m$, $m_1 = m_2 = 0.0115$ kg, $c_1 = c_2 = 1.45\ N\cdot s/m$, $A = 0.25$, $B = 0.05$, $\alpha = 0.18$, and $D_1 = D_2 = 1$ N^2/(rad/s). The state space in the FD method is discretized using a computational mesh of 35^4 nodes in the interval [-10, 10]×[-10, 10]×[-10, 10]×[-10, 10]. The initial distribution chosen is a zero mean, four dimensional Gaussian distribution with variance $\sigma_i^2 = 0.7$.The intensity of cubic nonlinearity, k_0, is chosen as 0.2 and time step used for the GEAR integration is 0.001. For three values of the normal load the mean square values of the displacement are compared with the results obtained from MCS, of six million realizations, in Fig. 9(a) with very good agreement between the two except at large times which may be due to the integration error.

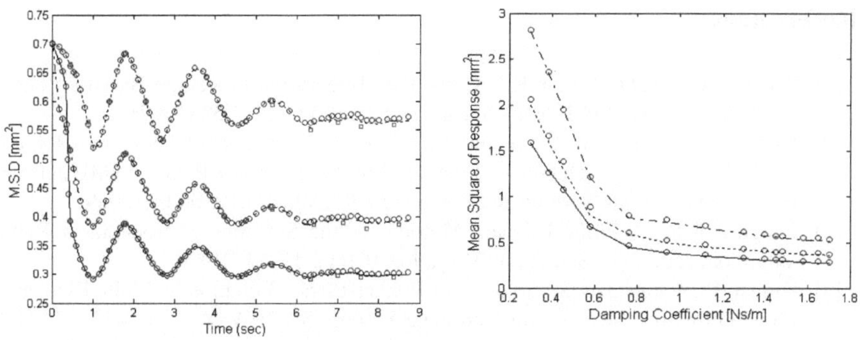

(a) (b)

Fig. 9 (a) Mean square of displacement, (b) Effects of damping coefficient on the mean square response. FD method for [F_N(N)—75,–·–·–25,------325]; °°°° MCS, ▪▪▪▪ FEM.

The effect of increasing the damping on the response statistics is studied by varying the damping coefficients c_1 and c_2 in the range of 0.3 to 1.75 N·m/s, The stationary mean square response at time t =24 sec is presented in Fig. 9(b) for different values of the damping coefficient. From the Fig. 9(b) it is seen that the mean square response of the blade to white noise excitation decreases with increased damping in the range of $c_1 = c_2 = 0.3$ to 1.4 Nm/s and, saturates beyond that value. It is also seen from the Fig. 9 that the FD results agree well with the MCS results.

5 Conclusions

A modified PI procedure is developed for the solution of the FP equation to predict the stochastic and chaotic response of number of nonlinear systems, subjected to external and parametric random excitations. The modified PI method is based on a non-Gaussian transition PDF which is the product of a Gaussian PDF and a series of Hermite polynomials. The examples considered show the efficacy of the proposed method. The method is useful in reducing the computational time and cost due to reduced number of time steps on the coarse grid. The paper also studies the stochastic jump, bifurcation, periodic and noise induced state change of PDF for nonlinear system subjected to combined harmonic and white noise excitation. A sixth order FD stencil scheme in conjunction with GEAR time integration is presented for the solution of four dimensional FP equation corresponding to a two DOF nonlinear system subjected to Gaussian white noise excitation representative of a tip-shrouded bladed-disk assembly. Effects of normal force and viscous damping on the mean square response of a blade are investigated.

References

[1] Yu, J.S., Cai, G.Q., Lin, Y.K.: A new Path Integration Procedure Based on Gauss – Legendre Scheme. Int. J. Non-Linear Mechanics 32, 759–768 (1997)

[2] Kumar, P., Narayanan, S.: Nonlinear Stochastic Dynamics, Chaos, and Reliability Analysis for a Single Degree of Freedom Model of a Rotor Blade. ASME Journal of Engineering for Gas Turbine and Power 130, 012506-1–012506-08 (2009)

[3] Langley, R.S.: A Finite Element Method for the Statistics of Non-Linear Random Vibration. Journal of Sound and Vibration 101(1), 41–54 (1985)

[4] Kumar, P., Narayanan, S.: Solution of Fokker-Planck Equation by Finite Element and Finite Difference Methods for Nonlinear System. SADHANA 31(4), 455–473 (2006)

[5] Wojtkiewicz, S.F., Bergman, L.A., Spencer Jr., B.F.: Robust Numerical Solution of the Fokker-Planck-Kolmogorov Equation for Two Dimensional Stochastic Dynamical Systems. Technical Report AAE 94-08, Department of Aeronautical and Astronautical Engineering, University of Illinois at Urbana-Champaign (1994)

[6] Masud, A., Bergman, L.A.: Application of Multi-Scale Finite Element Methods to the solution of the Fokker-Planck Equation. Computer Method in App. Mech. Engg. 194, 1513–1526 (2005)

[7] Yu, J.S., Lin, Y.K.: Numerical Path Integration of a Non-homogeneous Markov process. Int. J. Non-Linear Mechanics 39, 1493–1500 (2004)

[8] Kumar, P., Narayanan, S.: Modified Path Integral Solution of Fokker-Planck Equation: Response and Bifurcation of Nonlinear Systems. ASME Journal of Computational and Nonlinear Dynamics 05, 0110004-1:12 (2010)

[9] Zhang, D.S., Wei, G.W., Kouri, D.J., Hoffman, D.K.: Numerical Method for the Nonlinear Fokker-Planck Equation. Physical Review E 56(1), 1197–1206 (1997)

[10] Jung, P., Hänggi, P.: Invariant Measure of a Driven Nonlinear Oscillator with External Noise. Phys. Rev. Lett. 65, 3365–3368 (1990)

[11] Gear, C.W.: Numerical Initial Value Problems in Ordinary Differential Equation. Prentice-Hall, Englewood Cliffs (1971)

An Approximate Approach for Nonlinear System Evolutionary Response Spectrum Determination via Wavelets

P.D. Spanos[1] and I.A. Kougioumtzoglou[2]

[1] L.B. Ryon Chair in Engineering, Rice University, USA
[2] Department of Civil and Environmental Engineering, Rice University, USA

Abstract. A novel harmonic wavelet-based statistical linearization approach is proposed for determining the evolutionary power spectrum (EPS) of the response of nonlinear oscillators subject to stochastic excitation. Specifically, first a mathematically rigorous wavelet-based representation of non-stationary stochastic processes is presented. Next, a representation of the process corresponding to a specific scale and translation level is derived. This procedure leads to an EPS estimation approach which is applicable for estimating not only separable but non-separable in time and frequency EPS as well. Next, focusing on the case of the stochastic response of a nonlinear system and relying on the orthogonality properties of the developed representation an excitation-response EPS relationship is derived via statistical linearization. The approach involves the concept of assigning optimal and response dependent equivalent stiffness and damping elements corresponding to the specific frequency and time bands. This leads to an iterative determination of the EPS of the oscillator response. Pertinent Monte Carlo simulations demonstrate the reliability and versatility of the approach.

Keywords: Stochastic processes; Random vibration; Monte Carlo method; Nonlinear systems; Statistical linearization; Wavelets.

1 Introduction

Structural systems are often subject to stochastic excitations such as seismic motions, winds, and ocean waves which inherently possess the attribute of evolution in time. Therefore, representation of these phenomena by non-stationary stochastic processes is necessary to capture accurately the system behavior. Associated with the notion of a non-stationary stochastic process is the concept of the evolutionary power spectrum (EPS) (e.g. [1,2]). Several research efforts have focused on determining the response of systems under evolutionary excitation; see Kougioumtzoglou and Spanos [3] for a recent reference. Nevertheless, limited results exist in the context of a joint time-frequency response analysis.

Recently, wavelets have been successfully applied to study dynamic systems with time-varying characteristics and to develop system identification approaches.

W.Q. Zhu et al. (Eds.): Nonlinear Stochastic Dynamics & Control, IUTAM BOOK SERIES 29, pp. 87–96.
springerlink.com

A detailed presentation of engineering related applications can be found in review articles such as the one by Spanos and Failla [4]. In particular, harmonic wavelets have a critical presence in structural dynamics applications. In this paper, relying on the properties of the generalized harmonic wavelets and on a statistically rigorous approach to modeling non-stationary processes [2], a representation of the process corresponding to a specific scale and translation level is derived. Further, a novel statistical linearization approach is developed for the nonlinear response EPS determination. The advantages of this approach vis-à-vis the existing linearization schemes are emphasized and its reliability is verified by pertinent Monte Carlo studies.

2 Harmonic Wavelet-Based Stochastic Process Representation and EPS Estimation

2.1 *Harmonic Wavelet Transform*

The family of generalized harmonic wavelets uses two parameters (m,n) for the definition of the bandwidth at each scale level; see reference [5]. Generalized harmonic wavelets have a band-limited, box-shaped frequency spectrum. A wavelet of (m,n) scale and (k) position in time attains a representation in the frequency domain of the form

$$\Psi^{G}_{(m,n),k}(\omega)=\begin{cases}\dfrac{1}{(n-m)\Delta\omega}e^{-i\frac{\omega kT_o}{n-m}}, & m\Delta\omega\leq\omega<n\Delta\omega\\ 0, & otherwise\end{cases}, \tag{1}$$

where (m), (n) and (k) are considered to be positive integers and $(\Delta\omega=2\pi/T_o)$, where (T_o) is the total time duration of the signal under consideration. The inverse Fourier transform of Eq.(1) gives the time-domain representation of the wavelet which is equal to

$$\psi^{G}_{(m,n),k}(t)=\frac{e^{in\Delta\omega\left(t-\frac{kT_0}{n-m}\right)}-e^{im\Delta\omega\left(t-\frac{kT_0}{n-m}\right)}}{i(n-m)\Delta\omega\left(t-\dfrac{kT_0}{n-m}\right)}. \tag{2}$$

The continuous generalized harmonic wavelet transform (GHWT) is defined as

$$W^{G}_{(m,n),k} = \frac{n-m}{T_o} \int_{-\infty}^{\infty} f(t)\overline{\psi^{G}_{(m,n),k}(t)}dt, \tag{3}$$

and projects the $(f(t))$ on this wavelet basis; the bar over a symbol represents complex conjugation.

2.2 Locally Stationary Wavelet Process Representation

Most rigorous approaches to modeling non-stationary stochastic processes are based on extensions of the classical representation of stationary processes (e.g. [6]). Specifically, a zero mean stationary process $(f(t))$ can be represented as

$$f(t) = \int_{-\infty}^{\infty} A(\omega) e^{i\omega t} dZ(\omega), \tag{4}$$

where $(A(\omega))$ is a deterministic function and $(dZ(\omega))$ is a zero mean orthonormal increment stochastic process. The power spectrum of the process $(f(t))$ is then defined as $S_f(\omega) = |A(\omega)|^2$. Next, to develop models whose spectral content changes with time one option is to replace the amplitude $(A(\omega))$ in Eq.(4) by a time-varying version $(A(t,\omega))$. This leads to the slowly varying non-stationary processes as defined by Priestley [1]. Further, Nason et al. [2] introduced a novel representation of non-stationary stochastic processes in which the Fourier basis is replaced by a wavelet basis. The proposed process model is constructed to be locally stationary via constraints on the model coefficients, resulting in what is known as a locally stationary wavelet (LSW) process. This allows for defining a wavelet spectrum at a particular scale and location. According to the LSW process representation, the non-stationary process $(f(t))$ can be represented as

$$f(t) = \sum_{j}\sum_{k} w_{j,k}\psi_{j,k}(t)\xi_{j,k}, \tag{5}$$

where $(\xi_{j,k})$ is a stochastic orthonormal increment sequence; and $(\psi_{j,k}(t))$ is a non-decimated family of wavelets. According to the properties of the representation of Eq.(5), which are defined and proved in Nason et al. [2], the EPS at each time and scale is given by the equation $\left(S_{j,k}=\left|w_{j,k}\right|^2\right)$. Rewriting Eq.(5) for the case of the generalized harmonic wavelets yields

$$f(t)=\sum_{(m,n)}\sum_{k}\sqrt{S_{(m,n),k}}\,\psi^{G}_{(m,n),k}(t)\xi_{(m,n),k}\,, \tag{6}$$

where $\left(S_{(m,n),k}\right)$ is the evolutionary spectrum at scale (m,n) and translation (k). In conjunction with the preceding analysis consider a representation of a stochastic process $(f(t))$ in the form

$$f(t)=\sum_{(m,n)}\sum_{k}f_{(m,n),k}(t)\,, \tag{7}$$

where $\left(f_{(m,n),k}(t)\right)$ is the process at scale (m,n) and translation (k) given by

$$f_{(m,n),k}(t)=\int_{m\Delta\omega}^{n\Delta\omega}e^{i\omega\left(t-\frac{kT_0}{n-m}\right)}dZ_{(m,n),k}(\omega)\,, \tag{8}$$

where $dZ_{(m,n),k}(\omega)=\sqrt{S_{(m,n),k}}\left((n-m)\Delta\omega\right)\xi_{(m,n),k}$, with the properties $E\left(dZ_{(m,n),k}(\omega)\right)=0$, and $E\left(\left|dZ_{(m,n),k}(\omega)\right|^2\right)=S_{(m,n),k}(n-m)\Delta\omega$. In deriving Eq.(8), Eq.(2) has been taken into account. The similarity between Eq.(4) and Eq.(8) is obvious. This is not surprising considering the fact that the harmonic wavelet basis functions are essentially localized Fourier functions. Following a similar analysis as in reference [7], a spectral representation of $\left(f_{(m,n),k}(t)\right)$ of the form of Eq.(8) involving an adequately large number of N harmonics of constant amplitudes and of random phases can be cast in the form

$$f_{(m,n),k}(t)=\sqrt{4S_{(m,n),k}d\omega}\sum_{l=0}^{N-1}\cos\left((\omega_l)\left(t-\frac{kT_0}{n-m}\right)+\phi_l\right), \tag{9}$$

Note that this representation is defined in the intervals $(m\Delta\omega\leq\omega_l<n\Delta\omega)$ and $\left(\frac{k}{n-m}T_o\leq t<\frac{k+1}{n-m}T_o\right)$ with $\left(d\omega=\frac{(n-m)\Delta\omega}{\mathrm{N}}\right)$, and $(\omega_l=m\Delta\omega+ld\omega)$.

2.3 EPS Estimation

Early attempts of applications of wavelets in spectral estimation for vibration problems include the work by Basu and Gupta [8] where they related the mean square value of the wavelet transform at different scales and the time-dependent spectral content of the process (see also reference [9]). Spanos and Failla [10] followed an alternative approach developing relationships between the EPS and the wavelet coefficients in context with the theory of non-stationary processes as proposed by Priestley [1]. In the same context, the developments of the preceding section are used herein to derive a scheme for estimations of both separable and non-separable EPS for stochastic processes. According to Newland [5] the equation

$$\int_{-\infty}^{\infty} |f(t)|^2 dt = 2\sum_{m,n}\sum_{k}\left(\frac{T_o}{n-m}\left|W^G_{(m,n),k}\right|^2\right), \tag{10}$$

holds true. Moreover, considering Parseval's theorem yields

$$2\pi \int_{-\infty}^{\infty} E\left[|F(\omega)|^2\right] d\omega = \int_{-\infty}^{\infty} E\left[|f(t)|^2\right] dt. \tag{11}$$

Combining Eqs.(10) and (11) and considering the non-overlapping character of the different energy bands, it can be argued that the estimation of the EPS $\left(S(\omega,t)\right)$ can be obtained as

$$S(\omega_i,t_i) = \frac{T_o}{2\pi(n-m)} E\left[\left|W^G_{(m,n),k}\right|^2\right], \quad m\Delta\omega \le \omega_i < n\Delta\omega, \quad \frac{kT_o}{n-m_o} \le t_i < \frac{(k+1)T_o}{n-m}. \tag{12}$$

To demonstrate the accuracy of the proposed estimation the non-separable spectrum of the form

$$S(\omega,t) = S_2\left(\frac{\omega}{5\pi}\right)^2 e^{-0.15t} t^2 e^{-\left(\frac{\omega}{5\pi}\right)^2 t}, \quad t \ge 0, \quad -\infty < \omega < \infty. \tag{13}$$

where $\left(S_2 = 1\right)$, is considered next. This spectrum comprises some of the predominant features of seismic shaking, such as decreasing of the dominant frequency with time. Realization records compatible with Eq.(13) are produced using the concept of spectral representation of a stochastic process (e.g. [7]). In Fig.(1) the target and the EPS estimate are plotted at two time instants $(t = 1.6\,\text{sec})$ and $(t = 4.1\,\text{sec})$.

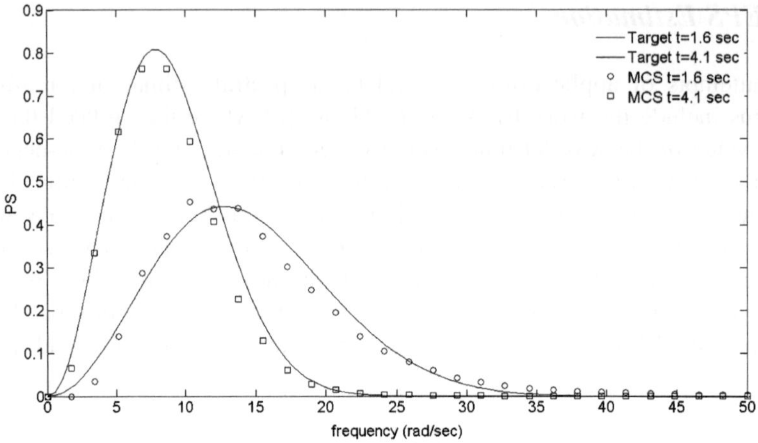

Fig. 1 Non-separable power spectrum $(S_2 = 1)$ at $(t = 1.6\,\text{sec})$ and $(t = 4.1\,\text{sec})$. Comparison between MCS data (500 realizations) and the target spectrum.

The two EPS are in good agreement with each other verifying the reliability of the generalized harmonic wavelet-based EPS estimation approach for physically realistic versions of EPS with time-varying frequency content.

3 Harmonic Wavelet-Based Statistical Linearization

3.1 Nonlinear Response EPS Determination

Early attempts towards developing a linearization approach include the work by Basu and Gupta [11], where the equivalent linear element is essentially averaged over the different wavelet scales leading to a linear time-variant (LTV) system. It can be argued that this approach negates in a sense the joint time-frequency representation capabilities of the wavelet transform. In this section, relying on the LSW representation developed in section 2 a statistical linearization approach is presented for each scale and translation level. In this regard, the equivalent stiffness and damping elements corresponding to the specific frequency and time band are employed to evaluate the response EPS in an iterative manner. Consider next a nonlinear single-degree-of-freedom system whose motion is governed by the differential equation

$$\ddot{x} + 2\zeta_0\omega_0\dot{x} + \omega_0^2 x + \varepsilon h[x, \dot{x}] = w(t)\,. \tag{14}$$

where (ζ_0) is the ratio of critical damping; (ω_0) is the natural frequency of the corresponding linear oscillator; $(w(t))$ represents a Gaussian, zero-mean non-stationary stochastic process possessing an EPS $S(\omega,t)$; (ε) denotes the degree of nonlinearity; and $(h[x,\dot{x}])$ is an arbitrary nonlinear function which depends on the response displacement and velocity. Replacing next the nonlinear Eq.(14) with the equivalent linear

$$\ddot{x} + 2\zeta_{eq}\omega_{eq}\dot{x} + \omega_{eq}^2 x = w(t)\,. \tag{15}$$

and performing minimization of the error in a mean square sense (e.g. [12]) yields

$$E\left(\dot{x}G[x,\dot{x}]\right) - 2\zeta_{eq}\omega_{eq}E\left(\dot{x}^2\right) - \omega_{eq}^2 E\left(x\dot{x}\right) = 0\,, \tag{16}$$

and

$$E\left(xG[x,\dot{x}]\right) - 2\zeta_{eq}\omega_{eq}E\left(x\dot{x}\right) - \omega_{eq}^2 E\left(x^2\right) = 0\,, \tag{17}$$

where $G[x,\dot{x}] = 2\zeta_0\omega_0\dot{x} + \omega_0^2 x + \varepsilon h[x,\dot{x}]$. Substituting next Eq.(7) into Eqs.(16-17) and considering Eq.(9) it can be readily seen that the only terms which survive from the expectation operation are those which correspond to the same scale (frequency) (m,n) and translation (time) (k) intervals. The expectation of the rest of the terms is equal to zero due to the independence of the random phase variables. This yields

$$2\zeta_{eq,(m,n),k}\,\omega_{eq,(m,n),k} = 2\zeta_0\omega_0 + \varepsilon E\left(\frac{\partial h\left(x_{(m,n),k}, \dot{x}_{(m,n),k}\right)}{\partial \dot{x}_{(m,n),k}}\right), \tag{18}$$

and

$$\omega_{eq,(m,n),k}^2 = \omega_0^2 + \varepsilon E\left(\frac{\partial h\left(x_{(m,n),k}, \dot{x}_{(m,n),k}\right)}{\partial x_{(m,n),k}}\right), \tag{19}$$

where the standard Gaussian assumption for the response processes has been invoked (e.g. [12]). It is important to note that the equivalent elements (Eqs.(18-19)) are treated as frequency and time dependent. In other words, Eqs.(18-19) are valid in the intervals $(m\Delta\omega \le \omega_l < n\Delta\omega)$ and $\left(\frac{k}{n-m}T_o \le t < \frac{k+1}{n-m}T_o\right)$. Obviously, the evaluation of the equivalent stiffness and damping elements depends on the response statistics. Thus, additional spectral input-output relationships must be considered to yield a system of simultaneous nonlinear equations which can be solved iteratively. To this aim, raising Eq.(15) to the second power, applying the expectation operator and taking into account the representation of Eq.(9) and the orthogonality properties of the monochromatic functions yields

$$S_{(m,n),k}^x = \frac{\left(\sum_{l=0}^{N-1}\left(\frac{1}{2}\right)\right)S_{(m,n),k}^w}{\sum_{l=0}^{N-1}\left(\frac{1}{2}\omega_l^4\right) - 2\omega_{eq,(m,n),k}^2\sum_{l=0}^{N-1}\left(\frac{1}{2}\omega_l^2\right) + \omega_{eq,(m,n),k}^4\sum_{l=0}^{N-1}\left(\frac{1}{2}\right) + \left(2\zeta_{eq,(m,n),k}\,\omega_{eq,(m,n),k}\right)^2\sum_{l=0}^{N-1}\left(\frac{1}{2}\omega_l^2\right)}, \tag{20}$$

It can be readily seen that for each scale (m,n) and translation level (k), the unknowns $\left(\omega^2_{eq,(m,n),k}\right)$, $\left(2\zeta_{eq,(m,n),k}\,\omega_{eq,(m,n),k}\right)$ and $\left(S^x_{(m,n),k}\right)$ can be obtained by solving Eqs.(18-20). To this aim, an iterative scheme can be adopted. Note that the proposed statistical linearization approach can be the basis for performing a joint time-frequency response analysis and is readily applicable for both separable and non-separable excitation EPS.

3.2 Duffing Oscillator Application

To assess the accuracy of the approach, a Duffing oscillator of the form

$$\ddot{x}+2\zeta_0\omega_0\dot{x}+\omega_0^2x+\varepsilon\omega_0^2x^3=w(t),\quad \varepsilon>0\,, \tag{21}$$

is considered. Applying Eqs. (18-19) and taking into account Eq.(9) yields $2\zeta_{eq,(m,n),k}\,\omega_{eq,(m,n),k}=2\zeta_0\omega_0$, and $\omega^2_{eq,(m,n),k}=\omega_0^2\left(1+3\varepsilon\left(4S^x_{(m,n),k}d\omega\sum_{l=0}^{N-1}\left(\frac{1}{2}\right)\right)\right)$. The parameters values $\left(\omega_0=10rad/\sec,\zeta=0.1,\varepsilon=10\right)$ are chosen for the Duffing oscillator of Eq.(21). The response EPS of the Duffing oscillator is calculated using the non-separable excitation EPS of Eq.(13). Comparisons of the EPS with MCS data at distinct time instants (Fig.(2)) demonstrate the accuracy of the approach even for this high degree of nonlinearity.

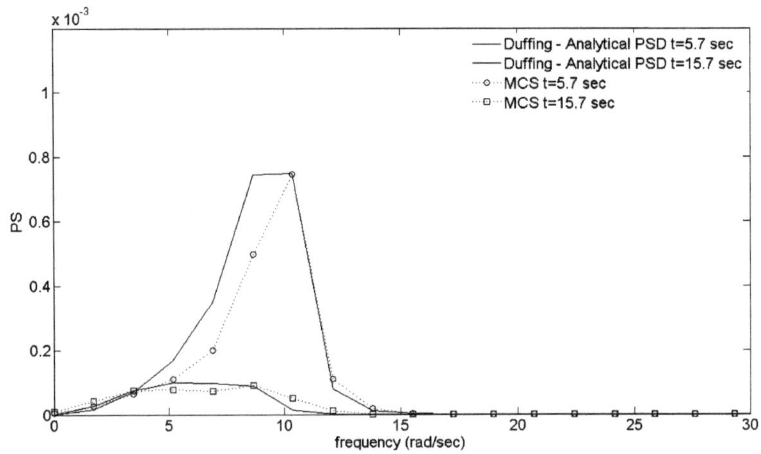

Fig. 2 Response EPS of a Duffing oscillator $\left(\omega_0=10rad/\sec,\zeta=0.1,\varepsilon=10\right)$ at $(t=5.7\,\sec)$ and $(t=15.7\,\sec)$ under a non-separable evolutionary spectrum $(S_2=1)$. Comparison between MCS data (500 realizations) and the analytical approach.

4 Conclusions

In this paper, a novel wavelet-based statistical linearization approach has been proposed for determining the EPS of the response of nonlinear oscillators under evolutionary stochastic excitation. The approach has been developed in conjunction with the family of harmonic wavelets. Specifically, relying on the properties of the generalized harmonic wavelet and exploiting the concept of a locally stationary wavelet-based representation of stochastic processes, a process corresponding to a specific scale and translation level has been defined. In this context, first an EPS estimation approach has been presented in conjunction with the proposed process representation and has been applied in estimating both separable and non-separable EPS. Thus, the capacity of the approach to capture successfully the time-varying frequency content of quite complex non-stationary stochastic phenomena is demonstrated. Finally, in context with the statistical linearization approach, excitation-response relationships have been obtained by employing the novel concept of wavelet scale (frequency) and wavelet translation level (time) dependent equivalent stiffness and damping elements. The resulting nonlinear system of equations has been solved iteratively to determine the response EPS. In this manner, a joint time-frequency response analysis has been achieved. Pertinent Monte Carlo simulations have demonstrated the accuracy of the approach.

References

1. Priestley, M.B.: Evolutionary spectra and non-stationary processes. Journal of the Royal Statistical Society 27, 204–237 (1965)
2. Nason, G.P., von Sachs, R., Kroisand, G.: Wavelet processes and adaptive estimation of evolutionary wavelet spectra. Journal of the Royal Statistical Society 62, 271–292 (2000)
3. Kougioumtzoglou, I.A., Spanos, P.D.: An approximate approach for nonlinear system response determination under evolutionary stochastic excitation. Current Science, Indian Academy of Sciences 97, 1203–1211 (2009)
4. Spanos, P.D., Failla, G.: Wavelets: Theoretical concepts and vibrations related applications. The Shock and Vibration Digest. 37, 359–375 (2005)
5. Newland, D.E.: Harmonic and musical wavelets. Proceedings of the Royal Society London A 444, 605–620 (1994)
6. Cramer, H., Leadbetter, M.R.: Stationary and related stochastic processes. Wiley, New York (1967)
7. Liang, J., Chaudhuri, S.R., Shinozuka, M.: Simulation of non-stationary stochastic processes by spectral representation. Journal of Engineering Mechanics 133, 616–627 (2007)
8. Basu, B., Gupta, V.K.: Seismic response of SDOF systems by wavelet modeling of non-stationary processes. Journal of Engineering Mechanics 124, 1142–1150 (1998)
9. Spanos, P.D., Tezcan, J., Tratskas, P.: Stochastic processes evolutionary spectrum estimation via harmonic wavelets. Computer Methods in Applied Mechanics and Engineering 194, 1367–1383 (2005)

10. Spanos, P.D., Failla, G.: Evolutionary spectra estimation using wavelets. Journal of Engineering Mechanics 130, 952–960 (2004)
11. Basu, B., Gupta, V.K.: On equivalent linearization using wavelet transform. Journal of Vibration and Acoustics 121, 429–432 (1999)
12. Roberts, J.B., Spanos, P.D.: Random Vibration and Statistical Linearization. Dover Publications, New York (2003)

On Information/Entropy Flow in Stochastic Dynamical Systems

K. Sobczyk and P. Hołobut

Institute of Fundamantal Technological Research, Polish Academy of Sciences,
Pawińskiego 5b, 02-106 Warsaw, Poland

Abstract. The objective of this paper is to show how some basic informational quality measures (such as entropy and relative entropy / Kullback divergence) of stochastic dynamical systems depend on the system properties and characteristics of the external/internal randomness. First, the Shannon entropy flow in dynamic systems with random initial states is considered with emphasis on the effects of the system properties. Next, we quantify the influence of random external noise as well as the parametric randomness on the entropy and on the Kullback-Leibler relative entropy of the system. The analysis is illustrated by specific dynamical systems for which the entropy change in time is presented graphically.

Keywords: stochastic systems, Shannon entropy, relative entropy, entropy production.

1 Introduction

Although the contemporary stochastic dynamics of linear and nonlinear systems has gained a high level of maturity (cf. Lin, Cai [7], Sobczyk [11]), it seems that in its applicatory aspects, it still concentrates primarily on the reliability-type properties. It is no doubt, that the information-theoretic approach to the system dynamics may seriously enrich the existing applied analysis of stochastic systems. Especially, the advantages of the potential inherent in information theory seems to be relevant to various complex biological and social systems.

Theoretic information reasoning is well known in physics, but in this field the basic focus is on the relationships of statistical entropy with nonequilibrium thermodynamics. A good example is the famous Boltzmann H-theorem, and the second law of thermodynamics. The language of information theory turned out to be useful to quantifying chaos, system predictability and self-organization (cf. Haken [5], Ebeling [3]). It is also very relevant to data processing and statistical inference including inference on the behaviour of dynamical systems. For variety of real systems an important question is, for example: how does the Shannon entropy flow through a dynamic system, and how does it depend on the system characteristics and properties of random disturbances? Although in the existing literature there are a number of contributions on information and dynamics (cf. Sobczyk [12] and references therein, Garbaczewski [4], Munakata and Igarashi [8]) the subject still calls for further insight. This paper is an attempt in this direction.

W.Q. Zhu et al. (Eds.): Nonlinear Stochastic Dynamics & Control, IUTAM BOOK SERIES 29, pp. 97–106.
springerlink.com

2 Entropy and Relative Entropy for SDS

Real stochastic dynamical systems (SDS) are usually modelled by stochastic differential equations which, generally, can be represented as

$$\frac{\mathrm{d}\mathbf{Y}(t)}{\mathrm{d}t} = \mathbf{F}[\mathbf{Y}(t), t; \mathbf{X}(t,\gamma)], \qquad \mathbf{Y}(t_0) = \mathbf{Y}_0 \tag{1}$$

where $\mathbf{Y}(t) = [Y_1(t), \dots, Y_n(t)]$ is unknown stochastic process, t denotes time ($t \in T$), and $\mathbf{X}(t,\gamma)$ is a given stochastic process ($\gamma \in \Gamma$, where Γ is the space of elementary events on which the probability is defined (cf. Sobczyk [11])). Function $\mathbf{F}(\mathbf{y},t,\mathbf{x})$ is a given sufficiently smooth function characterizing the physical nature of the system considered. We are interested in characterization of the process $\mathbf{Y}(t)$ on the basis of information/data about $\mathbf{F}$, $\mathbf{Y}_0$, and $\mathbf{X}$. Since $\mathbf{Y}(t)$ is a stochastic process we are interested in the probability density $f(\mathbf{y},t)$ of $\mathbf{Y}$ for each $t \in T$ (more completely, as is known, process $\mathbf{Y}(t)$ is described by the family of all joint probability densities over sets $\{t_1, t_2, \dots, t_k\}$ of $t \in T$). Here, we are interested in the behaviour of the following functional defined on $f(\mathbf{y},t)$:

$$H_t(f) = H_{\mathbf{Y}}(f;t) = -\int f(\mathbf{y},t) \log f(\mathbf{y},t) \mathrm{d}\mathbf{y} = -\langle \log f(\mathbf{y},t) \rangle_f \tag{2}$$

where the bracket $\langle \cdot \rangle$ denotes the average value (with respect to density f) of the quantity indicated, and the integration is extended over the support of $f(\mathbf{y},t)$. The quantity $H_t(f)$ is the Shannon entropy of $\mathbf{Y}(t)$ at time t. It quantifies, for each t, the overall randomness of $\mathbf{Y}(t)$; we may also say that $H_{\mathbf{Y}}(t)$ quantifies the information content I_t of $\mathbf{Y}(t)$ at time t (cf. Cover, Thomas [2]).

Another measure/functional assigned to process $\mathbf{Y}(t)$ which can be used for quantifying the information transfer in stochastic system (1) is the *relative entropy* (or, the *Kullback-Leibler* divergence measure). Let $f(\mathbf{y},t)$ and $q(\mathbf{y},t)$ be two probability densities with the same support and such that $f(\mathbf{y},t)$ is the density of our main interest (often unknown explicitly), whereas $q(\mathbf{y},t)$ is some *prior* or *reference* density associated with the system considered. We wish to quantify the informational difference between f and q. The relative entropy between f and q is defined as

$$H_t(f,q) = \int f(\mathbf{y},t) \log \frac{f(\mathbf{y},t)}{q(\mathbf{y},t)} \mathrm{d}\mathbf{y} = \left\langle \log \frac{f(\mathbf{y},t)}{q(\mathbf{y},t)} \right\rangle_f \tag{3}$$

with the convention that $0 \log(0/q) = 0$ and $f \log(f/0) = \infty$.

The relative entropy $H_t(f,q) \geqslant 0$ with equality if and only if densities f and q are identical. The quantity (3) is not a distance measure in the sense of metric (it is not symmetric and the triangle inequality does not hold) but it is widely accepted as an informational indicator of similarity or divergence of two densities f and q. When density $f(\mathbf{y},t)$ in (2) or both $f(\mathbf{y},t)$ and $q(\mathbf{y},t)$ in (3) are Gaussian then $H_t(f)$ and $H_t(f,q)$ have analytical expressions in terms of the corresponding covariance matrices (for $n = 1$, in terms of variances). When $\log = \log_2$ the units of $H_t(f)$ and $H_t(f,q)$ are bits.

In the next sections we will investigate the properties of the measures (2) and (3) for various forms of system (1).

3 Entropy in Liouvillian Systems

The simplest case of random dynamics occurs when randomness is caused by the statistical nature of the initial states. So, in model (1) there is no stochastic process in the system equations ($\mathbf{X}(t,\gamma) \equiv 0$); only the initial condition $\mathbf{Y}_0$ is random with given probability density $f_0(\mathbf{y})$. During the motion the initial density $f_0(\mathbf{Y})$ evolves to a new density $f(\mathbf{y},t)$ for $t > t_0$. This means that also the Shannon entropy of the system passes from H_0 to $H_{\mathbf{Y}}(t)$. This is a Liouvillian flow since (as it is known – cf. Sobczyk [11]) the probability density $f(\mathbf{y},t)$ satisfies the Liouville's equation (cf. Nemytskii and Stepanov [9])

$$\frac{\partial f}{\partial t} + \sum_{i=1}^{n} \frac{\partial}{\partial y_i} [f(\mathbf{Y},t)F_i(\mathbf{y},t)] = 0 \tag{4}$$

with $f(\mathbf{y},t_0) = f_0(\mathbf{y})$.

Making use of equation (4) we can derive an evolution equation for the Shannon entropy functional (2). After differentiation of $H_t(f)$ with respect to time and making the appropriate transformations (with the assumption that fF_i tend to zero at infinity) one obtains (cf. Sobczyk and Hołobut [13])

$$\frac{\mathrm{d}H_t(f)}{\mathrm{d}t} = \sum_{i=1}^{n} \left\langle \frac{\partial F_i}{\partial y_i} \right\rangle_f = \langle \mathrm{div}_{\mathbf{y}} \mathbf{F} \rangle_f \tag{5}$$

Equation (5) says that the sign of the entropy rate of the system considered depends on the sign of the divergence of the vector field $\mathbf{F}(\mathbf{y},t)$ governing the system. In contrast with standard thermodynamical intuition (and the second law of thermodynamics) the information entropy can grow with time only if the dynamic deterministic system (1), when $\mathbf{X}(t) \equiv 0$, has positive mean flow divergence. For the Hamiltonian systems (i.e. described by Hamilton's equations) commonly studied in statistical physics $\mathrm{div}_{\mathbf{Y}}\mathbf{F} = 0$, so $\mathrm{d}H/\mathrm{d}t = 0$ which means that entropy $H_{\mathbf{Y}}(t)$ is conserved during the motion, i.e. $H_{\mathbf{Y}}(t) = H_0$, $t \geqslant t_0$. For other systems we may observe the entropy decrease or increase with time. The simplest example is: let $\dot{Y}(t) = -Y(t)$, $\mathrm{div}F = -1$, therefore $\dot{H}(t) = -1$ which means that $H_Y(t) = H_0 - (t - t_0)$, $t \geqslant t_0$ (entropy decreases with time – the system is more and more predictable).

It is worth noticing that the entropy concept used here in the context of dynamical systems is the entropy defined by (2) and commonly known as the statistical or Shannon entropy (which, in fact, is the Boltzmann entropy with opposite sign – introduced in 1872). In the qualitative theory of dynamical systems a useful measure for the chaotic nature of motion of a system the analogous (but different) Kolmogorov-Sinai entropy is used.

4 Entropy Rate of Systems with External Random Excitation

It is of interest to know how a random external noise influences the entropy flow in the system in question.

Let us consider a system governed by a special case of equation (1), namely by the following Langevin-type vectorial equation

$$\frac{\mathrm{d}\mathbf{Y}(t)}{\mathrm{d}t} = \mathbf{F}(\mathbf{Y},t) + \sigma\xi(t), \qquad t \in [t_0, \infty) \tag{6}$$

where $\mathbf{Y}(t)$ is an unknown stochastic n-dimensional vector process, $\xi(t)$ is a given vector stochastic white noise excitation with the intensity matrix $\sigma = \{\sigma_{ij}\}$. Under known assumptions the probability density $f(\mathbf{y},t|\mathbf{y}_0,t_0) = f(\mathbf{y},t)$ satisfies the following Fokker-Planck-Kolmogorov equation

$$\frac{\partial f}{\partial t} + \sum_{i=1}^{n} \frac{\partial}{\partial y_i}[F_i(\mathbf{y},t)f(\mathbf{y},t)] - \frac{1}{2}\sum_{i,j=1}^{n} b_{ij}\frac{\partial^2 f}{\partial y_i \partial y_j} = 0 \tag{7}$$

where $b_{ij} = \sum_{r=1}^{n} \sigma_{ir}\sigma_{jr}$ (that is the matrix $\mathbf{B} = \{b_{ij}\} = \sigma\sigma^{\mathrm{T}}$).

What is the rate of the Shannon entropy flow in system (6) with its probability density governed by eq.(7)? According to the definition of $H(t)$ – eq.(2)

$$\frac{\mathrm{d}H}{\mathrm{d}t} = -\frac{\mathrm{d}}{\mathrm{d}t}\int f \ln f \mathrm{d}\mathbf{y} = -\int \frac{\partial f}{\partial t}(\ln f + 1)\mathrm{d}\mathbf{y} \tag{8}$$

The second term on the right hand side is (due to the normalization condition) equal to zero, and $\partial f/\partial t$ (due to F-P-K equation (7)) consists of two components, i.e.

$$A_L = -\mathrm{div}(\mathbf{F}f) \quad , \quad A_D = \frac{1}{2}\sum_{i,j} b_{ij}\frac{\partial^2 f}{\partial y_i \partial y_j}. \tag{9}$$

Therefore

$$\frac{\mathrm{d}H}{\mathrm{d}t} = \dot{H}_L + \dot{H}_D \tag{10}$$

where $\dot{H}_L$ – the Liouvillian rate is represented by formula (5). After suitable transformations, the second component $\dot{H}_D$ – the diffusion entropy rate is as follows

$$\begin{aligned} \frac{\mathrm{d}H_D}{\mathrm{d}t} &= -\int \frac{1}{2}\sum_{i,j} b_{ij}\frac{\partial^2 f}{\partial y_i \partial y_j}\ln f \mathrm{d}\mathbf{y} \\ &= \frac{1}{2}\sum_{i,j} b_{ij}\left\langle \frac{\partial \ln f}{\partial y_i}\frac{\partial \ln f}{\partial y_j}\right\rangle_f \end{aligned} \tag{11}$$

The above entropy rate is positive since the diffusion matrix $\{b_{ij}\}$ is positive definite. This indicates that the total entropy rate $\dot{H} = \dot{H}_L + \dot{H}_D$ may have positive or negative signs (with value of zero when random noise is absent and the system is divergenceless). The quantity $\dot{H}_D$ expressed by formula (11) can be interpreted as the information entropy production (by a random excitation). The quantity $\dot{H}_L$ characterizes the system entropy due to the process inside the system.

Example 1

As an illustration of the entropy change in a stochastic dynamical system let us consider a simple process governed by the equation

$$\dot{Y}(t) + aY(t) = \xi(t), \qquad a > 0 \tag{12}$$

and $\xi(t)$ is the Gaussian white noise, such that

$$\langle \xi(t) \rangle = 0, \qquad \langle \xi(t_1)\xi(t_2) \rangle = \sqrt{2D}\delta(t_2 - t_1) \tag{13}$$

Therefore the drift and diffusion coefficients are

$$F(y) = -ay, \qquad \sigma = \sqrt{2D} \tag{14}$$

and equation (7) is

$$\frac{\partial f}{\partial t} = -a\frac{\partial}{\partial y}[yf(y,t)] + D\frac{\partial^2 f}{\partial y^2} \tag{15}$$

with $f(y,0) = f_0(y)$ – being a Gaussian density with mean m_0 and variance σ_0^2. Since the system is linear the normality of this initial (Gaussian) density is preserved, and we have $N_Y\left(m_Y(t), \sigma_Y^2(t)\right)$

$$m_Y(t) = \langle Y(t) \rangle = m_0 \mathrm{e}^{-at}, \qquad \sigma_Y^2(t) = \sigma_0^2 \mathrm{e}^{-2at} + \frac{D}{2a}(1 - \mathrm{e}^{-2at}) \tag{16}$$

Therefore the density $f(y,t)$ and entropy $H_Y(t)$ are, respectively

$$f(y,t) = \frac{1}{[2\pi\sigma_Y^2(t)]^{1/2}} \exp\left[-\frac{(y - m_0\mathrm{e}^{-at})^2}{2\sigma_Y^2(t)}\right] \tag{17}$$

$$H_Y(t) = \frac{1}{2}\log[2\pi \mathrm{e} \sigma_Y^2(t)] \tag{18}$$

$$\dot{H}_Y(t) = \frac{a(D - 2a\sigma_0^2)\mathrm{e}^{-2at}}{D - (D - 2a\sigma_0^2)\mathrm{e}^{-2at}} \log \mathrm{e} \tag{19}$$

It is seen that increase of entropy ($\dot{H}_Y(t) > 0$) takes place if $\sigma_0^2 < D/2a$ whereas the entropy decreases when $\sigma_0^2 > D/2a$. Keeping in mind that in the case without noise ($D = 0$) we have Liouvillian system for which $\dot{H}_Y(t) = -a \log \mathrm{e}$, this decrease of entropy is "inverted" to its growth by noise if $\sigma_0^2 < D/2a$.

Figure 1 illustrates the entropy change given by equations (18), (19). For such numerical values of σ_0^2, a, and D that $\sigma_0^2 < D/2a$ ($\sigma_0 = 2$, $a = 0.1$, $D = 10$) we observe the monotonic growth of $H_Y(t)$, whereas for $\sigma_0 = 2$, $a = 0.3$, $D = 0.5$, we have $\sigma_0^2 > D/2a$ and the entropy decreases in time.

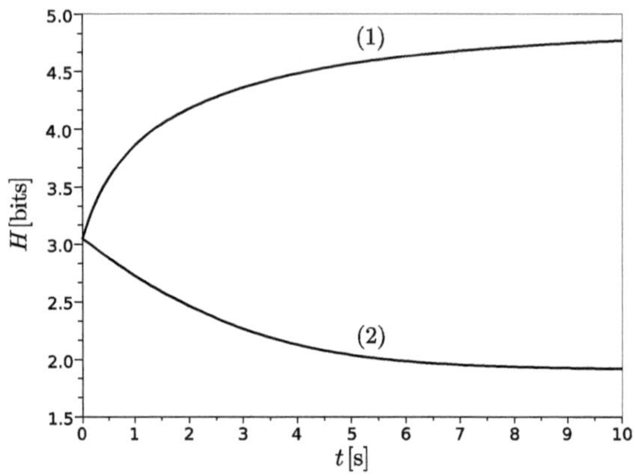

Fig. 1 $H_Y(t)$, as given by eq.(18), for $\sigma_0 = 2, a = 0.1, D = 10$ – curve (1), and $\sigma_0 = 2, a = 0.3, D = 0.5$ – curve (2)

Example 2

Let us take now a linear harmonic oscillator

$$\ddot{Y}(t) + 2\omega_0\zeta\dot{Y}(t) + \omega_0^2 Y(t) = \xi(t) \qquad (20)$$
$$Y(0) = 0, \qquad \dot{Y}(0) = 0$$

where ω_0 is the natural frequency of the system, ζ is the damping ratio, and $\xi(t)$ is a Gaussian white noise with constant spectral density g_0. As it is well known the nonstationary probability density of the response is Gaussian $N(0, \sigma_Y^2(t))$, where

$$\sigma_Y^2(t) = \frac{\pi g_0}{2\zeta\omega_0^3}\{1 - \lambda_0^{-2}\exp(-2\omega_0\zeta t)[\lambda_0^2 + 2\omega_0^2\zeta^2\sin^2\lambda_0 t + \omega_0\lambda_0\zeta\sin 2\lambda_0 t]\} \qquad (21)$$

where $\lambda_0 = \omega_0(1-\zeta^2)^{1/2}$. The entropy $H_Y(t)$ is in this case proportional to $\log\sigma^2(t)$ and its variability in time is shown in Fig. 2.

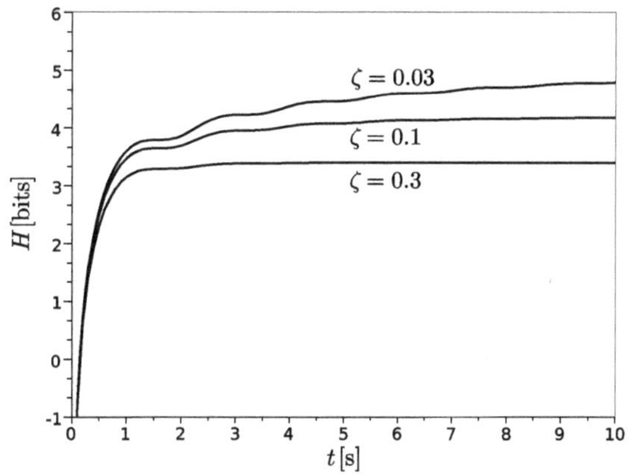

Fig. 2 $H_Y(t)$ of the harmonic oscillator (20), for $g_0 = 10$, $\omega_0 = 2$, and several values of ζ

5 Relative Entropy for True and Idealized Models

5.1 Non-stationary Dynamics

Let us consider now a general form of dynamic model

$$\dot{\mathbf{Y}}(t) = \mathbf{F}(\mathbf{Y},t) + \mathbf{X}(t) \tag{22}$$

where $\mathbf{F}$ is, in general, a nonlinear vector-valued function with values in $\mathbb{R}^n$, $\mathbf{X}(t)$ is random, time-varying excitation. The initial condition $\mathbf{Y}(t_0) = \mathbf{Y}_0$ can be deterministic or random.

Let $f(\mathbf{y},t)$ represent a true probability density of $\mathbf{Y}(t)$ at time t. However, this density is not easily available (due to complexity of the system). Let $q(\mathbf{y},t)$ be another probability density associated with the system response which can be obtained under some simplifying hypotheses; it can be viewed as an "idealized approximation". The question which arises is: what is the information loss if we make the system predictions on the basis of q instead of f.

A measure which seems to be particularly well suited for answering the above question is relative entropy defined by (3). According to its meaning it quantifies the amount of information which distribution $f(\mathbf{y},t)$ provides about the system behaviour in excess to that given by the distribution $q(\mathbf{y},t)$. We may think of the probability density f as being non-stationary – compared to the stationary one, non-Gaussian – compared to the Gaussian approximation, or – generally – as the "realistic" in contrast to "idealized" one. In the existing studies, the relative entropy has mostly been used in the analysis of the convergence of non-stationary densities of diffusion Markov processes to the stationary ones (cf. Lasota and Mackey [6]); for example, it has been shown that if f and q are two solutions of the F-P-K equation, then $H_t(f,q)$ decays monotonically with time. Here we wish to concentrate primarily on the quantitative characterization of $H_t(f,q)$ with a special attention to the effect of system parameters.

5.2 Vibratory System with Time-Varying Parameters

To illustrate further the utility of the relative entropy measure to prediction of non-stationary processes let us consider a system analysed by G.J. Sissingh [10]:

$$\ddot{Y}(t) + c(t)\dot{Y}(t) + [p^2 + k(t)]Y(t) = \alpha(t)\xi(t) \qquad (23)$$
$$Y(0) = \dot{Y}(0) = 0$$

where $\xi(t)$ is a Gaussian white noise with zero mean and one-sided spectral density g_0, and

$$c(t) = \frac{\gamma}{8}\left(1 + \frac{4\mu}{3}\sin t\right)$$
$$k(t) = \frac{\gamma\mu}{6}\cos t - \frac{\mu^2\gamma}{8}\sin 2t \qquad (24)$$
$$\alpha(t) = \frac{\gamma}{6}\left(1 + \frac{3\mu}{2}\sin t\right)$$

and γ, μ are constants identified in rotor dynamics.

Since the time-varying coefficients (24) are deterministic, the response process $Y(t)$ is Gaussian. It has been shown (cf. Sissingh [10]) that for $p \gg 1$ the approximate solution (as a series solution in terms of power of p^{-2}) for the variance of the response has the form

$$\sigma_Y^2(t) = \frac{\pi g_0}{p^2} e^{-\eta(t)} \int_0^t e^{\eta(\tau)} \alpha^2(\tau) d\tau \qquad (25)$$

where

$$\eta(t) = \int_0^t c(\tau) d\tau = \frac{\gamma}{8} t + \frac{8\mu}{6}(1 - \cos t) \qquad (26)$$

Therefore, the Gaussian probability density $f(y,t)$ of $Y(t)$ is $N(0, \sigma_Y^2(t))$, where $\sigma_Y^2(t)$ is given by (25)-(26).

Let us take as a reference (or, idealized) system associated with oscillator (23) the system (23) in which coefficients are constant and given by:

$$c(t) = 2\zeta\omega_0 = \frac{\gamma}{8} \quad , \quad k(t) = 0 \quad , \quad p^2 = \omega_0^2 \quad , \quad \alpha(t) = \frac{\gamma}{6} = \alpha_0 \qquad (27)$$

In this case the response is governed by the Gaussian process with density $q(y,t) = N(0, \tilde{\sigma}_Y^2(t))$, where $\tilde{\sigma}_Y^2(t)$ obtained from the approximate solution (25)-(26)-(27) as a particular case, is

$$\tilde{\sigma}_Y^2(t) = \frac{\pi g_0 \alpha_0^2}{2\zeta\omega_0^3}\left(1 - e^{-2\zeta\omega_0 t}\right) \qquad (28)$$

Of course, we should keep in mind that the exact variance of the oscillator characterized by (27) is given by (21) with multiplier α_0^2.

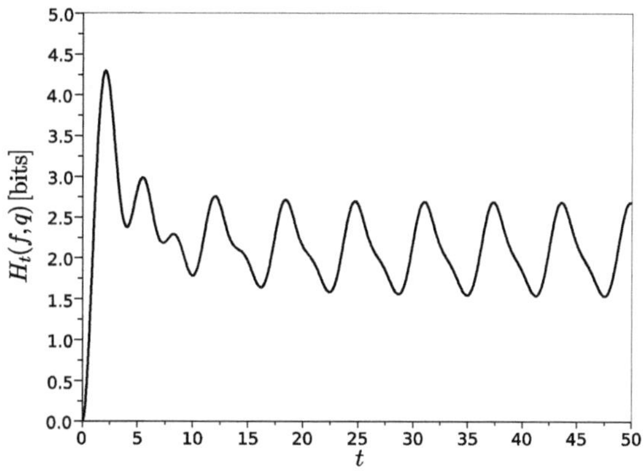

Fig. 3 $H_t(f,q)$ between the responses of oscillators with time-varying and constant parameters

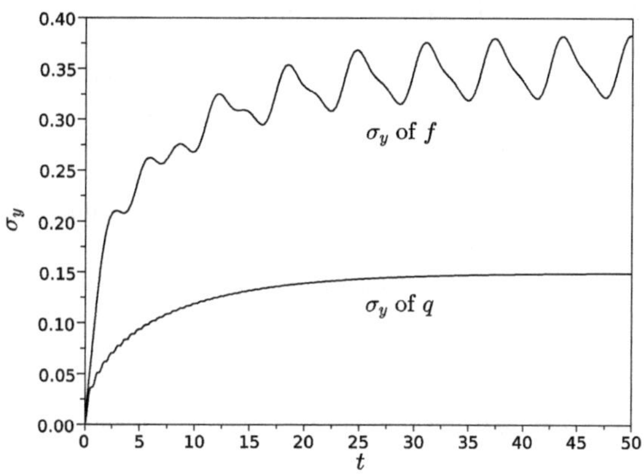

Fig. 4 Standard deviations of the responses of oscillators with time-varying and constant parameters

The relative entropy $H_t(f,q)$ has been calculated here on the basis of an appropriate formula for Gaussian densities. Its graphical illustration is shown in Fig. 3, whereas in Fig. 4 there are the plots of the standard deviations $\sigma_Y(t)$ determined by (25)-(26) and $\tilde{\sigma}_Y(t)$ from (28). The numerical values of parameters are taken as follows: $\gamma = 0.8$, $\mu = 2.0$, $p = 5.0$, $g_0 = 1$. It is seen that periodicity present in the standard deviation of the response results in the periodic-type variability of the relative entropy it time. Quantitatively, the informational difference between the presented responses can be regarded as essential.

6 Conclusions

In this paper a brief analysis of the entropy/information change in stochastic dynamical systems is presented with a special emphasis on the effect of the system parameters and intensities of random noise. More extensive treatment of the information dynamics problems (e.g. including effects of system uncertainty and the moment-based approximation) the reader can find in Sobczyk, Hołobut [13].

References

1. Barron, A.R., Shen, C.-H.: Approximation of the density functions by sequences of exponential families. Annals of Statistics 19(3), 1347–1369 (1991)
2. Cover, T.M., Thomas, J.A.: Elements of Information Theory. John Wiley, New York (1991)
3. Ebeling, W.: Entropy and information in processes of self-organization: uncertainty and predictability. Physica A 194, 563–575 (1993)
4. Garbaczewski, P.: Entropy methods in random motion. Journal of Statistical Physics 123, 315–355 (2006)
5. Haken, H.: Information and Self-Organization. Springer, Berlin (1988)
6. Lasota, A., Mackey, M.C.: Chaos, Fractals, and Noise. Springer, New York (1994)
7. Lin, Y.K., Cai, G.Q.: Probabilistic Structural Dynamics: Advanced Theory and Applications. McGraw-Hill, New York (1995)
8. Munakata, T., Igarashi, A.: Entropy and entropy production in simple stochastic models. Physical Review E 57(2), 1403–1409 (1998)
9. Nemytskii, V.V., Stepanov, V.V.: Qualitative Theory of Differential Equations. Princeton University Press, Princeton (1960)
10. Sissingh, G.J.: Dynamics of rotors operating on high advance ratios. Journal of Helicopter Society 13, 56–63 (1968)
11. Sobczyk, K.: Stochastic Differential Equations with Applications to Physics and Engineering. Kluwer, Dordrecht (1991)
12. Sobczyk, K.: Information dynamics: premises, challenges and results. Mechanical Systems & Signal Processing 15(3), 475–498 (2001)
13. Sobczyk, K., Hołobut, P. Information-theoretic approach to dynamics of stochastic systems (to be published)
14. Sobczyk, K., Trębicki, J.: Maximum entropy principle and nonlinear stochastic oscillators. Physica A 193, 448–468 (1993)

Finite Dimensional Markov Process Approximation for Time-Delayed Stochastic Dynamical Systems

Jian-Qiao Sun

School of Engineering, University of California, Merced, CA 95343, USA

Abstract. This paper presents a method of finite dimensional Markov process (FDMP) approximation for stochastic dynamical systems with time delay. The FDMP method preserves the standard state space format of the system, and allows us to apply all the existing methods and theories for analysis and control of stochastic dynamical systems. The paper presents the theoretical framework for stochastic dynamical systems with time delay based on the FDMP method, including the FPK equation, backward Kolmogorov equation, and reliability formulation. The work of this paper opens a door to various studies of stochastic dynamical systems with time delay.

Keywords: Time delay, stochastic systems, control, Markov approximation.

1 Introduction

Time delay comes from different sources and often leads to instability or poor performance in control systems. Other than a few exceptional cases, delay is undesired and control strategies to eliminate or minimize its unwanted effects have to be employed. Effects of time delay on the stability and performance of deterministic control systems have been a subject of many studies. For example, Yang and Wu [1] and Stepan [2] have studied structural systems with time delay. A powerful method using Chebyshev polynomial expansion to approximate general nonlinear functions of time has been developed to handle linear and nonlinear time-delayed dynamical systems with periodic coefficients [3, 4, 5, 6]. A study on stability and performance of feedback controls with multiple time delays is reported in [7] by considering the roots of the closed loop characteristic equation. For deterministic delayed linear systems a survey of recent methods for stability analysis is presented in [8].

There is a growing interest in the stochastic systems with time delay. An effective Monte Carlo simulation scheme that converges in a weak sense is presented by Kuchler and Platen [9]. Buckwar [10] studied numerical solutions of Itô type differential equations and their convergence where the system considered has time delay both in diffusion and drift terms. Guillouzic, L'Heureux and Longtin [11] studied first order delayed Itô differential equations using

W.Q. Zhu et al. (Eds.): Nonlinear Stochastic Dynamics & Control, IUTAM BOOK SERIES 29, pp. 107–116.
springerlink.com

a small delay approximation and obtained PDFs as well as the second order statistics analytically. Frank and Beek [12] obtained the PDFs using the FPK equation for linear delayed stochastic systems and studied the stability of fixed point solutions in biological systems. State feedback stabilization of nonlinear time delayed stochastic systems has been investigated by Fu, Tian and Shi [13] where a Lyapunov approach is used.

The delayed systems are studied using discretization techniques with an extended state vector. For example, Pinto and Goncalves [14] fully discretized a nonlinear SDOF system to study control problems with time delay. Klein and Ramirez [15] studied MDOF delayed optimal regulator controllers with a hybrid discretization technique where the state equation was partitioned into discrete and continuous portions. Another powerful discretization method is the semi discretization method. It is a well established method in the literature and used widely in structural and fluid mechanics applications. Recently, the method is applied to delayed deterministic systems by Insperger and Stepan [16]. They studied high dimensional multiple time delayed systems in [17]. The method can be extended to control systems with delayed feedback. We have studied the effect of various higher order approximations in semi discretization on the computational efficiency and accuracy. The merit of the semi-discretization method as introduced by Insperger and Stepan [16] lies in that it makes use of the exact solution of linear systems over a short time interval to construct the mapping of a finite dimensional state vector for the system with time delay. Recently, a method of continuous time approximation (CTA) of delayed dynamical systems was introduced [18]. The method can be viewed as an extension of the semi-discretization. It also discretizes the delayed portion of the response leading to a high and finite dimensional state space formulation of the time-delayed system in continuous time domain. The advantage of the CTA method lies in that the resulting finite dimensional state equations are in the standard state space form, making all the existing analysis methods and control design tools for linear and nonlinear dynamical systems amenable to the CTA method. The method can also handle multiple independent time delays in a natural way. In this paper, we apply the CTA method to study the responses of stochastic dynamical systems with time delay. The CTA method leads to a finite dimensional Markov process (FDMP) approximation of stochastic dynamical systems with time delay, whose responses are non-Markovian. For this reason, we refer to the current approach as the FDMP method for stochastic systems.

The paper is organized as follows. Section 2 introduces the FDMP method. Section 3 presents the theoretical framework for stochastic dynamical systems with time delay, including the FPK equation, the backward Kolmogorov equation, reliability theory and first-passage failure probability. Section 4 concludes the paper.

2 The FDMP Method

Consider a stochastic system in the Stratonovich sense,

$$\dot{\mathbf{x}} = \mathbf{f}\left(\mathbf{x}\left(t\right), \mathbf{x}\left(t-\tau\right), t\right) + \mathbf{G}\left(\mathbf{x}\left(t\right), \mathbf{x}\left(t-\tau\right), t\right) \mathbf{W}\left(t\right), \tag{1}$$

where $\mathbf{x} \in \mathbf{R}^n$, $\mathbf{W} \in \mathbf{R}^p$, $\mathbf{f}$ describes the system dynamics with time delay, and $\mathbf{G} = \{G_{ij}\}$ is a matrix determining the parametric and external random excitations. $W_i(t)$ are delta correlated Gaussian white noise processes with $E\left[W_i(t) W_j(t+T)\right] = 2\pi K_{ij} \delta(T)$. Equation (1) can be converted to the stochastic differential equations in the Itô sense,

$$d\mathbf{X} = \mathbf{m}\left(\mathbf{X}\left(t\right), \mathbf{X}\left(t-\tau\right), t\right) dt + \sigma\left(\mathbf{X}(t), \mathbf{X}\left(t-\tau\right), t\right) d\mathbf{B}\left(t\right), \tag{2}$$

where $\mathbf{m}$ is the drift vector including the Wong-Zakai correction term, and $\sigma\left(\mathbf{X}(t), \mathbf{X}\left(t-\tau\right), t\right)$ is the diffusion matrix given by

$$\sigma_{jl}\sigma_{kl}\left(\mathbf{X}(t), \mathbf{X}\left(t-\tau\right), t\right) = 2\pi K_{rs} G_{jr} G_{ks}, \tag{3}$$

where repeated indices imply summation. The Brownian motion $d\mathbf{B}(t)$ has the following properties

$$E\left[d\mathbf{B}\left(t\right)\right] = 0, \;\; E\left[dB_i\left(t_1\right) dB_j\left(t_2\right)\right] = \begin{cases} 0 & t_1 \neq t_2 \\ \delta_{ij} dt & t_1 = t_2 = t \end{cases}. \tag{4}$$

Note that the system lives in a state space with an infinite dimension and the state vector is given by $\left(\mathbf{X}\left(t\right), \mathbf{X}\left(t-t_1\right), 0 < t_1 \leq \tau\right)$. In general, $\mathbf{X}(t)$ is no longer a Markov process because it depends on its history. Following the idea of the semi-discretization method, we discretize the delayed part of the state vector $\left(\mathbf{X}\left(t-t_1\right), 0 < t_1 \leq \tau\right)$. Let N be integer such that $\Delta\tau = \tau/N$. $\tau_i = i\Delta\tau$ $(i = 1, 2, \cdots, N)$. Then, we introduce a finite difference approximation of the derivatives of $\left(\dot{\mathbf{X}}\left(t-\tau_i\right), 1 \leq i \leq N\right)$ as

$$\dot{\mathbf{X}}\left(t - j\Delta\tau\right) = \frac{1}{\Delta\tau}\left[\mathbf{X}\left(t-(j-1)\Delta\tau\right) - \mathbf{X}\left(t - j\Delta\tau\right)\right], \;\; 1 \leq j \leq N. \tag{5}$$

Note that higher order Runge-Kutta algorithms, Chebyshev nodes to replace the uniform sampled points τ_i and implicit-explicit methods can be applied leading to better approximation of $\dot{\mathbf{X}}\left(t - j\Delta\tau\right)$ and more accurate solutions overall in frequency and time domain [19,20,21,22,23,24,25].

Define a discrete vector as

$$\begin{aligned} \mathbf{Y}\left(t\right) &= \left[\mathbf{X}\left(t\right), \mathbf{X}\left(t-\Delta\tau\right), \mathbf{X}\left(t-2\Delta\tau\right), \cdots, \mathbf{X}\left(t-N\Delta\tau\right)\right]^T \\ &\equiv \left[\mathbf{Y}_1\left(t\right), \mathbf{Y}_2\left(t\right), \mathbf{Y}_3\left(t\right), \cdots, \mathbf{Y}_{N+1}\left(t\right)\right]^T. \end{aligned} \tag{6}$$

We obtain an Itô stochastic equation for the vector $\mathbf{Y}(t)$.

$$
\begin{aligned}
d\mathbf{Y}(t) &= \begin{bmatrix} \mathbf{m}\left(\mathbf{Y}_1(t), \mathbf{Y}_{N+1}(t), t\right) \\ \frac{1}{2\Delta\tau}\left[\mathbf{Y}_1(t) - \mathbf{Y}_3(t)\right] \\ \vdots \\ \frac{1}{\Delta\tau}\left[\mathbf{Y}_N(t) - \mathbf{Y}_{N+1}(t)\right] \end{bmatrix} dt + \begin{bmatrix} \sigma\left(\mathbf{Y}_1(t), \mathbf{Y}_{N+1}(t), t\right) \\ \mathbf{0} \\ \vdots \\ \mathbf{0} \end{bmatrix} d\mathbf{B}(t) \\
&\equiv \hat{\mathbf{m}}(\mathbf{Y}, t)dt + \hat{\sigma}(\mathbf{Y}, t)\, d\mathbf{B}(t) .
\end{aligned}
\tag{7}
$$

Note that for the indices $2 \le i \le N-1$, the central difference approximation of the derivatives is adopted. We have found that this substantially improves the accuracy of the approximate solution, and is used in all the numerical examples.

Remarks

Some remarks on the FDMP method are in order.

1. Equation (7) indicates that $\mathbf{Y}(t)$ is a Markov process when $d\mathbf{B}(t)$ is a Brownian motion [26]. The conditional probability density function of $\mathbf{Y}(t)$ satisfies a FPK equation as well as backward Kolmogorov equation. We shall study these equations later in the paper.
2. Discretization of the delayed time interval $(0, \tau)$ can be non-uniform. For systems with more than one time delays, we can make all the time delays to be the instances of the discretization. This enables the FDMP method to deal with multiple time delays in a consistent manner [18].
3. Note that Equation (5) introduces a time-domain approximation of the delayed system response in a similar manner to other numerical integration methods such as the Runge-Kutta method. The frequency domain properties of the approximation can be explicitly addressed as is the case in [20, 23, 24]. We have found that the FDMP method based on the finite difference scheme accurately computes the dominant poles of linear systems with time delay, hence, correctly predicts the stability of the system. When Chebyshev nodes or a low-pass filter based approximation are used to discretize the delayed response, many lower frequency poles can be predicted accurately.

3 Applications

One of the advantages of the FDMP method is that it keeps the system in a standard state space format. This allows us to extend all the existing methods and theories for analysis and control of stochastic systems to the ones with time delay. Here, we present the theory of stochastic systems within the framework of FDMP.

3.1 FPK Equation

The conditional probability density function $p_{\mathbf{Y}}(\mathbf{y}, t|\mathbf{y}_0, t_0)$ for the stochastic system (7) satisfies the FPK equation given by

$$\frac{\partial}{\partial t} p_{\mathbf{Y}}(\mathbf{y}, t|\mathbf{y}_0, t_0) = -\frac{\partial}{\partial y_j}[\hat{m}_j(\mathbf{y}, t) p_{\mathbf{Y}}] + \frac{\partial^2}{\partial y_j \partial y_k}\left[\frac{\hat{b}_{jk}(\mathbf{y}, t)}{2!} p_{\mathbf{Y}}\right], \quad (8)$$

where the index j runs from 1 to $m = n(N+1)$ and $\hat{b}_{jk} = \hat{\sigma}_{jl}\hat{\sigma}_{kl}$ or $\hat{\mathbf{b}} = \hat{\sigma}\hat{\sigma}^T$. The index l runs from 1 to p, the dimension of $d\mathbf{B}(t)$. The FPK equation is subject to an initial condition, for example,

$$p_{\mathbf{Y}}(\mathbf{y}, t|\mathbf{y}_0, t_0) = \delta(\mathbf{y} - \mathbf{y}_0). \quad (9)$$

Note that since the stochastic excitations only act on the vector $\mathbf{X}(t) = \mathbf{Y}_1(t)$, the second order derivatives of the FPK equation only involve the components of $\mathbf{Y}_1(t)$. Hence, there are only $n \times n$ diffusion terms, instead of $m \times m$. In other words, the time-delay within the FDMP method only affects the drift term of the FPK equation. This is also true with the backward Kolmogorov equation and its derivatives in the study of reliability and first-passage time probability.

3.2 Moment Equations

Recall that in the Itô sense, $dB_k(t)$ is defined as the forward difference and $\hat{\sigma}_{jk}(\mathbf{Y}, t)$ is independent of $dB_k(t)$. Also, $E[dB_k(t)] = 0$. Taking the mathematical expectation on both sides of Equation (7), we have

$$\frac{dE[Y_j(t)]}{dt} = E[\hat{m}_j(\mathbf{Y}, t)]. \quad (10)$$

Consider a function $F(\mathbf{Y}, t) = Y_j Y_k$. According to Itô's lemma, we have

$$d(Y_j Y_k) = \left(\hat{m}_j Y_k + \hat{m}_k Y_j + \frac{1}{2}\hat{b}_{jk}\right) dt + (\hat{\sigma}_{jl} Y_k + \hat{\sigma}_{kl} Y_j)\, dB_l(t). \quad (11)$$

Taking the expectation of the equation, we obtain the equation for the correlation function

$$\frac{dE[Y_j Y_k]}{dt} = E\left[\hat{m}_j Y_k + \hat{m}_k Y_j + \frac{1}{2}\hat{b}_{jk}\right]. \quad (12)$$

By following the same steps, we can construct differential equations governing the evolution of the moments of any order. Consider a linear example. The moment equations of the first and second orders are readily obtained in the matrix form,

$$\frac{d\mu_{\mathbf{Y}}}{dt} = \hat{\mathbf{A}}\mu_{\mathbf{Y}}, \tag{13}$$

$$\frac{d\mathbf{R}_{\mathbf{YY}}}{dt} = \hat{\mathbf{A}}\mathbf{R}_{\mathbf{YY}} + \mathbf{R}_{\mathbf{YY}}\hat{\mathbf{A}}^T + \frac{1}{2}\hat{\mathbf{b}}, \tag{14}$$

where $\mu_{\mathbf{Y}} = E[\mathbf{Y}]$ and $\mathbf{R}_{\mathbf{YY}} = E[\mathbf{YY}^T]$.

3.3 Reliability

The backward Kolmogorov equation for the process $\mathbf{Y}(t)$ can be derived as [27]

$$-\frac{\partial}{\partial t_0} p_{\mathbf{Y}}(\mathbf{y}, t|\mathbf{y}_0, t_0) = \hat{m}_j(\mathbf{y}_0, t_0)\frac{\partial p_{\mathbf{Y}}}{\partial y_{0j}} + \frac{\hat{b}_{jk}(\mathbf{y}_0, t_0)}{2!}\frac{\partial^2 p_{\mathbf{Y}}}{\partial y_{0j}\partial y_{0k}}. \tag{15}$$

Note that $t_0 \leq t < \infty$. Integrating Equation (15) with respect to the delayed components $(\mathbf{y}_2, \mathbf{y}_3, \cdots, \mathbf{y}_{N+1})$ of the state vector leads to the backward equation for the marginal probability density function.

$$-\frac{\partial}{\partial t_0} p_{\mathbf{X}}(\mathbf{x}, t|\mathbf{y}_0, t_0) = \hat{m}_j(\mathbf{y}_0, t_0)\frac{\partial p_{\mathbf{X}}}{\partial y_{0j}} + \frac{\hat{b}_{jk}(\mathbf{y}_0, t_0)}{2!}\frac{\partial^2 p_{\mathbf{X}}}{\partial y_{0j}\partial y_{0k}}. \tag{16}$$

An important application of the backward Kolmogorov equation is the reliability study. Consider the state vector $\mathbf{X}(t)$ of the original system. Let $\mathcal{S} \subseteq R^n$ be a domain in which the system is considered to be safe. Γ is the boundary of $\mathcal{S}$. Assume that all the components of $\mathbf{Y}(t_0) = \mathbf{y}_0$ lie inside $\mathcal{S}$ at time t_0. The probability that the system is still in the *safe domain* $\mathcal{S}$ at time t is given by

$$\begin{aligned} \mathcal{R}_{\mathcal{S}}(t, t_0, \mathbf{y}_0) &= P(t < T \cap \mathbf{X}(t) \in \mathcal{S}|\mathbf{Y}(t_0) = \mathbf{y}_0) \\ &= \int_{\mathcal{S}} p_{\mathbf{X}}(\mathbf{x}, t|\mathbf{y}_0, t_0)d\mathbf{x}, \end{aligned} \tag{17}$$

where T is the first time when $\mathbf{X}(t)$ crosses the boundary Γ. $\mathcal{R}_{\mathcal{S}}(t, t_0, \mathbf{y}_0)$ is also known as the reliability against the *first-passage failure* with respect to the safe domain $\mathcal{S}$.

Integrating Equation (15) over $\mathcal{S}$ with respect to $\mathbf{x}$, we obtain a partial differential equation of the reliability function $\mathcal{R}_{\mathcal{S}}(t, t_0, \mathbf{y}_0)$.

$$-\frac{\partial \mathcal{R}_{\mathcal{S}}(t, t_0, \mathbf{y}_0)}{\partial t_0} = \hat{m}_j(\mathbf{y}_0, t_0)\frac{\partial \mathcal{R}_{\mathcal{S}}(t, t_0, \mathbf{y}_0)}{\partial y_{0j}} + \frac{\hat{b}_{jk}(\mathbf{y}_0, t_0)}{2!}\frac{\partial^2 \mathcal{R}_{\mathcal{S}}(t, t_0, \mathbf{y}_0)}{\partial y_{0j}\partial y_{0k}}, \tag{18}$$

subject to the following initial and boundary conditions

$$\mathcal{R}_{\mathcal{S}}(t_0, t_0, \mathbf{y}_0) = 1, \ \mathbf{y}_{0i} \in \mathcal{S}, \ (1 \leq i \leq N+1), \tag{19}$$

$$\mathcal{R}_{\mathcal{S}}(t, t_0, \mathbf{y}_0) = 0, \ \mathbf{y}_{0i} \in \Gamma \text{ (for at least one } i). \tag{20}$$

3.4 First-Passage Time Probability

Denote the complement of $\mathcal{R}_{\mathcal{S}}(t, t_0, \mathbf{y}_0)$ as $F_{\mathcal{S}}(t, t_0, \mathbf{y}_0)$, which is the probability distribution function of the first-passage time. We have

$$F_{\mathcal{S}}(t, t_0, \mathbf{y}_0) = P(t \geq T | \mathbf{Y}(t_0) = \mathbf{y}_0) = 1 - \mathcal{R}_{\mathcal{S}}(t, t_0, \mathbf{y}_0). \tag{21}$$

Substituting this relationship to Equation (18), we obtain

$$-\frac{\partial F_{\mathcal{S}}(t, t_0, \mathbf{y}_0)}{\partial t_0} = \hat{m}_j(\mathbf{y}_0, t_0)\frac{\partial F_{\mathcal{S}}(t, t_0, \mathbf{y}_0)}{\partial y_{0j}} + \frac{\hat{b}_{jk}(\mathbf{y}_0, t_0)}{2!}\frac{\partial^2 F_{\mathcal{S}}(t, t_0, \mathbf{y}_0)}{\partial y_{0j}\partial y_{0k}}. \tag{22}$$

The probability density function of the first-passage time denoted by $p_T(t|\mathbf{y}_0, t_0)$ is given by

$$p_T(t|\mathbf{y}_0, t_0) = \frac{\partial F_{\mathcal{S}}(t, t_0, \mathbf{y}_0)}{\partial t} = -\frac{\partial \mathcal{R}_{\mathcal{S}}(t, t_0, \mathbf{y}_0)}{\partial t}. \tag{23}$$

Differentiating Equation (22) with respect to t, we yield the governing equation for $p_T(t|\mathbf{y}_0, t_0)$

$$-\frac{\partial p_T(t|\mathbf{y}_0, t_0)}{\partial t_0} = \hat{m}_j(\mathbf{y}_0, t_0)\frac{\partial p_T(t|\mathbf{y}_0, t_0)}{\partial y_{0j}} + \frac{\hat{b}_{jk}(\mathbf{y}_0, t_0)}{2!}\frac{\partial^2 p_T(t|\mathbf{y}_0, t_0)}{\partial y_{0j}\partial y_{0k}}. \tag{24}$$

Since, at a given time $t > t_0$ and when $\mathbf{y}_{0i} \in \Gamma$ (for at least one i), the reliability of the system vanishes $\mathcal{R}_{\mathcal{S}}(t, t_0, \mathbf{y}_0) = 0$, this suggests the boundary condition

$$p_T(t|\mathbf{y}_0, t_0) = 0, \; t > t_0, \;\; \mathbf{y}_{0i} \in \Gamma \text{ (for at least one } i). \tag{25}$$

Assume that initially, the system starts from a point in the safe domain with probability one, we have an initial condition

$$p_T(t_0|\mathbf{y}_0, t_0) = \delta(\mathbf{y}_0), \; \mathbf{y}_{0i} \in \mathcal{S}, \; (1 \leq i \leq N+1). \tag{26}$$

3.5 Pontryagin-Vitt Equations

The first-passage time is a random variable and its rth order moment can be defined as

$$M_r(\mathbf{y}_0, t_0) = E[(T - t_0)^r|\mathbf{y}_0, t_0] = \int_{t_0}^{\infty}(t - t_0)^r p_T(t|\mathbf{y}_0, t_0)\, dt. \tag{27}$$

From Equation (24), we obtain a set of integral-partial differential equations for the *moments of the first-passage time* as

$$-\int_{t_0}^{\infty}(t-t_0)^r \frac{\partial p_T(t|\mathbf{y}_0,t_0)}{\partial t_0}dt = \hat{m}_j(\mathbf{y}_0,t_0)\frac{\partial M_r(\mathbf{y}_0,t_0)}{\partial y_{0j}} + \frac{\hat{b}_{jk}(\mathbf{y}_0,t_0)}{2!}\frac{\partial^2 M_r(\mathbf{y}_0,t_0)}{\partial y_{0j}\partial y_{0k}}. \tag{28}$$

This equation is in general difficult to solve. Assume that $\mathbf{X}(t)$ is a stationary process such that $p_T(t|\mathbf{y}_0,t_0) = p_T(\tau|\mathbf{y}_0) = -\frac{\partial \mathcal{R}_S(\tau,\mathbf{y}_0)}{\partial \tau}$, $\hat{m}_j(\mathbf{y}_0,t_0) = \hat{m}_j(\mathbf{y}_0)$ and $\hat{b}_{jk}(\mathbf{y}_0,t_0) = \hat{b}_{jk}(\mathbf{y}_0)$ where $\tau = t - t_0$. Let

$$M_r(\mathbf{y}_0) = \int_0^{\infty} \tau^r p_T(\tau|\mathbf{y}_0)\,d\tau. \tag{29}$$

Assume that $lim_{\tau\to\infty}\tau^r p_T(\tau|\mathbf{y}_0) = 0$. We can derive a set of the *generalized Pontryagin-Vitt equations* for the moments of the first-passage time as

$$-rM_{r-1}(\mathbf{y}_0) = \hat{m}_j(\mathbf{y}_0)\frac{\partial M_r(\mathbf{y}_0)}{\partial y_{0j}} + \frac{\hat{b}_{jk}(\mathbf{y}_0)}{2!}\frac{\partial^2 M_r(\mathbf{y}_0)}{\partial y_{0j}\partial y_{0k}}. \tag{30}$$

All the moments satisfy the same boundary condition

$$M_r(\mathbf{y}_0) = 0, \quad \mathbf{y}_{0i} \in \Gamma \text{ (for at least one } i), \; r = 1,2,3,... \tag{31}$$

Note that $M_0(\mathbf{y}_0) = 1$ because $p_T(\tau|\mathbf{y}_0)$ is a probability density function of τ. Hence, the mean of the first-passage time satisfies the *Pontryagin-Vill* equation with $r = 1$,

$$-1 = \hat{m}_j(\mathbf{y}_0)\frac{\partial M_1(\mathbf{y}_0)}{\partial y_{0j}} + \frac{\hat{b}_{jk}(\mathbf{y}_0)}{2!}\frac{\partial^2 M_1(\mathbf{y}_0)}{\partial y_{0j}\partial y_{0k}}. \tag{32}$$

4 Conclusion

We have presented a method of finite dimensional Markov process approximation for stochastic dynamical systems with time delay. The FDMP method preserves the standard state space format of the system, and allows us to apply all the existing methods and theories for analysis and control of stochastic dynamical systems. We have presented the theoretical framework for stochastic dynamical systems with time delay based on the FDMP method, including the FPK equation, backward Kolmogorov equation, moment equations and reliability formulation. The present work opens a gate to various studies of stochastic dynamical systems with time delay.

References

1. Yang, B., Wu, X.: Modal expansion of structural systems with time delays. AIAA Journal 36(12), 2218–2224 (1998)
2. Stepan, G.: Delay-differential equation models for machine tool chatter. In: Moon, F.C. (ed.) Dynamics and Chaos in Manufacturing Processes, pp. 165–192. Wiley, New York (1998)
3. Deshmukh, V., Butcher, E.A., Bueler, E.: Dimensional reduction of nonlinear delay differential equations with periodic coefficients using Chebyshev spectral collocation. Nonlinear Dynamics 52(1-2), 137–149 (2008)
4. Deshmukh, V., Ma, H., Butchern, E.A.: Optimal control of parametrically excited linear delay differential systems via Chebyshev polynomials. Optimal Control Applications and Methods 27, 123–136 (2006)
5. Ma, H., Deshmukh, V., Butcher, E.A., Averina, V.: Delayed state feedback and chaos control for time periodic systems via a symbolic approach. Communications in Nonlinear Science and Numerical Simulation 10(5), 479–497 (2005)
6. Ma, H., Butcher, E.A., Bueler, E.: Chebyshev expansion of linear and piecewise linear dynamic systems with time delay and periodic coefficients under control excitations. Journal of Dynamic Systems, Measurement, and Control 125, 236–243 (2003)
7. Ali, M.S., Hou, Z.K., Noori, M.N.: Stability and performance of feedback control systems with time delays. Computers and Structures 66(2-3), 241–248 (1998)
8. Niculescu, S.I., Verriest, E.I., Dugard, L., Dion, J.M.: Stability of linear systems with delayed state: A guided tour. In: Proceedings of the IFAC Workshop: Linear Time Delay Systems, Grenoble, France, pp. 31–38 (1998)
9. Kuchler, U., Platen, E.: Weak discrete time approximation of stochastic differential equations with time delay. Mathematics and Computers in Simulation 59(6), 497–507 (2002)
10. Buckwar, E.: Introduction to the numerical analysis of stochastic delay differential equations. Journal of Computational and Applied Mathematics 125(1-2), 297–307 (2000)
11. Guillouzic, S., L'Heureux, I., Longtin, A.: Small delay approximation of stochastic delay differential equations. Physical Review E Statistical Physics, Plasmas, Fluids, and Related Interdisciplinary Topics 59(4), 3970–3982 (1999)
12. Frank, T.D., Beek, P.J.: Stationary solutions of linear stochastic delay differential equations: Applications to biological systems. Physical Review E Statistical Physics, Plasmas, Fluids, and Related Interdisciplinary Topics 64(2 I), 1–12 (2001)
13. Fu, Y., Tian, Z., Shi, S.: State feedback stabilization for a class of stochastic time-delay nonlinear systems. IEEE Transactions on Automatic Control 48(2), 282–286 (2003)
14. Pinto, O.C., Goncalves, P.B.: Control of structures with cubic and quadratic non-linearities with time delay consideration. Revista Brasileira de Ciencias Mecanicas/Journal of the Brazilian Society of Mechanical Sciences 24(2), 99–104 (2002)
15. Klein, E.J., Ramirez, W.F.: State controllability and optimal regulator control of time-delayed systems. International Journal of Control 74(3), 281–289 (2001)

16. Insperger, T., Stepan, G.: Semi-discretization of delayed dynamical systems. In: Proceedings of ASME 2001 Design Engineering Technical Conferences and Computers and Information in Engineering Conference, Pittsburgh, Pennsylvania (2001)
17. Insperger, T., Stepan, G.: Semi-discretization method for delayed systems. International Journal for Numerical Methods in Engineering 55(5), 503–518 (2002)
18. Sun, J.Q.: A method of continuous time approximation of delayed dynamical systems. Communications in Nonlinear Science and Numerical Simulation 14(4), 998–1007 (2009)
19. Bellen, A., Maset, S.: Numerical solution of constant coefficient linear delay differential equations as abstract Cauchy problems. Numerische Mathematik 84, 351–374 (2000)
20. Engelborghs, K., Roose, D.: On stability of LMS methods and characteristic roots of delay differential equations. IMA Journal of Numerical Analysis 40, 629–650 (2002)
21. Maset, S.: Numerical solution of retarded functional differential equations as abstract Cauchy problems. Journal of Computational and Applied Mathematics 161, 259–282 (2003)
22. Koto, T.: Method of lines approximations of delay differential equations. Computers and Mathematics with Applications 48, 45–59 (2004)
23. Breda, D., Maset, S., Vermiglio, R.: Computing the characteristic roots for delay differential equations. IMA Journal of Numerical Analysis 24, 1–19 (2004)
24. Breda, D., Maset, S., Vermiglio, R.: Pseudospectral differencing methods for characteristic roots of delay differential equations. SIAM Journal Science Computing 27, 482–495 (2005)
25. Koto, T.: Stability of implicit-explicit linear multistep methods for ordinary and delay differential equations. Frontiers of Mathematics in China 4, 113–129 (2009)
26. Lin, Y.K., Cai, G.Q.: Probabilistic Structural Dynamics - Advanced Theory and Applications. McGraw-Hill, New York (1995)
27. Sun, J.Q.: Stochastic Dynamics and Control. Elsevier Science, Ltd., Oxford (2006)

Response Analysis of Nonlinear Multi-degree of Freedom Systems to Non-Gaussian Random Excitations

Cho W. Solomon To

Department of Mechanical Engineering, University of Nebraska, Lincoln, NE 68588-0656, USA

Abstract. A novel approach for response analysis of multi-degree-of-freedom (mdof) nonlinear systems under non-Gaussian nonstationary random excitations is presented. It makes use of the stochastic central difference (SCD) method, time co-ordinate transformation (TCT), and adaptive time scheme (ATS). For tractability and simplicity of illustration, a two degree-of-freedom (dof) nonlinear asymmetric system under a non-Gaussian nonstationary random excitation is studied in the present investigation. Comparisons between computed results obtained for the system with Gaussian random excitations to those of the same system under non-Gaussian random excitations are made. It is concluded that the proposed approach is relatively very efficient, simple, and accurate for response analysis of mdof highly nonlinear systems under non-Gaussian nonstationary random excitations.

Keywords: Nonlinear, Multi-degree-of-freedom systems, Non-Gaussian excitations.

1 Introduction

For reasons of safety and economy, many modern structures and systems such as systems that house nuclear reactors, tall buildings, naval and aerospace installations, and their components have to be designed to withstand various intensive complicated loadings which can only be realistically modeled as random processes. Until very recently, these latter processes have generally been treated as Gaussian random processes. In practice, particularly in the designs and analysis of space shuttles and many other vehicles, there is a need to deal with the excitation processes as non-Gaussian ones. Specifically, the MIL-STD-810F DOD Test Method Standard [1], and Defence Standard 00-35 of the Ministry of Defence [2] require consideration of the non-Gaussian behavior in simulation and testing environments. Furthermore, in offshore structure design such as the tension-leg platform the response has been known to be non-Gaussian. Wind-generated waves in finite water depth have long been recognized as non-Gaussian random processes.

While analysis of non-Gaussian random processes has generated considerable amount of interests in recent years in the field of random vibration [3-5] it seems that an efficient and accurate analysis for nonlinear multi-degree-of-freedom (mdof) systems under non-Gaussian random excitations is not available. The

W.Q. Zhu et al. (Eds.): Nonlinear Stochastic Dynamics & Control, IUTAM BOOK SERIES 29, pp. 117–126.
springerlink.com

investigation being reported here was therefore concerned with providing a means of efficient and accurate analysis, and predicting responses of highly nonlinear mdof systems under non-Gaussian nonstationary random excitations. Owing to their abilities to provide efficient and accurate responses of mdof nonlinear systems and discretized nonlinear shell structures under Gaussian random excitations, the stochastic central difference (SCD) method [6], the time co-ordinate transformation (TCT) [7], and the adaptive time scheme (ATS) [7] were combined to form a novel procedure that was developed in the investigation. For tractability and simplicity of illustration, a two degree-of-freedom (dof) nonlinear asymmetric system under a non-Gaussian nonstationary random excitation is considered. This same nonlinear two dof system under a zero-mean Gaussian nonstationary random excitation was studied by the author and associate [7]. The latter excitation was modeled as a product of a deterministic modulating function and a zero-mean discrete Gaussian white noise (DGWN). Comparison between results obtained by the SCD method and those computed by the Monte Carlo simulation (MCS) was made and it was found to have excellent agreement [7]. Thus, the SCD method was further developed to be applied to nonlinear mdof systems under non-Gaussian nonstationary random excitations.

In Section 2, the SCD method for mdof nonlinear systems under non-stationary Gaussian random excitations is introduced. Section 3 is concerned with the SCD method for mdof nonlinear under non-Gaussian nonstationary random excitations. Computed results and comparisons between those obtained for systems under non-Gaussian nonstationary random excitations and those for systems under Gaussian nonstationary random excitations are included in Section 4. The final section, Section 5 includes concluding remarks.

2 Nonlinear Systems under Gaussian Excitations

For completeness and in order to provide a foundation for the formulation and presentation of the systems under non-Gaussian random excitations to be introduced in the following section, analysis of mdof nonlinear systems under Gaussian random excitations are consider in this section first.

The governing matrix equation of motion is given by

$$M\ddot{x} + C\dot{x} + Kx = P \tag{1}$$

where M, C and K are, respectively the nonlinear assembled mass, damping and stiffness matrices of the system; P is the external excitation vector which, in general, includes Gaussian nonstationary random forces. A typical example for P is the product of modulating forcing vector and a zero mean Gaussian white noise process.

Discretizing Eq. (1) in the time domain, one has

$$M_s\ddot{x}_s + C_s\dot{x}_s + K_s x_s = P_s, \tag{2}$$

where the subscript s denotes the time step; for instance, x_s is the value of x at time step t_s such that the time step size $\Delta t = t_{s+1} - t_s$ and $t_0 = 0$.

Following the steps in the derivation of recursive expression for the mean square matrix of generalized displacement vector in the SCD method [6], one obtains

$$\begin{aligned} R_{s+1} = &N_2 R_s N_2^T + N_3 R_{s-1} N_3^T + (\Delta t)^4 N_1 B_s N_1^T + N_2 D_s N_3^T \\ &+ N_3 D_s^T N_2^T + (\Delta t)^2 N_2 \langle x_s P_s^T \rangle N_1^T + (\Delta t)^2 N_1 \langle P_s x_s^T \rangle N_2^T \\ &+ (\Delta t)^2 N_3 \langle x_{s-1} P_s^T \rangle N_1^T + (\Delta t)^2 N_1 \langle P_s x_{s-1}^T \rangle N_3^T, \end{aligned} \tag{3}$$

in which

$$N_1 = \left[M_s + \tfrac{1}{2}(\Delta t) C_s\right]^{-1}, N_2 = N_1[2M_s - (\Delta t)^2 K_s],\ N_3 = N_1\left[\tfrac{1}{2}(\Delta t) C_s - M_s\right],$$

$$R_s = \langle x_s x_s^T \rangle,\ B_s = \langle P_s P_s^T \rangle,\ D_s = \langle x_s x_{s-1}^T \rangle, \tag{4a,b,c}$$

$$\langle x_{s+1} \rangle = (\Delta t)^2\, N_1 \langle P_s \rangle + N_2 \langle x_s \rangle + N_3 \langle x_{s-1} \rangle, \tag{4d}$$

where the angular brackets denote the mathematical expectation or ensemble aver age and the superscript T designates "the transpose of". Equation (3) contains the recursive covariance matrix. The terms containing $P_s x_s^T$, B_s, D_s and their transposes on the right-hand side (RHS) of Eq. (3) require further algebraic manipulation and expansion in the following.

Without loss of generality, the external random excitation vector P in Eq. (1) may be defined by

$$P = \Phi(t) w(t), \tag{5}$$

where $\Phi(t)$ is the vector of deterministic modulating functions, every element or entry of the deterministic modulating vector can be written as

$$\phi_i(t) = m_i(t) = E_{ri}(e^{-\alpha_{1i} t} - e^{-\alpha_{2i} t}) \tag{6}$$

in which α_{1i} , $i = 1, 2, 3, \ldots$, and α_{2i} are positive constants satisfying $\alpha_{1i} < \alpha_{2i}$, and E_{ri} is a constant applied to normalize $\phi_i(t)$ such that $max\{\phi_i(t)\} = 1.0$; $w(t)$ is the zero-mean Gaussian white noise whose discrete variance

$$\langle w_s^2(0) \rangle = 2\pi S_0 \delta(0) \tag{7}$$

where $\delta(.)$ is the Kronecker delta function such that $\delta(0) = 1$; S_0 is the spectral density of the discrete Gaussian white noise process. Of course, for stationary random excitations $\phi_i(t) = 1$ which are just special cases of Eq. (6).

Application of Eqs. (5) through (7) to Eq. (3) and after some algebraic manipulation, one can show that

$$R_{s+1} = R_s^{(1)} + R_s^{(2)} + R_s^{(3)}, \tag{8}$$

where

$$R_s^{(1)} = N_2\, R_s\, N_2^T + N_3\, R_{s-1}\, N_3^T, \quad R_s^{(2)} = N_2\, D_s\, N_3^T + N_3\, D_s^T N_2^T,$$
$$R_s^{(3)} = (\Delta t)^4\, N_1\, B_s\, N_1^T.$$

Equation (8) is the recursive mean square of generalized displacement vector of the nonlinear system under Gaussian nonstationary random excitations. It is noted that the parameter matrices N_1, N_2 and N_3 have to be updated at every time step since they are time-dependent for the nonlinear system. If the system is linear these parameter matrices are constant.

The expression for the recursive covariance matrix of the generalized displacement vector can be obtained from Eq. (8) as

$$U_{s+1} = U_s^{(1)} + U_s^{(2)} + U_s^{(3)}\,, \tag{9}$$

where

$$U_s^{(1)} = N_2\, U_s\, N_2^T + N_3\, U_{s-1}\, N_3^T\,, \qquad U_s^{(2)} = N_2\, A_s^m N_3^T + N_3\, (A_s^m)^T N_2^T\,,$$
$$U_s^{(3)} = (\Delta t)^4\ N_1\, B_s\, N_1^T,$$
$$U_s = R_s - \langle x_s\rangle\langle x_s\rangle^T\,,\ A_s^m = D_s - \langle x_s\rangle\langle x_{s-1}\rangle^T.$$

Before leaving this section it is noted that an efficient route of computing recursive covariances and mean squares of generalized displacement vector of the nonlinear system is to first apply Eq. (9) for the covariance matrix U_{s+1} and then evaluate the mean square matrix R_s and so on.

3 Nonlinear Systems under Non-Gaussian Excitations

The SCD method for the case with non-Gaussian random excitations, the non-Gaussian random excitation model, time co-ordinate transformation (TCT) technique, and an adaptive time scheme (ATS) are presented in the following subsections.

3.1 Recursive Expressions for Systems under Non-Gaussian Random Excitations

In this case, the recursive expression for the mean square matrix of displacements is identical to that given in Eq. (3) except that the third, sixth, seventh, eighth, and ninth terms on the RHS have to be evaluated differently. Unlike in the case of Gaussian random excitations these terms are not zero in general. Thus, the recursive mean square of the generalized displacement vector can be obtained as

$$R_{s+1} = R_s^{(1)} + R_s^{(2)} + R_s^{(n)}\,, \tag{10}$$

where the first two terms on the RHS are defined in Eq. (8) while the third term on the RHS is given by

$$\begin{aligned} R_s^{(n)} = {} & (\Delta t)^4 N_1 B_s N_1^T + (\Delta t)^2 N_2 \langle x_s P_s^T\rangle N_1^T \\ & + (\Delta t)^2 N_1 \langle P_s x_s^T\rangle N_2^T \\ & + (\Delta t)^2 N_3 \langle x_{s-1} P_s^T\rangle N_1^T + (\Delta t)^2 N_1 \langle P_s x_{s-1}^T\rangle N_3^T, \end{aligned} \tag{11}$$

in which now the random excitation vector P_s is non-Gaussian and the terms on the RHS have to be evaluated for a particular non-Gaussian excitation process.

For the covariance matrix, Eq. (9) can be similarly applied, of course, with due modification to the terms involving the excitation process.

3.2 Non-Gaussian Random Excitation Models

Various non-Gaussian random excitation processes can be found in [4,5], for example. However, for verification purpose and direct comparison to results of known nonlinear system under Gaussian random excitations the following specific non-Gaussian nonstationary random excitation process is employed. Consider

$$P_s = [P_{s1}\, P_{s2}\, P_{s3} \,...\, P_{sn}]^T, \tag{12}$$

where P_{si} is the i'th element of the non-Gaussian random excitation vector at time step t_s. In the present investigation,

$$P_{si} = P_{si}\,(t_s) = [m_i(t_s)]^2 \frac{w^2(t_s)}{\sqrt{6\pi S_0}}. \tag{13}$$

The corresponding discrete ensemble average is defined by

$$\langle P_{si} \rangle = \sqrt{2\pi S_o/3}\, m_i^2(t_s). \tag{14}$$

The corresponding discrete mean square is

$$\langle P_{si}^2(t_s) \rangle = 2\pi S_o\, m_i^4(t_s). \tag{15}$$

Finally, the corresponding discrete variance, however, is obtained as

$$\langle \sigma_{si}^2(t_s) \rangle = (4\pi S_o/3) m_i^4(t_s). \tag{16}$$

With the above results, Eq. (11) can be explicitly obtained, accordingly.

3.3 Time Co-ordinate Transformation

For stiff mdof nonlinear systems or systems with large numbers of dof, typically encountered in the finite element analysis (FEA), a computational strategy, known as the TCT technique has been proposed by the author [8] to deal with computational instability [9], and its use in the FEA has been demonstrated by the author and his associate [10], for example. Another advantage of applying the TCT technique in the FEA is the fact that the finite element size or mesh dimensions can be relatively much coarser for accurate response computations since the dimensionless highest natural frequency is always equal to unity. For completeness, it is briefly introduced in this sub-section.

In the TCT strategy it is assumed that the stiff linear counter part of the nonlinear system governed by Eq. (2) has its highest natural frequency Ω_s or simply writing as Ω. Within every time step the system can be considered as linear. Therefore, one can perform the following operation

$$M_{s\tau}\ddot{x}_{s\tau} + (C_{s\tau}/\Omega)\, \dot{x}_{s\tau} + (K_{s\tau}\,/\Omega^2)\, x_{s\tau} = P_{s\tau}/\Omega^2, \tag{17}$$

where the over-dot and double over-dot now denote, respectively the first and second order derivatives with respect to dimensionless time τ which is related to Ω by $\tau = \Omega t$. The second subscript τ refers to the quantity in the τ-domain. Thus, for example, $K_{s\tau}$ is the stiffness matrix at dimensionless time step τ_s, so that $K_{s\tau} = K_s/\Omega^2$, and $P_{s\tau} = P_s/\Omega^2$, and so on.

Equation (10) can now be applied to compute the recursive mean square matrix of the generalized displacement vector of the discretized system defined by Eq. (17) in the dimensionless time domain. It should be noted that the time step size Δt in Eq. (10) now should be replaced with the dimensionless time step size $\Delta\tau$ while the mass matrix M_s in Eq. (10) should be replaced with $M_{s\tau}$, for example.

Once the mean square and covariance matrices of displacements in the τ domain are obtained using Eqs. (10) and (9), they are converted back to the original t domain by the following relations [9]:

$$R_s = \Omega R_{s\tau}, \qquad U_s = \Omega U_{s\tau}. \qquad (18\text{a, b})$$

3.4 Adaptive Time Scheme

Three ATS have been studied by the author and his associate, and presented in [7] for the response computation and analysis of highly nonlinear mdof systems. In the latter reference it was found that in terms of computational effectiveness and accuracy, the SCD-TATS is a better strategy for stiff systems. This was further confirmed in [10]. Therefore, in the presently proposed approach only the SCD-TATS is adopted. The SCD-TATS means that in the application of the SCD method the TCT is applied once at the beginning of the computation process before the application of the ATS.

4 System with Asymmetric Nonlinear Stiffness

For tractability and simplicity of illustration, a two dof nonlinear asymmetric system under a non-Gaussian nonstationary random excitation is considered in this section. This same nonlinear two dof system under a zero-mean Gaussian nonstationary random excitation has been studied by the author and associate [7]. In the latter work the nonstationary random excitation was modeled as a product of a deterministic modulating function and a zero-mean discrete Gaussian white noise (DGWN). Comparison between results obtained by the SCD method and those computed by the MCS was made and it was found to have excellent agreement [7]. Therefore, responses of this system under Gaussian nonstationary random excitations are readily available for comparison. In Sub-section 4.1 system parameter matrices and other pertinent data are provided while computed results and comparison studies are included in Sub-section 4.2.

4.1 Two Dof System with Asymmetric Stiffness

The two dof nonlinear system considered here has previously been investigated by Kimura and Sakata [11], and the author and his associate [7,12]. It may be applied to study soil-structure interaction or a primary structure housing a secondary equipment under an earthquake excitation treated as a Gaussian nonstationary random process. The system parameter matrices are given as:

$$M = \begin{bmatrix} 1 & 0 \\ 0 & 1 \end{bmatrix}, \quad C = \begin{bmatrix} 2\zeta_1 W & -2\mu\zeta_2 \\ -2\zeta_1 W & 2(1+\mu)\zeta_2 \end{bmatrix}, \qquad (19a,b)$$

$$K = \begin{bmatrix} k_{11} & k_{12} \\ k_{21} & k_{22} \end{bmatrix}, \quad k_{11} = W^2, \quad k_{12} = -\mu + \mu\eta(x_2 - \mu\varepsilon x_2^2), \qquad (19c)$$

$$k_{21} = -W^2, \quad k_{22} = 1 + \mu + (1-\mu)\eta x_2 + (1+\mu)\varepsilon x_2^2,$$

$$\mu = \frac{m_{22}}{m_{11}}, \quad W = \frac{\omega_1}{\omega_2}, \quad \bar{\tau} = \omega_2 t, \quad \omega_1 = \sqrt{\frac{k_1}{m_{11}}}, \quad \omega_2 = \sqrt{\frac{k_2}{m_{22}}},$$

$$\frac{c_i}{m_{ii}} = 2\zeta_i\omega_i, \quad i = 1,2, \quad P = \begin{pmatrix} m_1(\bar{\tau})w(\bar{\tau}) \\ 0 \end{pmatrix}, \qquad (20)$$

η and ε are the nonlinear parameters of the system, k_1 and c_1 are the stiffness and damping constants between the first mass and base where the excitation is applied, k_2 and c_2 are the stiffness and damping constants between the first and second masses. Note that the above system nonlinear stiffness matrix is asymmetric.

For the system studied in the present investigation, the above parameters are: W=1, $\zeta_1 = \zeta_2 = 0.10$, $\mu = 1.0$, $S_o = 0.0012$, $\alpha_{11} = 0.125$, $\alpha_{21} = 0.250$, $E_{r1} = 4.0$, $\eta = -1.0$, and $\varepsilon = 1.5$, indicating the system is highly nonlinear.

4.2 Computed Results and Discussion

With the system parameters and nonstationary random excitations provided in the last sub-section and defined by Eqs.(6) and (13), the non-Gaussian nonstationary random excitation is presented in Figure 1. The computed mean square responses for the system under Gaussian and non-Gaussian nonstationary random excitations are included in Figures 2 through 4. Specifically, Figures 2 and 4 include results of Gaussian and non-Gaussian random responses of the first mass, and second mass of the system, respectively. Figure 3 includes results of ensemble average of x_1 and x_2. These figures clearly show that for the particular Gaussian and non-Gaussian random excitations considered in the present investigation the differences of the peak values of the mean squares of responses are not very much. It is also noted that the width of the plots are narrower for the non-Gaussian random excitation case. It is important to note that every MCS run requires approximately 90 times longer than the presently proposed approach for results of Gaussian random excitation. For systems with large numbers of dof the saving in computational time for the presented approach is believed to be much more substantial.

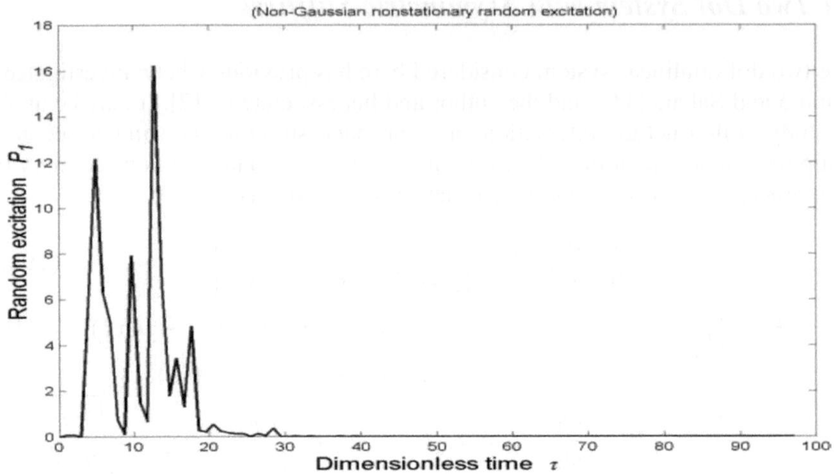

Fig. 1 Non-Gaussian nonstationary random excitation applied to first mass

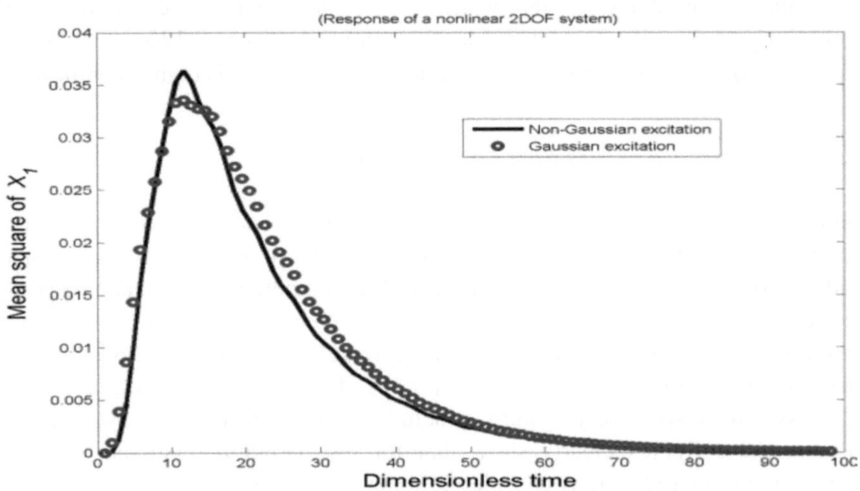

Fig. 2 Comparison of mean squares of displacement response x_1

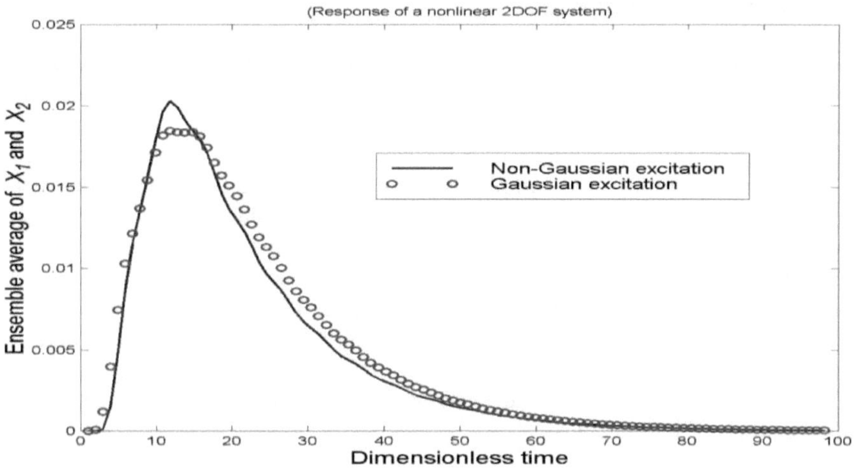

Fig. 3 Comparison of ensemble averages of x_1 and x_2

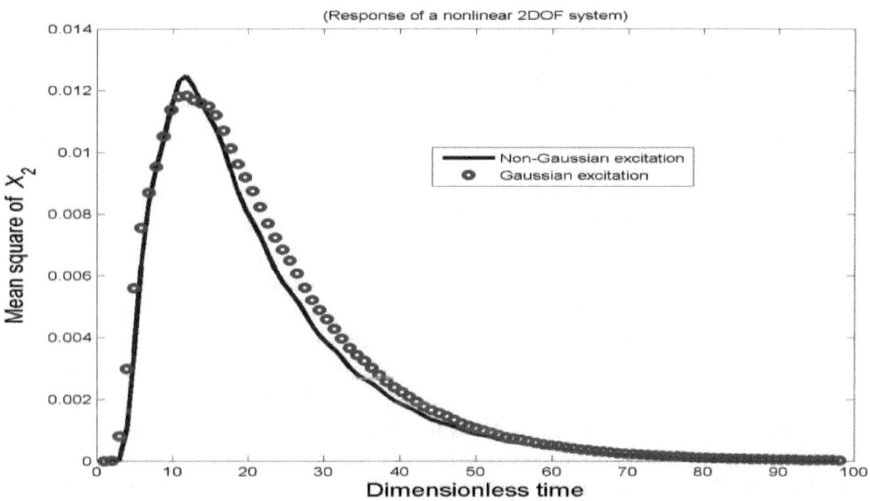

Fig. 4 Comparison of mean squares of displacement response x_2

4 Concluding Remarks

In the foregoing, a novel approach for response analysis of multi-degree-of-freedom (mdof) nonlinear systems under non-Gaussian nonstationary random excitations has been presented. It makes use of the stochastic central difference (SCD) method, time co-ordinate transformation (TCT), and adaptive time scheme (ATS). For tractability and simplicity of illustration, a two degree-of-freedom (dof) nonlinear asymmetric system under a non-Gaussian nonstationary random excitation was considered. Comparisons between computed results obtained for the system with Gaussian random excitations to those of the same system under non-Gaussian random excitations have been made. The proposed approach is very efficient compared with the MCS. It is relatively simple to implement and computed results are very accurate.

Computed results of nonlinear systems in the FEA will be obtained and published elsewhere in due course.

References

[1] MIL-STD-810F, Department of Defense Test Method Standard for Environmental Engineering Considerations and Laboratory Tests, USA, pp. 514.5-11 (2000)

[2] Defence Standard 00-35, Environmental Handbook for Defence Material, Part 5: Induced Mechanical Environments, Ministry of Defence, UK (1999)

[3] Steinwolf, A.: Shaker simulation of random vibrations with a high kurtosis value. Journal of the Institute of Environmental Sciences XL(3), 33–43 (1997)

[4] Grigoriu, M.: Applied Non-Gaussian Processes. Prentice-Hall, Englewood Cliffs (1995)

[5] Lutes, L.D., Sarkani, S.: Stochastic Analysis of Structural and Mechanical Vibrations. Prentice-Hall, Englewood Cliffs (1997)

[6] To, C.W.S.: On computational stochastic structural dynamics applying finite elements. Archives of Computational Methods in Engineering 8(1), 2–40 (2001)

[7] Liu, M.L., To, C.W.S.: Adaptive time schemes for responses of non-linear multi-degree-of-freedom systems under random excitations. Computers and Structures 52(3), 563–571 (1994)

[8] To, C.W.S.: Response of multi-degree-of-freedom systems with geometrical nonlinearity under random excitations by the stochastic central difference method. Paper presented at the Fourth Conference on Nonlinear Vibrations, Stability, and Dynamics of Structures and Mechanisms, Virginia Polytechnic Institute and State University, Blacksburg, Virginia, June 7-11 (1992)

[9] To, C.W.S.: Nonlinear Random Vibration: Computational Methods. Zip Publishing, Columbus (2010)

[10] To, C.W.S., Liu, M.L.: Large nonstationary random responses of shell structures with geometrical and material nonlinearities. Finite Elements in Analysis and Design 35, 59–77 (2000)

[11] Kimura, K., Sakata, M.: Nonstationary response analysis of a nonsymmetric nonlinear Multi-degree-of-freedom system to nonwhite random excitation. JSME International Journal 31(4), 690–697 (1988)

[12] To, C.W.S., Liu, M.L.: Recursive expressions for time dependent means and mean square responses of a multi-degree-of-freedom nonlinear system. Computers and Structures 48(6), 993–1000 (1993)

Stationary and Nontationary Response Probability Density Function of a Beam under Poisson White Noise

M. Vasta[1] and M. Di Paola[2]

[1] Engineering Department, University "G.D'Annunzio" of Chieti-Pescara, Viale Pindaro 42, 65100 Pescara, Italy
e-mail: mvasta@unich.it
[2] DISAG, Engineering Faculty, University of Palermo, Viale delle Scienze, 90128 Palermo, Italy

Abstract. In this paper an approximate explicit probability density function for the analysis of external oscillations of a linear and geometric nonlinear simply supported beam driven by random pulses is proposed. The adopted impulsive loading model is the Poisson White Noise , that is a process having Dirac's delta occurrences with random intensity distributed in time according to Poisson's law. The response probability density function can be obtained solving the related Kolmogorov-Feller (KF) integro-differential equation. An approximated solution, using path integral method, is derived transforming the KF equation to a first order partial differential equation. The method of characteristic is then applied to obtain an explicit solution. Different levels of approximation, depending on the physical assumption on the transition probability density function, are found and the solution for the response density is obtained as series expansion using convolution integrals.

Keywords: Poisson Pulses, Kolmogorov Feller Equation, Path Integral Method, Method of Characteristic, convolution integrals.

1 Introduction

In many civil and mechanical engineering applications, trains of pulses arriving at random times are used as ideal excitation. Such an idealization is suitable as an example for earthquakes [1], [2] and the dynamical behavior of a railway vehicle travelling over an imperfect track [3]. When the impulsive loads is the so called Poisson White Noise, that is a process having Dirac's delta occurrences with random intensity distributed in time according to Poisson's law, the probabilistic descriptors of the response can be obtained using either the stochastic integro-differential equation approach [4] or the differential approach [5]. Connection between the differential and the integro-differential approach has been recently found [10].

For the case of Gaussian White Noise input, an approximated response transition probability density function can be obtained solving the well known

W.Q. Zhu et al. (Eds.): Nonlinear Stochastic Dynamics & Control, IUTAM BOOK SERIES 29, pp. 127–136.
springerlink.com

Fokker-Planck equation. An approximated solution of the Fokker-Planck equation, for small values of the time step, is available in literature [6], [7]. This approximate solution was considered by many authors to solve engineering problems through path integral technique [8],[9]. For the case of Poisson White Noise input, the transition probability density function can be obtained as solution of an integro-differential equation [4], known as Kolmogorov-Feller (KF) equation. Exact stationary solutions for some particular classes of nonlinear systems excited by both external and parametric delta correlated excitations were obtained [11],[12]. A direct numerical solution of the KF equation is not an entirely straightforward matter, even for two-dimensional problems, and only approximated method have been derived [13],[14],[15]. An approximate technique, for deriving the response transition probability density function, solving the related KF equation, was proposed in [16], [21].

In this paper an approximate explicit response probability density function, for the analysis of external oscillations of a linear and geometric nonlinear simply supported beam driven by Poisson White Noise, is obtained as solution of the related KF equation. The approximated solution, using path integral method, is derived transforming the KF equation to a first order partial differential equation. The method of characteristic is then applied to obtain an explicit solution. Different levels of approximation, depending on the physical assumption on the transition probability density function, are found and the solution for the response density is obtained as series expansion using convolution integrals.

2 Linear Oscillations of a Beam under Transversal Poisson Pulses

Let us consider a simply supported beam subjected to a train of stationary Poisson pulses at midspan defined by

$$\xi(t)=\sum_{k=1}^{N(t)} A_k \delta(t-t_k) \tag{2}$$

where $\delta(\cdot)$ stand for a Dirac's delta. In eqn(1) the random instants t_k are independent of the intensities of the spikes A_k, having assigned probability density function $p_A(a)$ and assumed to be independent random variables, while $N(t)$ is an homogeneous Poisson counting process giving the total number of impulses occurring in the time interval [0,t) with mean arrival rate equal to λ.

The equation of motion of the beam can be written as follows

$$EJ\frac{\partial^4 w}{\partial x^4}+\mu\frac{\partial^2 w}{\partial t^2}+\rho\frac{\partial w}{\partial t}=\xi(t)\delta\left(x-\frac{l}{2}\right) \tag{3}$$

$w(x,t)$ being the vertical displacement of the beam. In Eq.(3) ρ is the beam density; J is the moment of inertia of the beam cross section; E is the Young's modulus; μ is a damping factor; l is the length of the beam.

Solution of eqn(3) can be obtained by mode superposition

$$w(x,t)=\sum_{n=1}^{\infty}\phi_n(x)Y_n(t) \tag{4}$$

In Eq.(4) $\phi_n(x)$ are the eigenfunctions, that for simply supported beams are given by $\phi_n(x)=\sin(n\pi x/l)$, and using the well known orthogonality properties of the eigenfunction basis the following set of stochastic uncoupled linear differential equations is obtained

$$\ddot{Y}_n + 2D\omega_n\dot{Y} + \omega_n^2 Y_n = P_n\xi(t) \quad n=1,2,\dots \tag{5}$$

having set

$$\omega_n = \left(\frac{\pi n}{l}\right)^2\sqrt{\frac{EJ}{\mu}}, \quad \frac{\rho}{\mu}=2D\omega_n, \quad P_n=\sin\left(\frac{n\pi}{2}\right)$$

For n=1 the following single degree of freedom equation of motion is obtained

$$\ddot{Y} + 2D\omega_1\dot{Y} + \omega_1^2 Y = \xi(t) \tag{6}$$

The probabilistic characterization of the response can be achieved, when truncating at the first mode, by considering the time evolution of the transition probability of the response

$$p(y,\dot{y};t+\tau/y',\dot{y}';t)=\frac{p(y,\dot{y},t+\tau;y',\dot{y}',t)}{p(y',\dot{y}',t)} \tag{7}$$

that respects the Kolmogorov-Feller integro-differential equation

$$\begin{aligned}\frac{\partial p}{\partial \tau} &= 2D\omega_1\frac{\partial(\dot{y}p)}{\partial \dot{y}} + \omega_1^2 y\frac{\partial p}{\partial \dot{y}} - \dot{y}\frac{\partial p}{\partial y} \\ &+ \lambda\int p(y,\dot{y}-a;t+\tau/y',\dot{y}';t)p_A(a)da - \lambda p\end{aligned} \tag{8}$$

The initial condition for Eq. (8) reads

$$p(y,\dot{y};t+\tau/y',\dot{y}';t)=\delta(y-y')\delta(\dot{y}-\dot{y}') \tag{9}$$

Exact stationary solutions (i.e. $\partial p/\partial\tau=0$) of Eq.(8) are known for particular classes of nonlinear systems [11], [12].

In order to find an approximate solution of Eq.(8), following Di Paola and Santoro [20], the following approximation hold for the transition density function up to order τ

$$p(y,\dot{y};t+\tau/y',\dot{y}';t)=\delta(y-y')[(1-\lambda\tau)\delta(\dot{y}-\dot{y}')+\lambda\tau p_A(\dot{y}-\dot{y}')] \tag{10}$$

It follows that the integral term appearing in Eq.(8) becomes

$$p(y,\dot{y};t+\tau / y',\dot{y}';t)=\delta(y-y')[(1-\lambda\tau)\delta(\dot{y}-a-\dot{y}') \\ +\lambda\tau p_A(\dot{y}-a-\dot{y}')] \tag{11}$$

Substituting Eq.(11) into Eq.(8) we obtain

$$\frac{\partial p}{\partial \tau}=2D\omega_1\frac{\partial(\dot{y}p)}{\partial \dot{y}}+\omega_1^2 y\frac{\partial p}{\partial \dot{y}}-\dot{y}\frac{\partial p}{\partial y}-\lambda p \\ +\lambda\int\delta(y-y')[(1-\lambda\tau)\delta(\dot{y}-a-\dot{y}')+\lambda\tau p_A(\dot{y}-a-\dot{y}')]p_A(a)da \tag{12}$$

Multiplying both members of Eq.(12) by the joint probability distribution $p_t=p(y',\dot{y}',t)$ and then integrating with respect of the initial variables, the following differential equation for the joint probability density function $p_{t+\tau}=p(y,\dot{y},t+\tau)$ up to order τ is obtained

$$\frac{\partial p_{t+\tau}}{\partial \tau}=2D\omega_1\frac{\partial(\dot{y}p_{t+\tau})}{\partial \dot{y}}+\omega_1^2 y\frac{\partial p_{t+\tau}}{\partial \dot{y}}-\dot{y}\frac{\partial p_{t+\tau}}{\partial y}-\lambda p_{t+\tau} \\ +\lambda G_t^{(0)}(y,\dot{y})+\lambda^2\tau\left(G_t^{(1)}(y,\dot{y})-G_t^{(0)}(y,\dot{y})\right) \tag{13}$$

where

$$G_t^{(0)}(y,\dot{y})=\int_{\mathbf{a}} p_t(y,\dot{y}-a;t)p_A(a)da \\ G_t^{(1)}(y,\dot{y})=\int_{\mathbf{a}} G_t^{(0)}(y,\dot{y}-a;t)p_A(a)da \tag{14}$$

Equation (13) is a linear partial differential equation that can be solved by the Method of Characteristics (see e.g. [17]). In order to construct the solution surface one solves the following system of ordinary differential equations on s

$$\begin{cases} \dfrac{d\tau}{ds}=1 \\ \dfrac{dy}{ds}=\dot{y} \\ \dfrac{d\dot{y}}{ds}=-\left(2D\omega_1\dot{y}+\omega_1^2 y\right) \\ \dfrac{dp}{ds}=\lambda G_t^{(0)}(y,\dot{y})+\lambda^2\tau\left(G_t^{(1)}(y,\dot{y})-G_t^{(0)}(y,\dot{y})\right)+(2D\omega_1-\lambda)p \end{cases} \tag{15}$$

Subjected to the following initial conditions

$$\begin{cases} \tau(0)=1 \\ y(0)=y' \\ \dot{y}(0)=\dot{y}' \\ p(0)=p_t(y',\dot{y}';t) \end{cases} \tag{16}$$

Solution of Eq.(15) takes the form

$$\begin{cases} \tau=s+C_1 \\ y(s)=C_2 e^{-D\omega_1 s}\cos(\overline{\omega}_1 s)+C_3 e^{-D\omega_1 s}\sin(\overline{\omega}_1 s) \\ p(s)=p(0)e^{cs}+\lambda G_t^{(0)}(y,\dot{y})s+\lambda^2\tau\left(G_t^{(1)}(y,\dot{y})-G_t^{(0)}(y,\dot{y})\right)s \end{cases} \tag{17}$$

having set

$$\overline{\omega}_1=\omega_1\sqrt{1-D^2},\quad c=2D\omega_1-\lambda$$

and because of the initial conditions Eq.(16)

$$C_1=0,\quad C_2=y',\quad C_3=\frac{\dot{y}'+D\omega_1 y'}{\overline{\omega}_1}$$

The solution found is finally

$$\begin{aligned} p(y,\dot{y};t+\tau)=\ & p_t(y'(y,\dot{y}),\dot{y}'(y,\dot{y});t)e^{c\tau} \\ & +\lambda\tau G_t^{(0)}(y,\dot{y})+\lambda^2\tau^2\left(G_t^{(1)}(y,\dot{y})-G_t^{(0)}(y,\dot{y})\right) \end{aligned} \tag{18}$$

Where for the lincar case the characteristic curves are explicitly given by

$$\begin{aligned} y'(y,\dot{y})&=\frac{e^{D\omega_1\tau}}{\overline{\omega}_1}[(\overline{\omega}_1\cos(\overline{\omega}_1\tau)-D\omega_1\sin(\overline{\omega}_1\tau))y-\sin(\overline{\omega}_1\tau)\dot{y}] \\ y'(y,\dot{y})&=\frac{e^{D\omega_1\tau}}{\overline{\omega}_1}[\overline{\omega}_1\sin(\overline{\omega}_1\tau)y+(\overline{\omega}_1\cos(\overline{\omega}_1\tau)+D\omega_1\sin(\overline{\omega}_1\tau))\dot{y}] \end{aligned} \tag{19}$$

3 Nonlinear Oscillations of a Beam under Transversal Poisson Pulses

Let us now consider the case of a geometric nonlinear simply supported beam subjected to a train of stationary Poisson pulses at mid span whose equations of motion are defined by

$$EJ\frac{\partial^4 w}{\partial x^4}+\mu\frac{\partial^2 w}{\partial t^2}+\rho\frac{\partial w}{\partial t}+\frac{1}{2l}\frac{\partial^2 w}{\partial x^2}\int_0^l\left(\frac{\partial w}{\partial x}\right)^2 dx=\xi(t)\delta\left(x-\frac{l}{2}\right) \tag{20}$$

By expanding the solution in linear mode superposition Eq.(4), the following set of nonlinear coupled stochastic differential equation of motion is obtained

$$\ddot{Y}_n+2D\omega_n\dot{Y}+\omega_n^2 Y_n+\sum_{m,p,q}^{n}\Gamma_{nmpq}Y_m Y_p Y_q=P_n\xi(t)\quad n=1,2,\ldots \tag{21}$$

where

$$\Gamma_{nmpq}=-\frac{1}{2l}\left[\int_0^l \phi_n'\phi_m' dx\right]\left[\int_0^l \phi_p'\phi_q' dx\right]$$

For the simple case $n=m=p=q=1$, by setting $\gamma=\Gamma_{1111}$, the following nonlinear stochastic differential equation is obtained

$$\ddot{Y}+2D\omega_1\dot{Y}+\omega_1^2 Y+\gamma Y^3=\xi(t) \tag{22}$$

The Kolmogorov-Feller equation associated with Eq.(22) reads

$$\begin{aligned}\frac{\partial p}{\partial \tau}&=2D\omega_1\frac{\partial(\dot{y}p)}{\partial \dot{y}}+\left(\omega_1^2 y+\gamma y^3\right)\frac{\partial p}{\partial \dot{y}}-\dot{y}\frac{\partial p}{\partial y}\\&+\lambda\int_{\mathbf{a}} p(y,\dot{y}-a;t+\tau / y',\dot{y}';t)p_A(a)da-\lambda p\end{aligned} \tag{23}$$

Following the same approach outlined for the linear case, the approximated solution of Eq.(23) by applying the Method of Characteristics is given by Eq.(18)

$$\begin{aligned}p(y,\dot{y};t+\tau)=&\,p_t\left(y'(y,\dot{y}),\dot{y}'(y,\dot{y});t\right)e^{c\tau}\\&+\lambda\tau G_t^{(0)}(y,\dot{y})+\lambda^2\tau^2\left(G_t^{(1)}(y,\dot{y})-G_t^{(0)}(y,\dot{y})\right)\end{aligned}$$

where the characteristic curves, because of the nonlinear term in Eq.(22), needs now to be evaluated by numerical integration solving the nonlinear system of ordinary differential equations

$$\begin{cases} \dfrac{dy}{d\tau} = \dot{y} \\ \dfrac{d\dot{y}}{d\tau} = -\left(2D\omega_1 \dot{y} + \omega_1^2 y + \gamma y^3\right) \end{cases} \tag{24}$$

Subjected to the initial conditions

$$\begin{cases} y(0) = y' \\ \dot{y}(0) = \dot{y}' \end{cases} \tag{25}$$

4 Solution of Higher Order τ

The solution found in the previous sections for linear and nonlinear systems is confined to the approximation of order τ of the transition pdf as given in Eq.(11). However, an approximate transition pdf of higher order can be defined, considering the probability of occurrence of n pulses in the time interval τ following the Poisson law

$$P[number\ of\ \ pulses\ in\ \tau \equiv n] = e^{-\lambda\tau} \frac{(\lambda\tau)^n}{n!} \tag{26}$$

The probability of having one spike of amplitude $(\dot{y} - \dot{y}')$, as stated in Eq.(11) is simply $p_A(\dot{y} - \dot{y}')$. The probability of having two spikes in τ of amplitude summation equal to $Y = (\dot{y} - \dot{y}')$, can be found as solution of the convolution equation

$$P[two\ spikes\ of\ \ amplitude\ summation\ Y\ in\ \tau] = \int_{\mathbf{a}} p_A(\dot{y} - \dot{y}' - a) p_A(a) da \tag{27}$$

and, for $n+1$ spikes arriving in τ of amplitude summation equal to$Y$

$$\begin{aligned} P[n\ spikes\ of\ \ amplitude\ summation\ Y\ in\ \tau] = & \int_{\mathbf{a}}(n)\int_{\mathbf{a}} p_A(a_1)\cdots p_A(a_n) \\ & \times p_A(\dot{y} - \dot{y}' - a_1 - \ldots - a_n) da_1 \cdots da_n \end{aligned} \tag{28}$$

It follows that combining Eqs.(26),(28) the following relationship hold for the transition pdf

$$
\begin{aligned}
p(y,\dot{y};t+\tau / y',\dot{y}';t) = \delta(y-y')e^{-\lambda\tau}\Big[&\delta(\dot{y}-\dot{y}') + \lambda\tau p_A(\dot{y}-\dot{y}') \\
&+ \frac{(\lambda\tau)^2}{2!}\int_{\mathbf{a}} p_A(\dot{y}-\dot{y}'-a)p_A(a)da \\
&+ \frac{(\lambda\tau)^3}{3!}\int_{\mathbf{a}}\int_{\mathbf{a}} p_A(\dot{y}-\dot{y}'-a_1-a_2)p_A(a_1)p_A(a_2)da_1 da_2 + \ldots\Big]
\end{aligned}
\tag{29}
$$

The transition pdf appearing in the integral term in Eq.(9), because of Eq.(29), can be written as follows (setting ρ=a)

$$
\begin{aligned}
p(y,\dot{y}-\rho;t+\tau / y',\dot{y}';t) = \delta(y-y')e^{-\lambda\tau}\Big[&\delta(\dot{y}-\dot{y}'-\rho) + \lambda\tau p_A(\dot{y}-\dot{y}'-\rho) \\
&+ \frac{(\lambda\tau)^2}{2!}\int_{\mathbf{a}} p_A(\dot{y}-\dot{y}'-\rho-a)p_A(a)da \\
&+ \frac{(\lambda\tau)^3}{3!}\int_{\mathbf{a}}\int_{\mathbf{a}} p_A(\dot{y}-\dot{y}'-\rho-a_1-a_2)p_A(a_1)p_A(a_2)da_1 da_2 + \ldots\Big]
\end{aligned}
\tag{30}
$$

Substituting Eq.(30) in Eq.(8), multiplying both members by the joint probability density $p_t = p(y',\dot{y}',t)$ and then integrating with respect of the initial variables, the following differential equation for the joint probability density function $p_{t+\tau} = p(y,\dot{y},t+\tau)$ up to order τ is obtained

$$
\begin{aligned}
\frac{\partial p_{t+\tau}}{\partial \tau} &= 2D\omega_1 \frac{\partial(\dot{y}p_{t+\tau})}{\partial \dot{y}} + \omega_1^2 y \frac{\partial p_{t+\tau}}{\partial \dot{y}} - \dot{y}\frac{\partial p_{t+\tau}}{\partial y} - \lambda p_{t+\tau} \\
&\quad + e^{-\lambda\tau}\sum_{j=0}^{\infty}\frac{\lambda^{j+1}\tau^j}{j!}G_t^{(j)}(y,\dot{y})
\end{aligned}
\tag{31}
$$

Solution of Eq.(31) can be explicitly derived by the method of characteristics, leading to the following formula for the joint pdf of displacement and velocity at $t+\tau$

$$
p(y,\dot{y};t+\tau) = e^{c\tau}p_t(y'(y,\dot{y}),\dot{y}'(y,\dot{y});t) + e^{-\lambda\tau}\sum_{j=0}^{\infty}\frac{(\lambda\tau)^{j+1}}{j!}G_t^{(j)}(y,\dot{y}) \tag{32}
$$

having set

$$
\begin{aligned}
G_t^{(0)}(y,\dot{y}) &= \int_{\mathbf{a}} p_t(y,\dot{y}-a;t)p_A(a)da \\
G_t^{(j)}(y,\dot{y}) &= \int_{\mathbf{a}} G_t^{(j-1)}(y,\dot{y}-a;t)p_A(a)da \qquad j=1,2,\ldots
\end{aligned}
\tag{33}
$$

It is worth noting that the transition pdf in Eq.(29) as well as the solution found in Eq.(32) recover to Eq.(11) and Eq.(18) if only terms up to order τ and τ^2 are considered respectively.

4 Conclusions

An approximate explicit response probability density function of a beam under external impulsive random Poisson excitation has obtained as approximate solution of the KF equation. It was shown that, using a physical-based transition pdf, the integro differential KF equation reduce to a first order partial differential equation, using path integral solution method for the integral term. Then, the Method of Characteristics has been applied to obtain the required solution. Different levels of approximation, depending on the physical assumption on the transition probability density function, were found and the solution for the response density has been obtained as series expansion using convolution integrals.

References

[1] Lin, Y.K.: Application of non-stationary shot noise in the study of system response to a class of non-stationary excitations. Journal of Applied Mechanics 30, 555–558 (1963)

[2] Lin, Y.K., Cai, G.Q.: Probabilistic Structural Dynamics: Advanced Theory and Applications. WilliamHill International edn. (1995)

[3] Tung, C.C.: Random response of highway bridges to vehicle loads. J. Engng. Mech. Div. 93, 79–94 (1967)

[4] Gihman, I.I., Skorohod, A.V.: Stochastic Differential Equations. Springer, Berlin (1972)

[5] Di Paola, M., Falsone, G.: Stochastic dynamics of non-linear systems driven by non-normal delta-correlated processes. Journal of Applied Mechanics 60, 141–148 (1993)

[6] Gardiner, C.W.: Handbook of Stochastic Methods. Springer, Berlin (1990)

[7] Risken, H.: The Fokker-Planck equation: methods of solution and applications. Springer, Berlin (1984)

[8] Feng, G.M., Wang, B., Lu, Y.F.: Path integral, functional method and stochastic dynamical systems. Probabilistic Engineering Mechanics 7, 149–157 (1992)

[9] Naess, A., Johnses, J.M.: Response statistics of nonlinear, compliant offshore structures by the path integral solution method. Probabilistic Engineering Mechanics 8, 91–106 (1993)

[10] Di Paola, M., Vasta, M.: Stochastic Integro-Differential and Differential Equations of Non Linear Systems Excited by Parametric Poisson Pulses. International Journal of Non-linear Mechanics 31, 855–862 (1997)

[11] Vasta, M.: Exact stationary solution for a class of non-linear systems driven by a non-normal delta-correlated process. Int. Journal of Non-linear Mechanics 30, 407–418 (1995)

[12] Proppe, C.: Exact stationary probability density functions for non-linear systems under Poisson white noise excitation. Int. Journal of Non-linear Mechanics 38, 557–564 (2003)

[13] Koyluoglu, H.U., Nielsen, R.K., Iwankievicz, R.: Response and reliability of Poisson-driven systems by path integration. Journal of Engineering Mechanics, 117–130 (1995)
[14] Iwankievicz, R., Nielsen, S.R.K.: Dynamic response of non-linear systems to Poisson distribuited random impulses. Journal of Sound and Vibration 156, 407–423 (1992)
[15] Roberts, J.B.: System response to random impulses. Journal of Sound and Vibration 24, 23–34 (1972)
[16] Vasta, M., Roberts, J.B.: An approximate transition probability density function for non-linear systems to impulsive loads. In: Proceedings of the Third International Conference on Computational Stochastic Mechanics (1998)
[17] Guenther, R.B., Lee, J.W.: Partial differential equations of mathematical physics and integral equations. Prentice-Hall, Englewood Cliffs (1988)
[18] Renger, A.: Eine dichtegleichung fur schwingungssysteme bei gleichzeitigen kontinuierlichen und diskreten stochastischen erregungen. ZAMM 59, 1–13 (1979)
[19] Nayfeh, A.H., Balachandran, B.: Applied Nonlinear Dynamics. Wiley Ed., New York (1995)
[20] Di Paola, M., Santoro, R.: Path integral solution for non-linear system enforced by Poisson White Noise. Probabilistic Engineering Mechanics 23, 164–169 (2008)
[21] Vasta, M., Luongo, A.: Dynamic Analysis of Linear and Nonlinear Oscillations of a Beam Under Axial and Transversal Random Poisson Pulses. Nonlinear Dynamics 36, 421–435 (2004)

Steady State Analysis of Stochastic Systems with Multiple Time Delays

W. Xu, C.Y. Sun and H.Q. Zhang

Department of Applied Mathematics, Northwestern Polytechnical University, Xi'an 710072, P.R. China

Abstract. In this paper, attention is focused on the steady state analysis of a class of nonlinear dynamic systems with multi-delayed feedbacks driven by multiplicative correlated Gaussian white noises. The Fokker-Planck equations for delayed variables are at first derived by Novikov's theorem. Then, under small delay assumption, the approximate stationary solutions are obtained by the probability density approach. As a special case, the effects of multi-delay feedbacks and the correlated additive and multiplicative Gaussian white noises on the response of a bistable system are considered. It is shown that the obtained analytical results are in good agreement with experimental results in Monte Carlo simulations.

Keywords: stochastic systems, multiple time delays, delay Fokker-Planck equation, Gaussian white noise.

1 Introduction

In the past few years, stochastic time-delayed dynamic systems have been extensively investigated, and the primary interests lie in revealing the interplay between stochastic excitations and time delay feedbacks, which have been proved capable of producing surprisingly rich phenomena. Resonance behavior in a quite simple stochastic model with time delay was observed both numerically and analytically [1]. Noise-induced resonance was studied in a single-mode semiconductor laser with weak optical feedback [2]. Analytical results were obtained for a prototypical bistable system with delayed feedback and the phenomena of coherence resonance was found [3]. Noise-induced coherence in bistable systems with multiple time delays was investigated by using the two-state approximation [4]. Hopf bifurcation analysis of a four-neuron network with multiple time delays was presented in [5]. Stochastic resonance in time-delayed bistable systems driven by weak periodic signal was investigated in [6].

As an important theoretic tool for learning about the response of stochastic dynamical systems, the Fokker-Planck equation has been studied extensively [7, 8]. In 1991, Fulinski etc [9] first introduced a correlation between a multiplicative noise and an additive noise. Then the effects of correlation between noise have attracted attention of many researchers [10, 11, 12].

The variety of applications demand a theory of stochastic time-delayed system. In this context, the delay Fokker-Planck equation was first introduced by Guilouzic

W.Q. Zhu et al. (Eds.): Nonlinear Stochastic Dynamics & Control, IUTAM BOOK SERIES 29, pp. 137–146.
springerlink.com

et al [13, 14]. Then Frank [15, 16] obtained the same result by using Novikov's theorem and compared the small delay approximation based on stochastic delay differential equation approach with the probability density approach. The impact of time delays on noise-induced transitions for a Hongler model was studied in [17, 18]. Delay Fokker-Planck equations have been obtained for stochastic systems with multiple delays by using Kramers-Moyal expansion technique [19]. In our previous work, a class of stochastic time-delayed systems with single time delay is studied [20].

In this paper, the delay Fokker-Planck equation is first obtained for a class of stochastic time-delayed systems driven by correlated Gaussian white noises with multiple time delays. As a kind of generalization, it matches the result derived by Frank [19] for some special cases. Then, the approximate stationary probability densities are obtained by using the small delay approximations based on a probability density approach. Numerical simulations are performed for a special case and it is shown that the analytical result is in good agreement with the Monte Carlo simulation.

2 Delay Fokker-Planck Equations

The class of stochastic time-delayed systems to be considered within this paper are of the form

$$\begin{aligned}\frac{dx(t)}{dt} &= h(x(t),x(t-\tau_1),\ldots,x(t-\tau_m)) + g_1(x(t),x(t-\tau_1),\ldots,x(t-\tau_m))\xi(t)\\ &\quad + g_2(x(t),x(t-\tau_1),\ldots,x(t-\tau_m))\eta(t).\end{aligned} \tag{1}$$

Further, $p(x,t) =< \delta(x-x(t)) >$ denotes the probability density of the stochastic process defined by Eq.(1). Differentiating $p(x,t)$ with respect to time yields

$$\begin{aligned}\frac{\partial}{\partial t}p(x,t) &= -< \frac{\partial}{\partial x}\delta(x-x(t))\frac{d}{dt}x(t) >\\ &= -\frac{\partial}{\partial x}\int h(x,x_{\tau_1},\ldots,x_{\tau_m})P(x,t;x_{\tau_1},t-\tau_1;\ldots;x_{\tau_m},t-\tau_m)dV_\tau\\ &\quad -\frac{\partial}{\partial x}\int g_1(x,x_{\tau_1},\ldots,x_{\tau_m})<\xi(t)\delta(x-x(t))\prod_{i=1}^{m}\delta(x_{\tau_i}-x(t-\tau_i))>dV_\tau\\ &\quad -\frac{\partial}{\partial x}\int g_2(x,x_{\tau_1},\ldots,x_{\tau_m})<\eta(t)\delta(x-x(t))\prod_{i=1}^{m}\delta(x_{\tau_i}-x(t-\tau_i))>dV_\tau,\end{aligned} \tag{2}$$

dV_τ corresponds to the M-dimensional differential $dV_\tau = dx_{\tau_1}\ldots dx_{\tau_m}$.

Next the Novikov's theorem is adopted to calculate the ensemble average $<\xi(t)\delta(x-x(t))\prod_{i=1}^{m}\delta(x_{\tau_i}-x(t-\tau_i))>$ and $<\eta(t)\delta(x-x(t))\prod_{i=1}^{m}\delta(x_{\tau_i}-x(t-\tau_i))>$, which reads

$$< \xi_k\phi(\xi_k,\xi_l) > = \int_0^t \gamma_{kl}(t,t') < \frac{\delta\phi(\xi_k,\xi_l)}{\delta\xi_{l'}(t')} > dt', \tag{3}$$

here ξ_k and ξ_l are Gaussian processes with correlation function γ_{kl}, $\phi(\xi_k,\xi_l)$ is an arbitrary functional of them. Thus

$$
\begin{aligned}
&< \xi(t)\delta(x-x(t))\prod_{i=1}^{m}\delta(x_{\tau_i}-x(t-\tau_i)) > \\
&= 2D\int_0^t \delta(t-t') < \frac{\delta(\delta(x-x(t))\prod_{i=1}^{m}\delta(x_{\tau_i}-x(t-\tau_i)))}{\delta x(t')}\frac{\delta x(t')}{\delta\xi(t')} > \\
&+2D\int_0^t \delta(t-t') < \frac{\delta(\delta(x-x(t))\prod_{i=1}^{m}\delta(x_{\tau_i}-x(t-\tau_i)))}{\delta x(t'-\tau_1)}\frac{\delta x(t'-\tau_1)}{\delta\xi(t')} > \\
&+\ldots \\
&+2D\int_0^t \delta(t-t') < \frac{\delta(\delta(x-x(t))\prod_{i=1}^{m}\delta(x_{\tau_i}-x(t-\tau_i)))}{\delta x(t'-\tau_m)}\frac{\delta x(t'-\tau_m)}{\delta\xi(t')} > \\
&+2\lambda\sqrt{DQ}\int_0^t \delta(t-t') < \frac{\delta(\delta(x-x(t))\prod_{i=1}^{m}\delta(x_{\tau_i}-x(t-\tau_i)))}{\delta x(t')}\frac{\delta x(t')}{\delta\eta(t')} > \\
&+2\lambda\sqrt{DQ}\int_0^t \delta(t-t') < \frac{\delta(\delta(x-x(t))\prod_{i=1}^{m}\delta(x_{\tau_i}-x(t-\tau_i)))}{\delta x(t'-\tau_1)}\frac{\delta x(t'-\tau_1)}{\delta\eta(t')} > \\
&+\ldots \\
&+2\lambda\sqrt{DQ}\int_0^t \delta(t-t') < \frac{\delta(\delta(x-x(t))\prod_{i=1}^{m}\delta(x_{\tau_i}-x(t-\tau_i)))}{\delta x(t'-\tau_m)}\frac{\delta x(t'-\tau_m)}{\delta\eta(t')} > .
\end{aligned}
\tag{4}
$$

Further, due to causality,

$$
\frac{\delta x(t'-\tau_i)}{\delta\xi(t')} = \frac{\delta x(t'-\tau_i)}{\delta\eta(t')} = 0, i = 1,\ldots,m. \tag{5}
$$

On the other hand, integrating Eq.(1) leads to

$$
\begin{aligned}
x(t) = x(0) &+ \int_0^t h(x(t),x(t-\tau_1),\ldots,x(t-\tau_m))ds \\
&+ \int_0^t g_1(x(t),x(t-\tau_1),\ldots,x(t-\tau_m))\xi(t)ds \\
&+ \int_0^t g_2(x(t),x(t-\tau_1),\ldots,x(t-\tau_m))\eta(t)ds,
\end{aligned}
\tag{6}
$$

and the functional derivative gives

$$
\frac{x(t)}{\xi(t)} = \frac{1}{2}g_1(x(t),x(t-\tau_1),\ldots,x(t-\tau_m)), \tag{7}
$$

$$
\frac{x(t)}{\eta(t)} = \frac{1}{2}g_2(x(t),x(t-\tau_1),\ldots,x(t-\tau_m)). \tag{8}
$$

Substituting Eq.(5), (7) and Eq.(8) into Eq.(4) yields

$$
\begin{aligned}
&< \xi(t)\delta(x-x(t))\prod_{i=1}^{m}\delta(x_{\tau_i}-x(t-\tau_i)) > \\
&= -D\frac{\partial}{\partial x}g_1(x,x_{\tau_1},\ldots,x_{\tau_m}) < \delta(x-x(t))\prod_{i=1}^{m}\delta(x_{\tau_i}-x(t-\tau_i)) > \\
&\quad -\lambda\sqrt{DQ}\frac{\partial}{\partial x}g_2(x,x_{\tau_1},\ldots,x_{\tau_m}) < \delta(x-x(t))\prod_{i=1}^{m}\delta(x_{\tau_i}-x(t-\tau_i)) > \\
&= -D\frac{\partial}{\partial x}g_1(x,x_{\tau_1},\ldots,x_{\tau_m})p(x,t;x_{\tau_1},t-\tau_1;\ldots;x_{\tau_m},t-\tau_m) \\
&\quad -\lambda\sqrt{DQ}\frac{\partial}{\partial x}g_2(x,x_{\tau_1},\ldots,x_{\tau_m})p(x,t;x_{\tau_1},t-\tau_1;\ldots;x_{\tau_m},t-\tau_m).
\end{aligned} \tag{9}
$$

Similarly,

$$
\begin{aligned}
&< \eta(t)\delta(x-x(t))\prod_{i=1}^{m}\delta(x_{\tau_i}-x(t-\tau_i)) > \\
&= -Q\frac{\partial}{\partial x}g_2(x,x_{\tau_1},\ldots,x_{\tau_m})p(x,t;x_{\tau_1},t-\tau_1;\ldots;x_{\tau_m},t-\tau_m) \\
&\quad -\lambda\sqrt{DQ}\frac{\partial}{\partial x}g_1(x,x_{\tau_1},\ldots,x_{\tau_m})p(x,t;x_{\tau_1},t-\tau_1;\ldots;x_{\tau_m},t-\tau_m).
\end{aligned} \tag{10}
$$

Inserting Eq.(9) and Eq.(10) into Eq.(2)leads to the delay Fokker-Planck equation

$$
\frac{\partial}{\partial t}p(x,t) = \hat{F}p(x,t) \tag{11}
$$

where

$$
\begin{aligned}
\hat{F} = \int [&-\frac{\partial}{\partial x}h(x,x_{\tau_1},\ldots,x_{\tau_m}) + D\frac{\partial}{\partial x}g_1(x,x_{\tau_1},\ldots,x_{\tau_m})\frac{\partial}{\partial x}g_1(x,x_{\tau_1},\ldots,x_{\tau_m})] \\
&+\lambda\sqrt{DQ}\frac{\partial}{\partial x}g_1(x,x_{\tau_1},\ldots,x_{\tau_m})\frac{\partial}{\partial x}g_2(x,x_{\tau_1},\ldots,x_{\tau_m})] \\
&+Q\frac{\partial}{\partial x}g_2(x,x_{\tau_1},\ldots,x_{\tau_m})\frac{\partial}{\partial x}g_2(x,x_{\tau_1},\ldots,x_{\tau_m})] \\
&+\lambda\sqrt{DQ}\frac{\partial}{\partial x}g_2(x,x_{\tau_1},\ldots,x_{\tau_m})\frac{\partial}{\partial x}g_1(x,x_{\tau_1},\ldots,x_{\tau_m})] \\
&p(x_{\tau_1},t-\tau_1;\ldots;x_{\tau_m},t-\tau_m|x,t)dV_{\tau}.
\end{aligned} \tag{12}
$$

When $\lambda = D = 0$, the result matches the delay Fokker-Planck equation derived by using the Kamers-Moyal expansion [19].

3 Approximate Stationary Solution

The perturbation theoretical approach proposed by Frank [15, 16] has been proven a useful tool for obtaining the approximate stationary solution of delay Fokker-Planck equation. The method can also be applied to Eq.(11). Like Ref [19], we can also illustrate this issue for a stochastic system in the form

$$\frac{dx(t)}{dt} = h^{(0)}(x(t)) + \sum_{i=1}^{n} h_i^{(1)}(x(t), x(t-\tau_i)) + g_1(x(t))\xi(t) + g_2(x(t))\eta(t). \tag{13}$$

Eq.(11) can be rewritten as

$$\begin{aligned}\frac{\partial}{\partial t}p(x,t) = &-\frac{\partial}{\partial x}[h^{(0)}(x) + \sum_{i=1}^{n}\int h_i^{(1)}(x(t),x(t-\tau_i))p^{(0)}(x_{\tau_i},t-\tau_i|x,t)dx_{\tau_i} \\ &+Dg_1\frac{\partial g_1}{\partial x} + \lambda\sqrt{DQ}g_2\frac{\partial g_1}{\partial x} + Qg_2\frac{\partial g_2}{\partial x} + \lambda\sqrt{DQ}g_1\frac{\partial g_2}{\partial x}]p(x,t) \\ &+\frac{\partial^2}{\partial x^2}[Dg_1^2 + 2\lambda\sqrt{DQ}g_1g_2 + Qg_2^2]p(x,t).\end{aligned} \tag{14}$$

Like Ref [19, 20], a perturbation theoretical technique is used and we have

$$\begin{aligned}&\frac{\partial}{\partial x}[h^{(0)}(x) + \sum_{i=1}^{n}\int h_i^{(1)}(x(t),x(t-\tau_i))p_{st}^{(0)}(x_{\tau_i},t-\tau_i|x,t)dx_{\tau_i} \\ &+Dg_1\frac{\partial g_1}{\partial x} + \lambda\sqrt{DQ}g_2\frac{\partial g_1}{\partial x} + Qg_2\frac{\partial g_2}{\partial x} + \lambda\sqrt{DQ}g_1\frac{\partial g_2}{\partial x}]p_{st}^{(1)}(x) \\ &= \frac{\partial^2}{\partial x^2}[Dg_1^2 + 2\lambda\sqrt{DQ}g_1g_2 + Qg_2^2]p_{st}^{(1)}(x),\end{aligned} \tag{15}$$

where $p_{st}^{(0)}(x_{\tau_i},t-\tau_i|x,t)$ is the zero order approximation of the stationary conditional distribution $p_{st}(x_{\tau_i},t-\tau_i|x,t)$, $p_{st}^{(1)}(x)$ is the first order approximation of the stationary probability density $p_{st}(x)$.

On the other hand, since $p_{st}^{(0)}(x_{\tau_i},t-\tau_i|x,t)$ is the conditional probability density of the unperturbed problem, using the short time propagator discussed in Ref [7, 8] gives

$$p_{st}^{(0)}(x_{\tau_i},t-\tau_i|x,t) = \sqrt{\frac{1}{4\pi R^{(0)}(x)\tau_i}}exp(-\frac{[x_{\tau_i} - x - f^{(0)}(x)\tau_i]^2}{4R^{(0)}(x)\tau_i}). \tag{16}$$

Substituting Eq.(16) into Eq.(15), the approximate stationary solution

$$p_{st}^{(1)}(x) = \frac{N}{R_{eff}}exp(\int \frac{f_{eff}(x')}{R_{eff}(x')}dx') \tag{17}$$

is derived with

$$f_{eff} = h^{(0)}(x) + \sum_{i=1}^{n} \sqrt{\frac{1}{4\pi R^{(0)}(x)\tau_i}} \int h_i^{(1)}(x(t), x(t-\tau_i))$$
$$*exp(-\frac{[x_{\tau_i} - x - f^{(0)}(x)\tau_i]^2}{4R^{(0)}(x)\tau_i})dx_{\tau_i}$$
$$+Dg_1\frac{\partial g_1}{\partial x} + \lambda\sqrt{DQ}g_2\frac{\partial g_1}{\partial x} + Qg_2\frac{\partial g_2}{\partial x} + \lambda\sqrt{DQ}g_1\frac{\partial g_2}{\partial x}, \quad (18)$$

$$R_{eff} = Dg_1^2 + 2\lambda\sqrt{DQ}g_1g_2 + Qg_2^2. \quad (19)$$

4 Case Study and Numerical Simulations

As a special case, we consider a bistable system with coupling between additive and multiplicative noise which can be written as

$$\frac{d}{dt}x(t) = x - x^3 + \varepsilon_1 x(t-\tau_1) + \varepsilon_2 x(t-\tau_2) + x\xi(x) + \eta(t). \quad (20)$$

The delay Fokker-Planck equation of the system is

$$\frac{\partial}{\partial t}p(x,t) = \int [-\frac{\partial}{\partial x}(x - x^3 + \varepsilon_1 x(t-\tau_1) + \varepsilon_2 x(t-\tau_2) + Dx + \lambda\sqrt{DQ})$$
$$+\frac{\partial^2}{\partial x^2}(Dx^2 + 2\lambda\sqrt{DQ}x + Q)]p(x_{\tau_1}, t-\tau_1; x_{\tau_2}, t-\tau_2|x,t)$$
$$dV_\tau p(x,t). \quad (21)$$

When $0 < \lambda < 1$, the approximate stationary probability density is

$$P_{st}(x) = \frac{N}{R(x)}\exp\{-\frac{1+\varepsilon_1\tau_1+\varepsilon_2\tau_2}{D}[\frac{1}{2}x^2 - cx + \frac{c^2+a-e}{2}ln|x^2+cx+e|$$
$$-\frac{c(c^2+a)-3ce-2b}{2\sqrt{e-\frac{c^2}{4}}}Arctg(\frac{1}{\sqrt{e-\frac{c^2}{4}}}(x+\frac{c}{2})), \quad (22)$$

where $R(x) = Dx^2 + 2\lambda\sqrt{DQ}x + Q, a = -(1+\varepsilon_1+\varepsilon_2+D), b = -\lambda\sqrt{DQ}, c = \frac{2\lambda\sqrt{DQ}}{D}, e = \frac{Q}{D}$. The critical curve separating the bimodal and unimodal regions is

$$\frac{1}{4}\lambda^2 DQ(1-\varepsilon_1\tau_1-\varepsilon_2\tau_2)^2 + \frac{1}{27}\frac{[D-\varepsilon\tau-(1+\varepsilon\tau)(1+\varepsilon)]^3}{1+\varepsilon\tau} = 0. \quad (23)$$

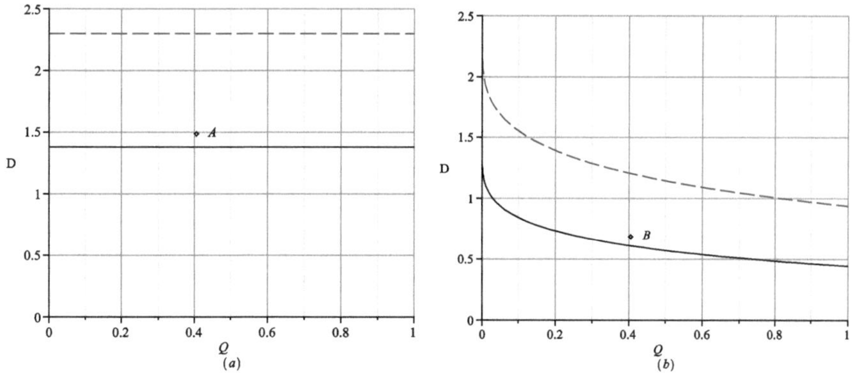

Fig. 1 The critical curves in the $Q-D$ parameter plane. (a) λ=0, ε_1=0.3, τ_1=0.1, ε_2=0, τ_2=0 (solid line), ε_2=0.2, τ_2=0.1 (dashed line). (b) λ=0.5, ε_1=0.3, τ_1=0.1, ε_2=0, τ_2=0 (solid line), ε_2=0.2, τ_2=0.1 (dashed line).

The critical curves are plotted in Fig.1. It can be seen that the unimodal region is increased while the bimodal region is decreased when the second time delay is increased. It can also be seen that when $\lambda = 0$, the critical curves are straight lines. The fact can be confirmed by examining Eq.(23).

Numerical simulation is performed to verify the validity of the analytical result and second-order Runge-Kutta algorithm is adopted (ensemble size $N = 2 \times 10^7$, single time step $\Delta t = 0.001$).

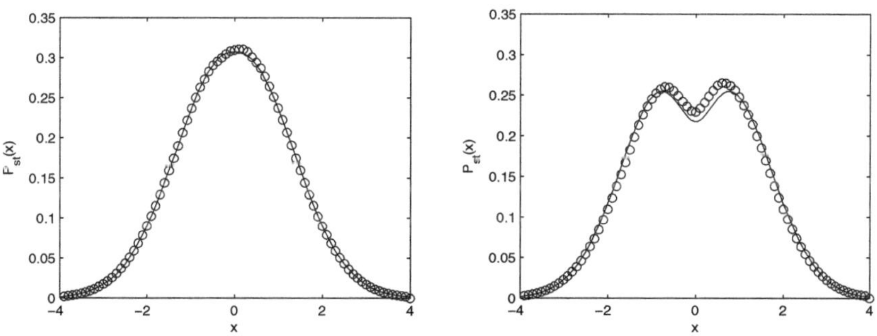

Fig. 2 The stationary probability density functions $P_{st}(x)$ relative to point A in Fig.1 are plotted('$-$' analytical results, '$\circ$' numerical simulations results). D=1.5, Q=0.4, λ=0, ε_1=0.3, τ_1=0.1, (a) ε_2=0, τ_2=0; (b) ε_2=0.4, τ_2=0.3.

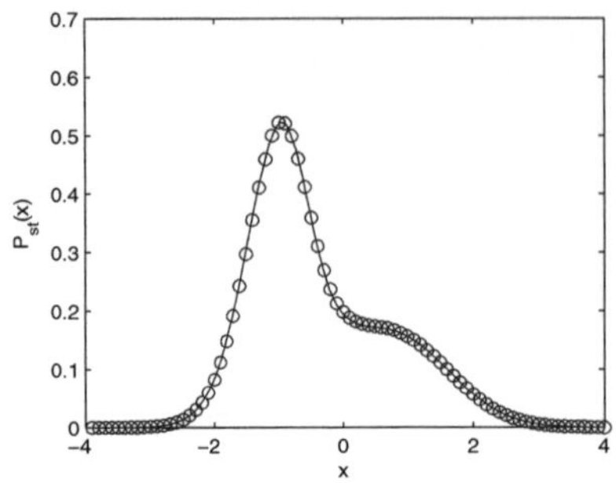

Fig. 3 The stationary probability density functions $P_{st}(x)$ relative to point B in Fig.1 are plotted('—' analytical results, 'o' numerical simulations results). D=0.7, Q=0.4, λ=0.5, ε_1=0.3, τ_1=0.1, (a) ε_2=0, τ_2=0; (b) ε_2=0.4, τ_2=0.3.

From Fig.2, it can be seen that $P_{st}(x)$ corresponding to the point A in Fig.1 experiences the transition from a bimodal to a unimodal structure when the second time delay is increased. Corresponding to the point B in Fig.1, Fig.3 show such a transition, too. We can also find that when the noises are uncorrelated, the stationary probability density function exhibits a symmetry bimodal structure. But when the noises are correlated, the symmetry bimodal structure is broken.

Based on the probability density derived above, further discussion about the mean first passage time (MFPT) can be performed. In this regard, interest is focused on in the effects of the two different time delays and the degree of correlation between noises on the transient property of the system. The switch time to escape from one stable state $-x_s$ to another stable state $+x_s$ is given by the equation

$$T_{+x_s}(-x_s) = \int_{-x_s}^{+x_s} \frac{dx}{R(x)p_{st}(x)} \int_{-\infty}^{x} p_{st}(y)dy, \tag{24}$$

through which MFPT can be evaluated analytically by inserting Eq.(22) into Eq.(24).

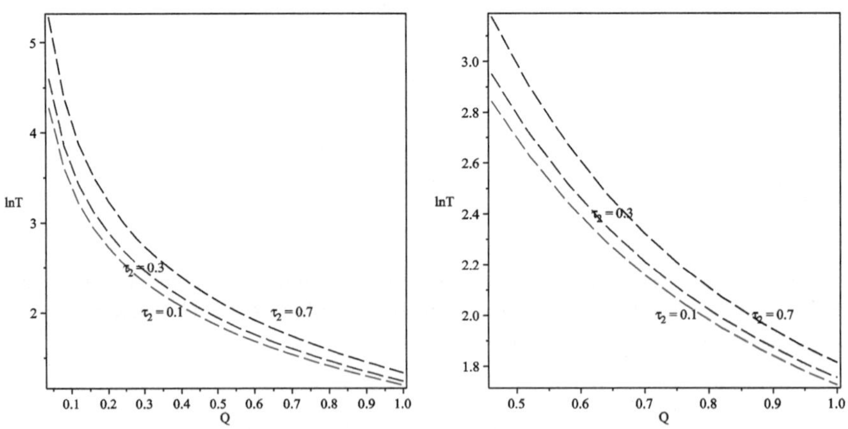

Fig. 4 The MFPT of the bistable system when parameters are fixed in the bimodal region are depicted. ε_1=0.3, τ_1=0.1, ε_2=0.4, (a) λ=0, D=1.5, (b) λ=0.5, D=0.7.

Shown in Fig.4 are results for various value of the parameter Q when one time delay is fixed to a certain value, the MFPT is seen increased with the other time delay in both symmetric and asymmetric cases.

The quantitative aspects in conjunction with the comparison with Monte Carlo simulation of the pervious section venders the proposed approach a viable tool for analyzing of stochastic systems with multiple time delays.

Acknowledgement. The authors are grateful for the support of the National Natural Science Foundation of China (Grant no.10872165).

References

1. Ohira, T., Sato, Y.: Resonance with Noise and Delay. Phys. Rev. Lett. 82, 2811–2815 (1999)
2. Masoller, C.: Distribution of Residence Times of Time-Delayed Bistable Systems Driven by Noise. Phys. Rev. Lett. 90, 20601 (2003); Noise-Induced Resonance in Delayed Feedback Systems. Phys. Rev. Lett. 88, 034102 (2002)
3. Tsimring, L.S., Pikovsky, A.: Noise-Induced Dynamics in Bistable Systems with Delay. Phys. Rev. Lett. 87, 250602 (2001)
4. Jiang, Y., Dong, S.H., Lozada-Cassou, M.: Noise-induced coherence in bistable systems with multiple time delays. Phys. Rev. E 69, 56225 (2004)
5. Mao, X.C., Hu, H.Y.: Hopf bifurcation analysis of a four-neuron network with multiple time delays. Nonlinear Dyn. 55, 95–112 (2009)
6. Wu, D., Zhu, S.Q.: Stochastic resonance in a bistable system with time-delayed feedback and non-Gaussian noise. Phys. Lett. A 363, 202–212 (2007)
7. Gardiner, C.W.: Handbook of Stochastic Methods. Springer, Berlin (1983)
8. Risken, H.: The Fokker-Planck Equation. Springer, Berlin (1983)
9. Fulinski, A., Telejko, T.: On the effect of interference of additive and multiplicative noises. Phys. Lett. A 152, 11–14 (1991)
10. Wu, D.J., Cao, L.: Bistable kinetic model driven by correlated noises: Steady-state analysis. Phys. Rev. E 50, 2496–2502 (1994)
11. Jia, Y., Li, J.R.: Stochastic system with colored correlation between white noise and colored noise. Physica A 252, 417–427 (1998)
12. Liang, G.Y., Cao, L., Wu, D.J.: Approximate Fokker-Planck equation of system driven by multiplicative colored noises with colored cross-correlation. Physica A 335, 371–384 (2004)
13. Guillouzic, S., Heureux, I.L., Longtin, A.: Small delay approximation of stochastic delay differential equations. Phys. Rev. E 59, 3970–3982 (1999)
14. Guillouzic, S., Heureux, I.L., Longtin, A.: Rate processes in a delayed, stochastically driven, and overdamped system. Phys. Rev. E 61, 4906–4914 (2000)
15. Frank, T.D.: Delay Fokker-Planck equations, perturbation theory, and data analysis for nonlinear stochastic systems with time delays. Phys. Rev. E 71, 31106 (2005)
16. Frank, T.D.: Delay Fokker-Planck equations, Novikovs theorem, and Boltzmann distributions as small delay approximations. Phys. Rev. E 72, 11112 (2005)
17. Frank, T.D., Patanarapeelert, K., Tang, I.M.: Delay- and noise-induced transitions: a case study for a Hongler model with time delay. Phys. Lett. A 339, 246–251 (2005)

18. Patanarapeelert, K., Frank, T.D., Friedrich, R., Tang, I.M.: On reducible nonlinear time-Delayed stochastic systems: Fluctuation-dissipation relations, transitions to bistability, and secondary transitions to non-stationarity. J. Phys. A: Math. Gen. 38, 10069–10083 (2005)
19. Frank, T.D.: Kramers-Moyal expansion for stochastic differential equations with single and multiple delays: Applications to financial physics and neurophysics. Phys. Lett. A 360, 552–562 (2007)
20. Zhang, H.Q., Xu, W., Xu, Y., Li, D.X.: Stochastic time-delayed systems driven by correlated noises: Steady-state analysis. Physica A 388, 3017–3023 (2009)

Stochastic Averaging of Strongly Nonlinear Oscillators under Poisson White Noise Excitation

Y. Zeng and W.Q. Zhu

Department of Mechanics, State Key Laboratory of Fluid Power Transmission and Control, Zhejiang University, Hangzhou, China

Abstract. A stochastic averaging method for single-degree-of-freedom (SDOF) strongly nonlinear oscillators under Poisson white noise excitation is proposed by using the so-called generalized harmonic functions. The stationary averaged generalized Fokker-Planck-Kolmogorov (GFPK) equation is solved by using the classical perturbation method. Then the procedure is applied to estimate the stationary probability density of response of a Duffing-van der Pol oscillator under Poisson white noise excitation. Theoretical results agree well with Monte Carlo simulations.

Keywords: Strongly nonlinear oscillator, Poisson white noise, stationary response, stochastic averaging.

1 Introduction

In stochastic dynamics, stochastic excitation is usually modeled as Gaussian white noise or filtered Gaussian white noise. Many analytical and numerical methods have been developed for predicting the response of such kind of stochastic dynamical systems. However, in real engineering there are many non-Gaussian random excitations. The representative examples of such random excitations are highway traffic loading, loadings caused by wind gusts associated with eddies, by atmospheric turbulence, or by buffeting of an airplane, and the loading acting on moving vehicles due to rough ground or imperfect track, etc. Therefore, many techniques for predicting the response of non-linear dynamical systems to non-Gaussian random excitation have been developed in recent years, including moment closure method [1], equivalent linearization method [2], perturbation method [3], cell mapping method [4], spectral finite difference method [5], characteristic function method [6], Wiener-Hermite expansion method [7] and Monte Carlo simulation method [8]. However there are several difficulties which restrict the extensive use of these techniques, e.g., complexity in calculation for high-dimensional nonlinear systems, much more time in repeating simulations for different parameter conditions, etc..

As a powerful technique, stochastic averaging methods based on diffusion approximation were extensively used in the prediction of response of various

W.Q. Zhu et al. (Eds.): Nonlinear Stochastic Dynamics & Control, IUTAM BOOK SERIES 29, pp. 147–155.
springerlink.com

nonlinear systems to Gaussian random excitation [9-12]. Through stochastic averaging, nonlinear systems can be simplified and approximate statistics of response can be more easily obtained from the averaged equations. However, for non-Gaussian random excitations the stochastic averaging based on diffusion approximation is not applicable and it is necessary to establish new stochastic averaging procedures. As the application of stochastic averaging theorems in predicting the response of nonlinear systems under non-Gaussian random excitations, stochastic averaging method of coefficients of GFPK equation for quasi linear systems under Poisson white noise excitation [13], standard stochastic averaging method for quasi linear systems under filtered Poisson white noise excitation [14], and stochastic averaging method for quasi-nonintegrable-Hamiltonian system under Poisson white noise excitation [15] have been developed by the authors successfully. In this paper, a new stochastic averaging procedure is proposed for predicting the response of SDOF strongly nonlinear oscillators under Poisson white noise excitation by using the so-called generalized harmonic functions. The approximate stationary solution of averaged GFPK equation is obtained by using the classical perturbation method. As an example, approximate probability density of stationary response of a Duffing-van der Pol oscillator under Poisson white noise excitation is obtained by using this new stochastic averaging method and confirmed by using Monte Carlo simulation.

2 Stochastic Averaging Method

Consider the following SDOF strongly nonlinear oscillator,

$$\begin{aligned} \dot{Q} &= P \\ \dot{P} &= -g(Q) - \varepsilon^2 f(Q,P) + \varepsilon \sum_{k=1}^{r} f_k(Q,P)\xi_k(t) \end{aligned} \tag{1}$$

where $g(Q)$ represents nonlinear restoring force, $\varepsilon^2 f(Q,P)$ represents lightly linear and(or) nonlinear damping, ε is a small positive parameter and $\xi_k(t)$, $k = 1,\cdots,r$ represent independent Poisson white noises which can be treated as formal derivatives of the following homogeneous compound Poisson processes:

$$\xi_k(t) \equiv \frac{dC_k(t)}{dt}, \quad k = 1,2,\cdots,r \tag{2}$$

$$C_k(t) = \sum_{i=1}^{N_k(t)} Y_{ki} U(t - t_{ki}) \tag{3}$$

in which, $N_k(t)$ is Poisson counting process with average arrival rate λ_k , $U(t-t_{ki})$ is unit step function at pulse arrival time t_{ki} and Y_{ki} is a random variable representing the intensity of the i -th impulse of ξ_k .

Because ε is small, the sample motion of system (1) will be nearly periodic and can be written as

$$Q = A\cos\Phi + B, P = -A\upsilon(A,\Phi)\sin\Phi, \Phi = \Psi + \Theta, \tag{4}$$

where Θ is initial phase angle, $\upsilon(A,\Phi)$ is instantaneous angle frequency, A is amplitude and B is symmetric center coordinate. $\sin\Phi$ and $\cos\Phi$ are called generalized harmonic functions. In Eq. (4), A , B , Φ , Ψ and Θ are random processes and

$$\upsilon(A,\Phi) = \frac{d\Psi}{dt} = \sqrt{\frac{2[V(A+B) - V(A\cos\Phi + B)]}{A^2 \sin^2\Phi}}, \tag{5}$$

$$V(A+B) = V(-A+B) = H, \tag{6}$$

$$H = \frac{1}{2}P^2 + V(Q), \tag{7}$$

$$V(Q) = \int_0^Q g(u)du. \tag{8}$$

Eq.(4) can be regarded as a set of random van der Pol transformations from Q , P to A , Θ . With the transformations accomplished, Stratonovich SDE of new state variables A , Θ are obtained from Eq.(1) as follows:

$$\begin{aligned} dA &= \varepsilon^2 m_1(A,\Phi)dt + \varepsilon \sum_{k=1}^{r} g_{1k}(A,\Phi) \circ dC_k(t), \\ d\Theta &= \varepsilon^2 m_2(A,\Phi)dt + \varepsilon \sum_{k=1}^{r} g_{2k}(A,\Phi) \circ dC_k(t), \end{aligned} \tag{9}$$

where

$$m_1(A,\Theta) = -A\upsilon f' \sin\Phi, \tag{10}$$

$$m_2(A,\Theta) = -\upsilon f'(\cos\Phi + R), \tag{11}$$

$$g_{1k}(A,\Theta) = -A\upsilon f_k' \sin\Phi, \tag{12}$$

$$g_{2k}(A,\Theta) = -(\cos\Phi + R)\upsilon f_k', \tag{13}$$

$$f' = \frac{f(A\cos\Phi + B, -A\upsilon(A,\Phi)\sin\Phi)}{g(A+B)(1+R)}, \tag{14}$$

$$f_k' = \frac{f_k(A\cos\Phi + B, -A\upsilon(A,\Phi)\sin\Phi)}{g(A+B)(1+R)}, \tag{15}$$

$$R = \frac{dB}{dA} = \frac{g(-A+B)+g(A+B)}{g(-A+B)-g(A+B)}. \tag{16}$$

By using the transformation from Stratonovich SDE to Itô SDE proposed by Di Paola and Falsone [16, 17], Itô SDE of A, Θ can be obtained as follows:

$$\begin{aligned} dA &= \varepsilon^2 m_1(A,\Theta)dt + \sum_{s=1}^{\infty}\sum_{k=1}^{r} \frac{\varepsilon^s}{s!} G^{(s)}_{1\underbrace{kk\cdots k}_{s-fold}} (dC_k)^s, \\ d\Theta &= \varepsilon^2 m_2(A,\Theta)dt + \sum_{s=1}^{\infty}\sum_{k=1}^{r} \frac{\varepsilon^s}{s!} G^{(s)}_{2\underbrace{kk\cdots k}_{s-fold}} (dC_k)^s, \end{aligned} \tag{17}$$

in which,

$$\begin{aligned} G^{(s)}_{i\underbrace{kk\cdots k}_{s\sim fold}} &= g_{1k}(A,\Theta)\frac{\partial}{\partial A} G^{(s-1)}_{i\underbrace{kk\cdots k}_{(s-1)\sim fold}} + g_{2k}(A,\Theta)\frac{\partial}{\partial \Theta} G^{(s-1)}_{i\underbrace{kk\cdots k}_{(s-1)\sim fold}}, \\ G^{(1)}_{ik} &= g_{ik}(A,\Theta),\ i=1,2;\ k=1,2,\cdots r. \end{aligned} \tag{18}$$

By using probability evolution equation of Markov process, GFPK equation is obtained from Itô SDE (17) as follows:

$$\begin{aligned}\frac{\partial\rho}{\partial t}=&\sum_{i=1}^{\infty}\frac{(-1)^i}{i!}\frac{\partial^i\left(M_{1i}\rho\right)}{\partial a^i}+\sum_{j=1}^{\infty}\frac{(-1)^j}{j!}\frac{\partial^j\left(M_{2j}\rho\right)}{\partial\theta^j}\\&+\sum_{i=1}^{\infty}\sum_{j=1}^{\infty}\frac{(-1)^{i+j}}{i!j!}\frac{\partial^{i+j}\left(N_{ij}\rho\right)}{\partial a^i\partial\theta^j},\end{aligned}\tag{19}$$

where $\rho=\rho(a,\theta,t)$ and

$$M_{ij}=\begin{cases}\varepsilon^2 m_i+\sum\limits_{s=1}^{\infty}\sum\limits_{k=1}^{r}G^{(s)}_{i\underbrace{kk\cdots k}_{s\sim fold}}\dfrac{\varepsilon^s\lambda_k\left\langle Y_k^s\right\rangle}{s!}, & j=1\\ \sum\limits_{s_1=1}^{\infty}\sum\limits_{s_2=1}^{\infty}\cdots\sum\limits_{s_j=1}^{\infty}\sum\limits_{k=1}^{r}G^{(s_1)}_{i\underbrace{kk\cdots k}_{s_1\sim fold}}G^{(s_2)}_{i\underbrace{kk\cdots k}_{s_2\sim fold}}\cdots G^{(s_j)}_{i\underbrace{kk\cdots k}_{s_j\sim fold}}\\ \times\dfrac{\varepsilon^{s_1+\cdots+s_j}\lambda_k\left\langle Y_k^{s_1+\cdots+s_j}\right\rangle}{s_1!s_2!\cdots s_j!}, & j\ge 2\end{cases}\tag{20}$$

$$\begin{aligned}N_{ij}=&\sum_{s_1=1}^{\infty}\cdots\sum_{s_i=1}^{\infty}\sum_{r_1=1}^{\infty}\cdots\sum_{r_j=1}^{\infty}\sum_{k=1}^{r}G^{(s_1)}_{1\underbrace{kk\cdots k}_{s_1\sim fold}}\cdots G^{(s_i)}_{1\underbrace{kk\cdots k}_{s_i\sim fold}}G^{(r_1)}_{2\underbrace{kk\cdots k}_{r_1\sim fold}}\cdots G^{(r_j)}_{2\underbrace{kk\cdots k}_{r_j\sim fold}}\\&\times\frac{\varepsilon^{s_1+\cdots+s_i+r_1+\cdots+r_j}\lambda_k\left\langle Y_k^{s_1+\cdots+s_i+r_1+\cdots+r_j}\right\rangle}{s_1!\cdots s_j!r_1!\cdots r_j!},\ i,j=1,2,\cdots,\infty.\end{aligned}\tag{21}$$

in which, $\langle\bullet\rangle$ denotes mathematical expectation.

Obviously, coefficients of GFPK Eq. (19) are functions of A and Θ. By using Eqs. (4), (5) and (17), it is found Φ is a rapid variable, i.e.,

$$d\Phi=\upsilon(A,\Phi)dt+\varepsilon^2 m_2(A,\Theta)dt+\sum_{s=1}^{\infty}\sum_{k=1}^{r}\frac{\varepsilon^s}{s!}G^{(s)}_{2\underbrace{kk\cdots k}_{s-fold}}\left(dC_k\right)^s\tag{22}$$

Therefore, time averaging to GFPK Eq.(19) can be replaced by space averaging with respect to Φ, averaged GFPK equation is obtained as follows:

$$\frac{\partial \rho}{\partial t} = \sum_{i=1}^{\infty} \frac{(-1)^i}{i!} \frac{\partial^i \left(\bar{M}_{1i}\rho\right)}{\partial a^i}, \tag{23}$$

where $\rho = \rho(a,t)$ and

$$\bar{M}_{ij} = \lim_{T\to\infty} \frac{1}{T} \int_0^T M_{ij}(a,\varphi)dt = \frac{1}{2\pi} \int_0^{2\pi} M_{ij}(a,\varphi)d\varphi. \tag{24}$$

Because all the terms in the right side of averaged GFPK Eq. (23) can be rearranged with respect to the order of ε, following perturbation expansion of stationary solution can be made:

$$\rho(a) = \rho_0(a) + \varepsilon\rho_1(a) + \varepsilon^2\rho_2(a) + \cdots. \tag{25}$$

Substituting Eq.(25) into Eq.(23) and using the classical perturbation method, approximate stationary solution of $\rho(a)$ can be obtained. And then, approximate stationary probability density $\rho(a,\theta)$ can be derived as follows:

$$\rho(a,\theta) = \frac{1}{2\pi}\rho(a). \tag{26}$$

Using the transformations in Eq.(4), the joint probability density of Q and P is obtained as follows:

$$\rho(q,p) = \left|\frac{\partial(a,\theta)}{\partial(q,p)}\right| \rho(a,\theta) = \frac{1}{2\pi}\left|\frac{\partial(a,\theta)}{\partial(q,p)}\right| \rho(a)\Big|_{a=V^{-1}(p^2/2+V(q))-b}. \tag{27}$$

and marginal probability density of Q can be obtained in the following:

$$\rho(q) = \int_{-\infty}^{\infty} \rho(q,p)dp. \tag{28}$$

In the following section, intensity of random impulses of Poisson white noise is assumed to be Gaussian distributed with zero mean for illustrative purpose, i.e., $\langle Y^4\rangle = 3\langle Y^2\rangle^2$.

3 Example

Consider a Duffing-van der Pol oscillator subject to external excitation of Poisson white noise in the form:

$$\begin{aligned} \dot{Q} &= P \\ \dot{P} &= -(wQ + \alpha Q^3) - \varepsilon^2(\beta Q^2 - 1)P + \varepsilon\xi(t) \end{aligned} \tag{29}$$

Substituting Eq. (29) into the Stratonovich and Itô SDEs and averaged GFPK equation in section 2 and solving stationary averaged GFPK equation by using the proposed perturbation method, approximate stationary probability density $\rho(a)$ and stationary marginal probability density $\rho(q)$ are obtained and presented separately in Figures 1 and 2 with parameters: $w = 1.0$, $\alpha = 1.0$, $\beta = 1.0$, $\varepsilon = 0.4$, $\lambda = 1.0$, and $\langle Y^2 \rangle = 2.0$.

It is shown in Figure 1 that approximate stationary probability density $\rho(a)$ agrees well with results from Monte Carlo simulation and is better than the Gaussian approximate solution which is obtained under the assumption that the random excitation is a Gaussian white noise with the same intensity of Poisson white noise in Eq. (29). It is also shown in Figure 2 that stationary marginal probability density $\rho(q)$ obtained by using the proposed method agrees well with results from Monte Carlo simulation.

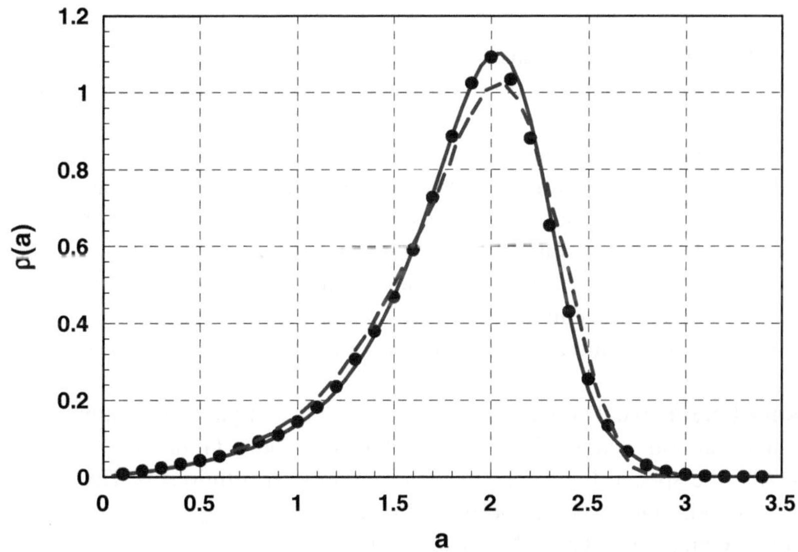

Fig. 1 Stationary probability density $\rho(a)$ of Duffing-van der Pol oscillator: ——, approximate stationary solution obtain by the proposed method; ---, Gaussian approximate solution; •, results from Monte Carlo simulation

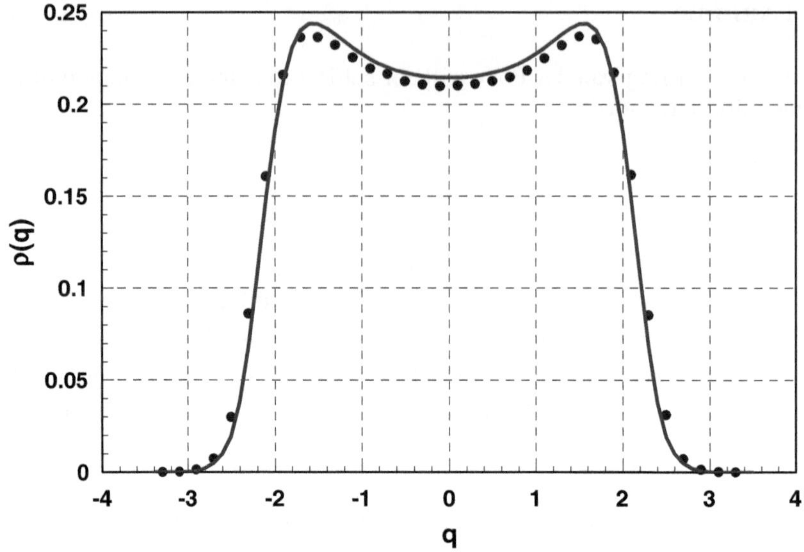

Fig. 2 Stationary marginal probability density $\rho(q)$ of Duffing-van der Pol oscillator: ——, approximate stationary solution obtain by the proposed method; •, results from Monte Carlo simulation

4 Conclusions

In the present paper, a stochastic averaging procedure for predicting the response of SDOF strongly nonlinear oscillator under Poisson white noise excitation is proposed by using the so-called generalized harmonic functions and a classical perturbation method is introduced to solve approximate stationary probability density form corresponding averaged GFPK equation. Using this stochastic averaging method, approximate stationary density of amplitude and stationary marginal probability density of displacement of Duffing-van der Pol oscillator are obtained and confirmed by comparing with results from Monte Carlo simulation. This stochastic averaging procedure is the basis of further research on stability and reliability of strongly nonlinear oscillators under Poisson white noise excitation.

Acknowledgments. The work reported in the present paper was supported by the National Natural Science Foundation of China under Grant No. 10772159 and 1093209, the special Fund for Doctor Program in Institutions of Higher Learning of China under Grant No. 20060335125, and the Zhejiang Natural Science Foundation of China under Grant No. Y7080070.

References

1. Di Paola, M., Falsone, G.: Stochastic dynamics of MDOF structural systems under non-normal filtered inputs. Probab. Eng. Mech. 9, 265–272 (1994)
2. Grigoriu, M.: Equivalent linearization for Poisson white-noise input. Probab. Eng. Mech. 10(1), 45–51 (1995)
3. Cai, G.Q., Lin, Y.K.: Response distribution of nonlinear-systems excited by non-Gaussian impulsive noise. Int. J. Non-Linear Mech. 27(6), 955–967 (1992)
4. Koyluoglu, H.U., Nielsen, S.R.K., et al.: Fast cell-to-cell mapping (path integration) for nonlinear white-noise and Poisson driven systems. Struct. Saf. 17(3), 151–165 (1995)
5. Wojtkiewicz, S.F., Johnson, E.A., et al.: Response of stochastic dynamical systems driven by additive Gaussian and Poisson white noise: Solution of a forward generalized Kolmogorov equation by a spectral finite difference method. Computer Methods in Appl. Mech. Eng. 168(1-4), 73–89 (1999)
6. Grigoriu, M.: Characteristic function equations for the state of dynamic systems with Gaussian, Poisson, and Levy white noise. Probab. Eng. Mech. 19(4), 449–461 (2004)
7. Orabi, I.I., Ahmadi, G.: Response of the Duffing oscillator to a non-Gaussian random-excitation. J. Appl. Mech.-Trans. ASME 55(3), 740–743 (1988)
8. Grigoriu, M.: Linear and nonlinear-systems with non-Gaussian white-noise input. Probab. Eng. Mech. 10(3), 171–179 (1995)
9. Stratonovich, R.L.: Topics in the Theory of Random Noise, vol. 1. Gordon and Breach, New York (1963)
10. Roberts, J.B., Spanos, P.D.: Stochastic averaging: an approximate method of solving random vibration problems. Int. J. Non-Linear Mech. 21, 111–134 (1986)
11. Zhu, W.Q.: Stochastic averaging methods in random vibration. Appl. Mech. Rev. 41(5), 189–199 (1988)
12. Zhu, W.Q.: Nonlinear stochastic dynamics and control in Hamiltonian formulation. Appl. Mech. Rev. 59(1-6), 230–248 (2006)
13. Zeng, Y., Zhu, W.Q.: Stochastic averaging of quasi linear systems driven by Poisson white noise. Probab. Eng. Mech. 25(1), 99–107 (2010)
14. Zeng, Y., Zhu, W.Q.: Stochastic averaging of n-dimensional non-linear dynamical systems subject to non-Gaussian wide-band random excitations. Int. J. Non-Linear Mech. 45(5), 572–586 (2010)
15. Zeng, Y., Zhu, W.Q.: Stochastic averaging of quasi-nonintegrable-Hamiltonian systems under Poisson white noise excitation. J. Appl. Mech.-Trans. ASME (accepted)
16. Di Paola, M., Falsone, G.: Stochastic dynamics of nonlinear systems driven by non-normal delta-correlated processes. J. Appl. Mech.-Trans. ASME 60, 141–148 (1993)
17. Di Paola, M., Falsone, G.: Itô and Stratonovich integrals for delta-correlated processes. Probab. Eng. Mech. 8, 197–208 (1993)

Part 2
Stability, Bifurcation and Chaos of Nonlinear Stochastic Systems

Marginal Instability and Intermittency in Stochastic Systems

M.F. Dimentberg[1], A. Hera[2], and A. Naess[3]

[1] Mechanical Engineering Department, Worcester Polytechnic Institute, 100 Institute Road, Worcester, MA 01609, USA
[2] Information Technology Division, Worcester Polytechnic Institute, 100 Institute Road, Worcester, MA 01609, USA
[3] Centre for Ships and Ocean Structures and Department of Mathematical Sciences, Norwegian University of Science and Technology, NO-7491 Trondheim, Norway

Abstract. Dynamic systems with lumped parameters which experience random temporal variations are considered. The variations may "smear" boundary between the system's states which are dynamically stable and unstable in the classical sense. The system's response within such a "twilight zone" of marginal instability is found to be of an intermittent nature, with alternating periods of zero or almost zero response and rare short outbreaks. As long as it may be impractical to preclude completely such outbreaks for a designed system, the corresponding response should be analyzed to evaluate the system's reliability.

Results of such analyses are presented separately for cases of slow and rapid parameter variations. Linear models of the systems are studied in the former case using parabolic approximation for the variations in the vicinity of their peaks together with Krylov-Bogoliubov averaging for the transient response. This results in a solution for the response probability density function (PDF). The analysis is also used to derive on-line identification procedure for the system from its observed response with set of rare outbreaks. Potential examples of applications include 1D and 2D short-term galloping of elastically suspended bodies in cross-flow of fluid with random temporal variations of flow speed; bundles of heat exchanger tubes in cross-flow with potential for flutter-type instability; and rotating shafts.

The case of rapid broadband parameter variations is studied using theory of Markov processes. The system is assumed to operate beyond its stochastic instability threshold – although only slightly – and its nonlinear model is used accordingly. The analysis is based on solution of the Fokker-Planck-Kolmogorov (FPK) partial differential equation for stationary PDF of the response. Several such PDFs are analyzed; they are found to have integrable singularities at the origin indicating an intermittent nature of the response. One of potential applications is population dynamics where behaviour of predator-prey (or parasite-host) pair in random environment is studied using extended stochastic Lotka-Volterra model. The analysis provides potential for probabilistic predictions of response outbreaks, in particular for the cases of intermittency (like the notorious case of seven outbreaks in budworms (forest parasites) in eastern Canada since 1710).

Keywords: marginal instability, excursions, probability density function, intermittency.

W.Q. Zhu et al. (Eds.): Nonlinear Stochastic Dynamics & Control, IUTAM BOOK SERIES 29, pp. 159–170.
springerlink.com

1 Introduction

Classical definitions of stability and instability deal with long-term behavior of dynamic systems, that is, behavior as time $t \to \infty$. These definitions are known to be not perfectly appropriate for applications with limited service life (such as missiles, projectiles, etc.) which may sometimes be qualified as acceptable in spite of being unstable in the classical sense. Design of such marginally unstable systems may be based on analysis of their transient response within limited service life.

These classical definitions may also prove to be not perfectly adequate for some dynamic systems that may be intended for long-term operation. Such systems are designed, as a rule, to operate within their stability domain in the classical sense as long as their "nominal" design parameters are considered. However, if the parameters may experience random temporal variations around their "nominal" or expected values, the system may become "marginally unstable" within the "smeared" stability boundary. Whenever complete elimination of this kind of response may lead to impossible or impractical design the corresponding short-time outbreaks in response should be analyzed to evaluate the system's reliability with respect to, say, first-passage failure and/or of low-cycle fatigue. Relevant dynamic studies may also be of importance for *interpretation* of measured response signals.

Results of such analyses are presented separately for cases of slow and rapid parameter variations. Linear models of the systems are studied in the former case using parabolic approximation (PA) for the variations in the vicinity of their peaks together with Krylov-Bogoliubov (KB) averaging for the transient response. This results in a solution for the probability density function (PDF) of the response in terms of that of the bifurcation parameter. The analysis is also used to derive procedure for on-line system identification from its observed intermittent response.

The case of rapid broadband parameter variations is studied using theory of Markov processes. The system is assumed to operate beyond its stochastic instability threshold but only slightly and its nonlinear model is used accordingly. The analysis is based on solution of the FPK equation for stationary response PDF. Several such PDFs are analyzed with integrable singularities at their origins indicating an intermittent nature of the response. One of the potential applications is population dynamics where behaviour of predator-prey pair in random environment is studied using extended stochastic Lotka-Volterra model.

2 Systems with Slow Variations of Parameters

Linear systems with slow stationary random temporal variations of parameter(s) are considered which operate *within classical stability domain* for the mean or "nominal" system. However, any brief excursion beyond the instability threshold may lead to growth of the system's response. The growth is assumed to be limited as long as the system quickly returns back into the stability domain. The response would be seen then as a set of spontaneous brief outbreaks alternating with intervals of zero response.

The Slepian model [12] of a stationary zero-mean random process *g(t)* with unit standard deviation implies parabolic approximation (PA) [14] in the vicinity of its peak –during upcrossing given level u that starts at time instant $t = 0$ namely

$$\begin{gathered} g(t/u) \cong u + (1/u)\left(\varsigma t - \lambda^2 t^2/2\right) \text{ so that } g(t) \cong u + \varsigma t - (u/2)(\lambda t)^2 \\ \text{for } t \in \left[0, 2\varsigma/\lambda^2 u\right] \text{ and } \max_t g(t) = g\left(\varsigma/\lambda^2 u\right) = g_p = u + \varsigma^2/2\lambda^2 u. \end{gathered} \tag{1}$$

Subscript "*p*" is used for peak values of random processes, ς is random slope of *g(t)* at the instant of upcrossing and $\lambda^2 = \sigma_{\dot{g}}^2 = \int_{-\infty}^{\infty} \omega^2 \Phi_{gg}(\omega) d\omega$ where $\Phi_{gg}(\omega)$ is power spectral density (PSD) of *g(t)* so that λ is a mean frequency of *g(t)*. Thus, according to the equation (1) random process *g(t)* is regarded as deterministic within the high-level excursion of duration $\tau_f = \lambda t_f = 2\varsigma/\lambda u$ above level u, depending just on its initial slope ς at the instant of upcrossing. This slope is regarded as a random *variable* for the excursion; in particular, it has the Rayleigh PDF in case of a Gaussian *g(t)*[14] . This probabilistic description is used together with the solution for the transient response within the instability domain.

The first example is a SDOF system with randomly varying damping – say, Den-Hartog model of 1D galloping [2] under variable windspeed

$$\ddot{X} + 2(\alpha - q(t))\dot{X} + \Omega^2 X = 0 \text{ so that } q(t) = \sigma_q \cdot g(t) \text{ and } u = \alpha/\sigma_q . \tag{2}$$

Substituting the PA (1) into equation (2) reduces the latter to an ordinary differential equation (ODE) with a single random parameter ς. This ODE for a certain representative crossing should be integrated starting from the instant of upcrossing t_u until instant of peak of *X(t)* for a given outbreak. Here it can be done analytically using the KB-averaging over the response period for a quite common case of a lightly damped system (2) with slow temporal variations of the damping coefficient: $|\alpha - q(t)| << \Omega$, $\lambda << \Omega$ [1]. The method leads to first-order ODE for slowly varying amplitude $A(t) = \sqrt{X^2 + \dot{X}^2/\Omega^2}$ which has the solution

$$A(\tau) = A_0 \exp f(\tau); f(\tau) = (\sigma_q/\lambda)\left[(\varsigma/\lambda)(\tau^2/2) - u\tau^3/6\right], \tau = \lambda(t - t_u). \tag{3}$$

The peak amplitude of the response as attained at $\tau_f = 2\varsigma/\lambda u$ is

$$A_p = A(\tau_f) = A_0 \exp(2\delta),\ \delta = (\sigma_q/3\lambda u^2)(\varsigma/\lambda)^3 . \tag{4}$$

This solution together with equation (1) define, in parametric form, relation between $\bar{A}_p = A_p/A_0$ and g_p. Let $\bar{A}_p = h(g_p)$ for $g_p \geq u$. Then the function inverse to h (denoted by superscript "-1") can be obtained as

$$g_p = u + (\varsigma/\lambda)^2 (1/2u) = h^{-1}(\bar{A}_p) = u + (1/2u)\left[(3\lambda u^2/2\sigma_q)\ln \bar{A}_p\right]^{2/3}. \quad (5)$$

These relations open way to predicting reliability for the system (2) based on relevant statistics of *g(t)*. Thus, the first-passage problem for *A(t)* with barrier A_* is reduced to that for *g(t)* with barrier $g_* = h^{-1}(\bar{A}_*)$ as evaluated by equation (5). Furthermore, the PDF of *g(t)* can be used to obtain the PDF of $\bar{A}_p$ as

$$p(\bar{A}_p) = p_g\left(h^{-1}(\bar{A}_p)\right)\cdot\left|dh^{-1}/d\bar{A}_p\right|. \quad (6)$$

This PDF is non-zero for $\bar{A}_p \geq 1$ rather than for $\bar{A}_p \geq 0$ and is normalized not to unity but to $\Pr ob\{g_p > u\}$; according to the equations (5) and (6) it has a singularity at $\bar{A}_p = 1$.

Figure 1 illustrates response sample of the system (2) with $\alpha = 0.16, \Omega = 2$ which contains one excursion of the apparent damping into negative domain (see dash-dot curve of *q(t)*) with the corresponding response outbreak. To guarantee nonzero response during the short-term instability a small zero-mean stationary broadband random process had been added to the RHS of the equation (2); one can see that the corresponding subcritical response is really very small.

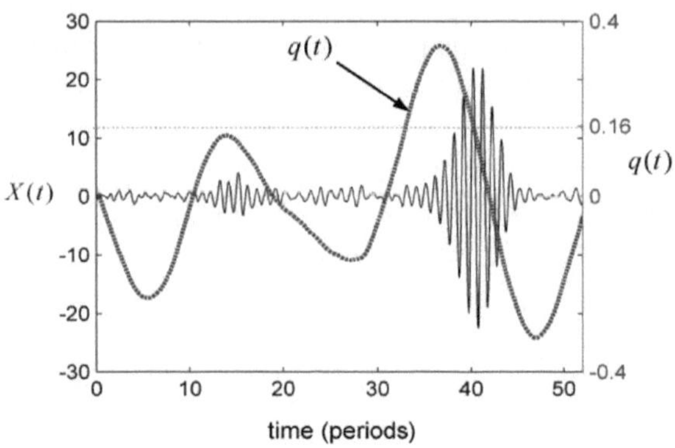

Fig. 1 Response sample with "outbreak" (solid line) of a SDOF system with apparent linear viscous damping 0.16 – q(t); sample of random process q(t) is shown by dotted line

In Fig. 2 the PDF $p(\bar{A}_p)$ for the case of Gaussian *g(t)* is compared with Monte-Carlo simulation (where values of A_0 were measured for each upcrossing). The corresponding prediction of the PDF of the actual (nonscaled) response amplitude and/or its peaks may be improved if PDF $p(A_0)$ of the random variable A_0 is known [8].

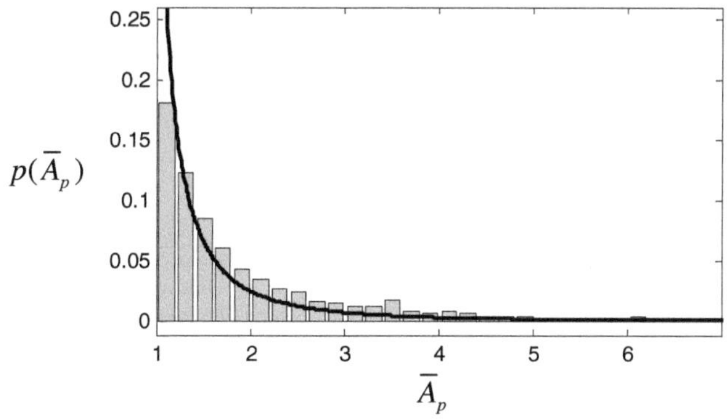

Fig. 2 Theoretical PDF of scaled amplitude $\bar{A}_p = A_p / A_0$ and corresponding histogram as obtained from sample of X(t)

The simplest approach for getting rough estimates for A_0 is just to ignore the parameter(s) variations in prediction of small-level steady-state response [7, 8]. However, such a crude approximation may be unconservative. To improve its accuracy a model with deterministic parameter variations and external random excitation is considered.

$$\ddot{X} + 2(\alpha - q(t))\dot{X} + \Omega^2 X = \varsigma(t) \text{ with } q(t) = \beta \sin \lambda t \text{ and } \alpha, \beta, q, \lambda << \Omega. \quad (7)$$

To this model the method of moments can be applied together with stochastic averaging [4, 13] resulting in three ODEs for second-order moments $D_{cc,ss} = \langle X^2_{c,s} \rangle$, $D_{cs} = \langle X_{cs} \rangle$ where

$$X_c(t) = X \cos \Omega t - (\dot{X}/\Omega) \sin \Omega t, X_s(t) = X \sin \Omega t + (\dot{X}/\Omega) \cos \Omega t.$$

Numerical solutions to this ODE set have been obtained within $[0, t_u]$ with initial conditions (ICs) corresponding to steady-state response of a system with constant parameters: $D_{cc}(0) = D_{cc}(0) = D_\varsigma / 4\alpha\Omega^2, D_{cs}(0) = 0$. The results provide ICs for the following analytical solution as based on the PA for $q(t) = \beta \sin \lambda t$ within $[t_u, t_f], \lambda t_u = \sin^{-1}(\alpha/\beta), \lambda t_f = \pi - \sin^{-1}(\alpha/\beta)$

$$D_{cc,ss}(t_f) = D_{cc,ss}(t_u) \cdot \exp\left\{(4\alpha/3\lambda)\left[(\beta/\alpha)^2 - 1\right]^{3/2}\right\}. \quad (8)$$

The results may be compared with "stationary approximation" whereby the simple ICs are enforced as $D_{cc,ss}(t_u) = D_\varsigma / 4\alpha\Omega^2$. Figure 3 illustrates the corresponding correction factor for ***peak*** values of $D_+(t) = D_{cc}(t) + D_{ss}(t)$ as established upon identifying β as peak value of *g(t)*:

$$K = 1/\int_1^{\infty} \frac{\sqrt{D_+(z)_{\text{numerical}}}}{\sqrt{D_+(z)_{\text{stationary}}}} p(z|*)dz,\ p(z|*) = u^2 z \cdot \exp(-u^2 z^2/2)/P_* \text{ for } z > 1$$

where $z = \beta/\alpha = g_p/u,\ g_p = q_p/\sigma_q, u = \alpha/\sigma_q$.

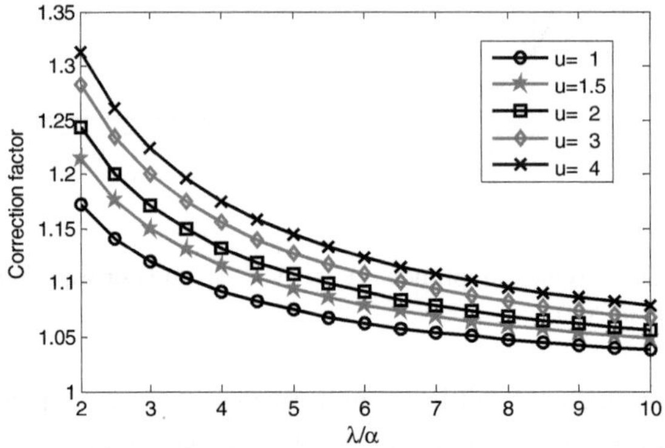

Fig. 3 Correction factor K as function of λ/α and u.

These correction factors may be used to obtain improved estimates for initial values of the response amplitude at each starting instant of the response outbreak due to short-term instability. Namely, the estimate of a steady-state rms value for the system without parameter variations may be directly multiplied by K. As could be expected, the correction is seen to be reduced with increasing λ.

The equation (4) is convenient for evaluating the system's properties from its measured (on-line!) response with outbreaks as one shown in Fig. 1. To this end one can use peak amplitudes A_p, as attained at instants $\tau_f = 2\varsigma/\lambda u$ in the local time frames and corresponding amplitudes A_i at inflexion points of the curve $\ln A(\tau)$. From the equation (4) $A_i = A(\tau_i) = A_0 \exp\delta$, so that $A_p/A_i = \exp\delta$ and $A_0 = A_i^2/A_p$ Thus, for each one of the observed response outbreaks one can identify in a global time frame the instants $t_f = t_u + \tau_f/\lambda$ and $t_i = t_u + \tau_i/\lambda$ which correspond to peak and inflexion-point amplitudes A_p and A_i respectively; the instants of upcrossings can also be identified as $t_u = t_f - 2(t_f - t_i) = 2t_i - t_f$. The frequency λ may now be obtained by averaging time difference $t_f - t_i$ over all observed outbreaks of response. The identification procedure as described in details in [7] relies upon averaging $\ln(A_p/A_i) = \exp\delta$ over all observed outbreaks. It provides on-line estimates both for the mean apparent damping coefficient – which may be regarded as a nominal stability margin – and for standard deviation and mean frequency of its random temporal variations.

The described analytical solutions can be extended to certain TDOF systems with certain "symmetry" which permit to "wrap up" two equations of motion into a single complex equation [6, 8]. Thus, translational and tilting oscillations of the Jeffcott rotor had been considered in [6] and [8] respectively.

In the general case of TDOF system two coupled response amplitudes remain after KB-averaging for the case of lightly damped system thereby requiring numerical integration for the two ODEs of slow motion from starting point of the response outbreak till the instant when both response variables pass their peaks. Then, as long as relation is established (numerically) between peak value(s) of response(s) and that of scaled zero-mean part *g(t)* of the bifurcation parameter the basic procedure can be applied for predicting response PDF. Several examples of such analysis as presented in [8] are:

- Rotating shaft with anisotropic stiffness in translational vibrations;
- TDOF flutter of a tube row in heat exchanger in cross-flow of fluid;
- Two-dimensional galloping of a rigid body in a fluid flow (case of full 2×2 damping matrix).

3 Systems with Rapid Variations of Parameters

Systems with broadband stationary random temporal variations of parameters are considered here which may be described by the theory of Markov processes. They have clearly defined boundaries corresponding to various definitions of stochastic stability for the system's linear part [13]. They operate *within the domain of stochastic instability* - although close to the corresponding instability threshold - where the instability may be called marginal indeed. Therefore, adequate modelling requires the system's *nonlinearity* to be accounted for. The response is found to be of the intermittent nature indeed in such cases. The analysis provides the potential for predicting the response PDF's through solution of the stationary Fokker-Planck-Kolmogorov (FPK) partial differential equation. Several such stationary PDF's are analyzed; all of them are found to possess an integrable singularity at the origin, whereas the response itself does exhibit the intermittency indeed. Common characteristic features of these solutions are also certain other typical patterns of a stationary intermittent response; for example, if amplitude *A(t)* has mean <*A*> and standard deviation σ_A then typically a small relative stay time of *A(t)* above <*A*> is observed, and also σ_A >> <*A*>.

Stationary PDF of response amplitude with singularity at zero has been known for a long time for a SDOF system with small nonlinear damping [14] but correlation between the singularity and intermittency has been established by numerical simulation only recently [11]. Extensive study of intermittency in distributed-parameter system through their response moments see in [10].

Solutions to the FPK equations for response PDFs are presented here for two cases of potential intermittency in systems with impacts. They have different types of damping than in [11, 14] which may effectively restrict response growth in case of stochastic instability. The first is that of a single-barrier SDOF system with inelastic impacts as governed by the equation of motion between impacts for its displacement *Y(t)*

$$\ddot{Y} + 2\alpha\dot{Y} + \Omega^2 Y\left[1+\xi(t)\right] = 0 \text{ for } Y > -h, \tag{9}$$

and impact condition at the barrier at $Y = -h$ (with subscripts "plus" and "minus" corresponding to rebound and impact velocities respectively)

$$\dot{Y}_+ = -r\dot{Y}_- \text{ where } \dot{Y}_\pm = \dot{Y}(t_* \pm 0), Y(t_*) = -h \text{ and } 0 < r \le 1. \tag{10}$$

Here r is a restitution factor, $\xi(t)$ is a stationary zero-mean Gaussian white noise with intensity D_ξ. Stationary PDF of response energy $H(t) = (1/2)(\dot{X}^2 + \Omega^2 X^2)$ had been obtained in [3] for the system (9), (10) by a quasiconservative stochastic averaging [4, 12]. It does exist (is normable) if and only if $\delta < 1$ and $\Omega(1-r)/\pi\alpha > (1-\delta)/\delta$ where $\delta = \alpha/2B_\xi$ and is described by different analytical expressions for $H < H_h$ and for $H > H_h$ where $H_h = \Omega^2 h^2/2$ is the system's potential energy at the barrier. The first of these expressions which may be regarded as the conditional PDF $p_c(H) = p(H|H < H_h)$ normalized within $[0, H_h]$ is

$$p_c(H) = \left[(1-\delta)/H_h\right](H/H_h)^{-\delta}. \tag{11}$$

The (conditional) mean and relative stay time above this mean of the response energy may be found from (11) as

$$\langle H\rangle = H_h\left[(1-\delta)/(2-\delta)\right], \lambda_H = \int_{\langle H\rangle}^{H_h} p_c(H)dH \Big/ \int_0^{\langle H\rangle} p_c(H)dH = \left(\frac{2-\delta}{1-\delta}\right)^{1-\delta} - 1.$$

Thus $\lambda_H << 1$ for small $1-\delta$ clearly indicating that the response should be intermittent indeed in this case. The intermittency may also be seen from expected time $\langle T\rangle$ for reaching mean energy from initial state with $H < \langle H\rangle$. The latter may be adopted as an index of expected period between outbreaks and can be obtained from solution to the corresponding Pontryagin equation for *T(H)* - one with given initial state *H* [13]. Averaging this *T(H)* over *H* using the conditional PDF (11) yields

$$\alpha\langle T\rangle = \alpha\int_0^{\langle H\rangle} T(H)p_c(H)dH = (\delta/2)(1-\delta)^{-(1+\delta)}(2-\delta)^{-(1-\delta)} \cong 1/\left[2(1-\delta)\right]. \tag{12}$$

This expected period between outbreaks is clearly increasing with reducing $1-\delta$ The other case of intermittency is considered for vibroimpact system with double-sided barrier and *linear* damping. The equation of motion between impacts is

$$\ddot{Y}+2\alpha\dot{Y}\left[1+\eta(t)\right]+\Omega^2 Y\left[1+\xi(t)\right]=0 \text{ for } -h<Y<h\,, \tag{13}$$

where $\xi(t),\eta(t)$ are stationary zero-mean independent Gaussian white noises with intensities $D_\xi, D_\eta=(2\alpha/\Omega)^2 D_\xi$ respectively. Impact condition (10) with $r = 1$ is now imposed both at $Y = -h$ and $Y = +h$. Under the above relation between intensities of excitations the FPK equation for the joint PDF of response displacement and velocity has an **exact** stationary solution [4]

$$w(y,\dot{y})=C/\left(y^2+\dot{y}^2/\Omega^2\right)^\delta \text{ where } \delta=2\alpha/D_\xi\Omega^2+1/2\,. \tag{15}$$

Without barriers this solution is not normable. However, nonlinearity due to the double-sided barrier may restrict response growth even in the absence of nonlinear damping if $1/2<\delta<1$. Specifically, integrating PDF (15) over $\dot{y}$ provides PDF $p(y)$ of the displacement

$$p(y)=\frac{1-\delta}{h(y/h)^{2\delta-1}} \text{ for } -h<y<h \tag{16}$$

with integrable singularity at the origin if $\delta>1/2$. Relative stay time $\lambda_{|Y|}$ of the magnitude of $Y(t)$ above its mean may be considered as an index of intermittency and it is seen to approach zero with $1-\delta\to 0$ thereby indicating intermittency:

$$\langle|Y|\rangle=2\int_0^h yp(y)dy=\frac{2(1-\delta)}{3-2\delta}\cdot h,\ P=\Pr ob\{Y<|Y|\}=2\int_0^{\langle|Y|\rangle} p(y)dy= \\ \left[2(1-\delta)/(3-2\delta)\right]^{2(1-\delta)} \text{ and } \lambda_{|Y|}=(1-P)/P. \tag{17}$$

Finally, consider extended stochastic Lotka-Volterra (L-V) model [4, 5] of interacting populations of the predator-prey or parasite-host type

$$\dot{u}=-mu+k\beta uv,\ \dot{v}=\alpha v\left[1+\xi(t)\right]-\beta uv-\gamma v^2\,. \tag{18}$$

Here $u(t)$ and $v(t)$are population sizes of predators (or parasites) and preys (or hosts) respectively, whereas $\xi(t)$ is a zero-mean Gaussian random white noise with intensity D. The system (18) has asymptotically stable equilibrium point $u_0=(\alpha-\gamma m/k\beta)/\beta; v_0=m/k\beta$ and an additional equilibrium state $u_*=0,\ v_*=\alpha/\gamma$ which is unstable if $v_o<v_*$ or $\gamma<\gamma_*=\alpha k\beta/m$. At the bifurcation point $\gamma=\gamma_*$ these two equilibrium states merge, with the state $u_*=0, v_*=\alpha/\gamma$ becoming stable for $\gamma>\gamma_*$. Physical meaning of these transformations: beyond this transcritical bifurcation point growth of the preys' population is bounded by its interspecies competition rather than by predators' activity, whereas the latter become extinct because of food shortage. Joint stationary PDF $w(u, v)$ of population sizes as derived in [4, 5] from exact solution to the corresponding FPK equation is

$$w(u,v) = p(\Delta u/k)\cdot p(\Delta v),\ \text{where}\ p(z) = z^{z_o-1}e^{-z}/\Gamma(z_o)\ \text{and}\ \Delta = 2\gamma/D\alpha^2 \quad (19)$$

where Γ is gamma-function and $z = \Delta u/k$, $z_0 = \Delta u_0/k$ and $z = \Delta v$, $z_0 = \Delta v_0$ for stationary PDFs of $u(t)$ and $v(t)$, respectively. Thus, both steady-state population sizes $u(t)$ and $v(t)$ are independent gamma-distributed random variables with the following mean values and standard deviations

$$\langle u\rangle = u_0 = (\alpha - \gamma m/k\beta)/\beta = (\alpha/\beta)(1 - v_o/v_*), \langle v\rangle = v_0 = m/k\beta$$
$$\text{and}\ \sigma_u = \sqrt{(u-\langle u\rangle)^2} = \sqrt{u_o k/\Delta}, \sigma_v = \sqrt{(v-\langle v\rangle)^2} = \sqrt{v_o/\Delta}. \quad (20)$$

Both these PDFs do exist provided that $\gamma < \gamma_*$. It can be seen from the equations (19) that intermittent behavior should be expected for $v(t)$ ($u(t)$) if $\Delta v_0 << 1$ ($\Delta u_0/k << 1$), so that $\sigma_v >> v_o (\sigma_u >> u_o)$, with rare and short pulse-like intensive outbreaks in $v(t)$ ($u(t)$) and low-level oscillations between the pulses. Relevant indices of intermittency such as small relative stay time above the mean and large values of the ratio standard deviation/mean may also be used. Figure 4 illustrates such a behavior of predators for small $1 - v_o/v_* = 1 - \gamma/\gamma_*$. Cases with intermittent behavior of both predators and preys are illustrated in [5, 9].

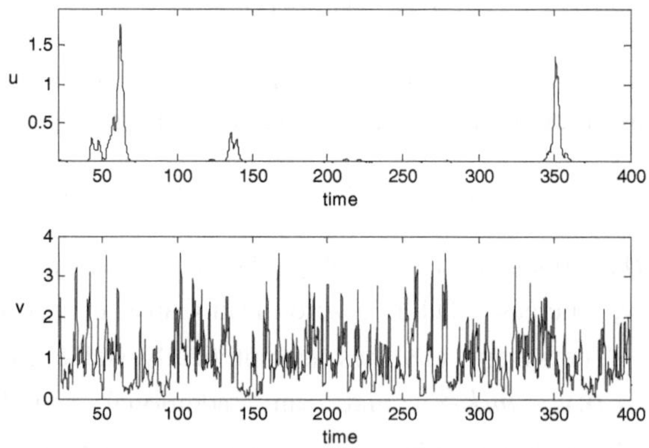

Fig. 4 Samples of $u(t)$ and $v(t)$ as obtained from numerical simulation for the system (18) with m=1, α=1, β=1, k=1, γ=0.98, D=1. It illustrates an intermittency in predators only (u_0=.01 and Δ=1.98). All quantities are non-dimensional.

4 Conclusions

Temporal random variations of parameters in dynamic systems may "smear" classical neutral stability boundaries. The system's response within such a "twilight zone" of marginal instability is found to be of an intermittent nature, with alternating periods of zero (or almost zero) response and rare short outbreaks. As long as it may be impractical to preclude completely such outbreaks for a designed system its response should be analyzed to evaluate reliability, in particular, to predict response PDF and/or to solve first-passage problem.

Procedures and results of such analyses are presented for cases of slow and rapid parameter variations. Linear models may be adequate in the former case where using parabolic approximation for the variations together with transient analysis yields response PDF. The analysis is also used to derive on-line identification procedure for the system from its observed response with set of rare outbreaks. Potential applications include 1D and 2D short-term galloping of suspended bodies in fluid cross-flow; heat exchanger tubes in cross-flow with potential flutter-type instability; rotating shafts.

The case of broadband parameter variations is studied using theory of Markov processes with the system operating beyond its stochastic instability threshold – although only slightly – and its nonlinear model is used accordingly. The analysis is based on solution of the FPK equation for stationary PDF of the response. Several such PDFs are analyzed; they are found to have integrable singularities at the origin indicating an intermittent nature of the response. One of the potential applications is population dynamics where behaviour of predator-prey pair in random environment is studied using extended stochastic L-V model. The analysis provides potential for probabilistic predictions of response outbreaks, in particular for the cases of intermittency (like the notorious case of seven outbreaks in budworms (forest parasites) in eastern Canada since 1710).

References

[1] Bogoliubov, N.N., Mitropol'sky, Y.A.: Asymptotic Methods in the Theory of Nonlinear Oscillations. Gordon&Breach, New York (1961)
[2] Den Hartog, J.P.: Mechanical Vibrations, 4th edn. Dover, New York (1985)
[3] Dimentberg, M.F., Menyailov, A.: Response of a Single-Mass Vibroimpact System to White-Noise Random Excitation. ZAMM 59, 709–716 (1979)
[4] Dimentberg, M.F.: Statistical Dynamics of Nonlinear and Time-Varying Systems. Research Studies Press, Taunton (1988)
[5] Dimentberg, M.F.: Lotka-Volterra system in a random environment. Physical Review E 65, 36204 (2002)
[6] Dimentberg, M.: Vibration of a Rotating Shaft with Randomly Varying Internal Damping. Journal of Sound and Vibration 285, 759–765 (2005)
[7] Dimentberg, M., Naess, A.: Short-term dynamic instability of a system with randomly varying damping. Journal of Vibration and Control 12, #5, 527–536 (2006)

[8] Dimentberg, M., Hera, A., Naess, A.: Marginal Instability and Intermittency in Stochastic Systems. Part I – Systems with Slow Random Variations of Parameters. Journal of Applied Mechanics 75(#4), 041002-1–041002-8 (2008)
[9] Dimentberg, M., Hera, A., Naess, A.: Marginal Instability and Intermittency in Stochastic Systems. Part II – Systems with Rapid Random Variations of Parameters. Journal of Applied Mechanics 76(#3), 031002-1–031002-08 (2009)
[10] Gartner, J., Molchanov, S.A.: Parabolic Problems for the Anderson Model. I. Intermittency and Related Topics. Commun. Math. Phys. 132, 613–655 (1990)
[11] Ibrahim, R.: Stabilization and stochastic bifurcation, with application to nonlinear ocean structures. In: Shlesinger, M.F., Swean, T. (eds.) Stochastically Excited Nonlinear Ocean Structures, pp. 1–52. World Scientific, Singapore (1998)
[12] Leadbetter, M.R., Lindgren, G., Rootzén, H.: Extremes and Related Properties of Random Sequences and Processes. Springer, New York (1983)
[13] Lin, Y.K., Cai, G.Q.: Probabilistic Structural Dynamics. McGraw Hill, New York (1995)
[14] Stratonovich, R.L.: Topics in the Theory of Random Noise, vol. II. Gordon&Breach, New York (1967)

A Practical Strategy to Study Stochastic Chaos

T. Fang[1], C.L. Wu[2], and X.L. Yang[3]

[1] Dept. of Eng. Mech., Northwestern Polytechnical University, Xian 710072, China
[2] Aircraft Strength Research Institute of China, Xian 710065, China
[3] College of Math. and Info. Sci., Shaanxi Normal University, Xian 710062, China

Abstract. Stochastic chaos often appears in random nonlinear dynamic systems. Based on orthogonal polynomial approximation and ergodic theorem, a practical strategy for studying stochastic chaos is proposed and illustrated by a Duffing oscillator with bounded random parameter and driven by an ergodic random excitation. By the proposed method we have studied stochastic chaos and its control, and synchronization in typical nonlinear dynamical systems. Some representative results are reported.

Keywords: Stochastic chaos, Orthogonal polynomial approximation, Chaos control, Synchronization.

1 Introduction

Chaos is a specific type of motion featuring a sensitive dependence on initial conditions [1-3]. The repellence between any two adjacent chaotic responses causes not only the sensitive dependence on initial conditions but also the hold of at least one positive Top Lyapunov Exponent (TLE) for chaos. Systems with random parameters or under random excitations are called stochastic systems. Chaotic phenomena in stochastic systems are called stochastic chaos, which not only reflects the intrinsic randomness of the nonlinear system but also the random effects of the random parameter or/and the random excitation. Hence, stochastic chaos also features at least one positive TLE. For analysis of random phenomena, one used to look for the probability density function (PDF) of random responses. However, the PDF information is not favorable to study repellency of the neighboring chaotic responses, nor to calculate the related TLE; so we would rather study stochastic chaos through its sample responses than its PDF. Moreover, any sample of a mature stochastic chaos should be a deterministic one, so we need not to supply any additional definition on stochastic chaos. We mainly concern two basic kinds of nonlinear stochastic systems: one with random variables as its parameters and one with ergodic random processes as excitations. To solve the stochastic chaos problems in the first type of systems, we transform the stochastic system into an equivalent deterministic one using the orthogonal polynomial approximation. For the second type one, we replace the ergodic random excitations by their representative deterministic samples. Having transformed the stochastic chaos problem into the deterministic one of equivalent systems, we can use all available effective methods for further chaos analysis. In this paper, we aim to review the state of art of studying stochastic chaos by the proposed strategy.

W.Q. Zhu et al. (Eds.): Nonlinear Stochastic Dynamics & Control, IUTAM BOOK SERIES 29, pp. 171–181.
springerlink.com

2 Implement of Orthogonal Polynomial Approximation

Taking a low dimensional system for example, the two kinds of stochastic systems can be described in a unified form as follows:

$$\ddot{x} + a\dot{x} + bx + f(x,\dot{x},\varepsilon/0) = F_1(t) + F_2(t)\,, \quad a>0 \tag{1}$$

where a and b are constants; the assumption $a > 0$ ensures system (1) is dissipative; $F_1(t)$ is a deterministic excitation, $F_2(t)$ is an ergodic random one; $f(x,\dot{x},\varepsilon/0)$ is a polynomial function of x and $\dot{x}$, and ε is a bounded random parameter. It is assumed that $F_2(t)$ and ε are mutually independent. Express ε as $\varepsilon = \bar{\varepsilon} + \delta u$, where $\bar{\varepsilon}$ is the mean value, δ is an intensity coefficient, and u is a random variable defined on $[-1, 1]$ with an arch-like PDF:

$$p(u) = \begin{cases} (2/\pi)\sqrt{1-u^2} & as \quad |u| \le 1 \\ 0 & as \quad |u| > 1 \end{cases}$$

which well matches the second kind Chebyshev polynomials $U_i(u)$ $(i = 0, 1, \cdots)$, due to the orthogonal property: $\int_{-1}^{1} p(u)U_i(u)U_j(u)du = \delta_{ij}$. The recurrent formula for Chebyshev polynomials of the second kind is

$$2uU_i(u) = U_{i-1}(u) + U_{i+1}(u) \tag{2}$$

Due to orthogonality of Chebyshev polynomials, any measurable function $\phi(u) \subset L^2$ can be expressed in the series form: $\phi(u) = \sum_{i=0}^{\infty} c_i U_i(u)$. Hence, the response $x(t,u)$ of system (1) can be approximately expressed as

$$x(t,u) = \sum_{i=0}^{N} x_i(t)U_i(u) \tag{3}$$

The implement of orthogonal polynomial approximation for Eq. (1) to study stochastic chaos can be illustrated for any specific $f(x,\dot{x},\varepsilon/0)$, say $f = \varepsilon x^3$. In this case, Eq. (1) is reduced to

$$\ddot{x} + a\dot{x} + bx + \varepsilon x^3 = F_1(t) + F_2(t), \quad a > 0 \tag{4}$$

Substituting the expression of ε and Eq. (3) into Eq. (4), we have

$$\begin{aligned}\left(\frac{\mathrm{d}^2}{\mathrm{d}t^2}+a\frac{\mathrm{d}}{\mathrm{d}t}+b\right)\sum_{i=0}^{N}x_i(t)U_i(u)+\bar{\varepsilon}\left(\sum_{i=0}^{N}x_i(t)U_i(u)\right)^3\\ +\delta u\left(\sum_{i=0}^{N}x_i(t)U_i(u)\right)^3=F_1(t)+F_2(t)\end{aligned} \tag{5}$$

The nonlinear cubic term in Eq. (5) can be reduced into a linear combination of basic polynomial functions. Multiplying both sides of Eq. (5) by $U_i(u)$ in sequence and taking expectation with respect to u, we obtain

$$\begin{cases}\left(\dfrac{\mathrm{d}^2}{\mathrm{d}t^2}+a\dfrac{\mathrm{d}}{\mathrm{d}t}+b\right)x_0(t)+\bar{\varepsilon}X_0(t)+\dfrac{1}{2}\delta X_1(t)=F_1(t)+F_2(t),\\ \left(\dfrac{\mathrm{d}^2}{\mathrm{d}t^2}+a\dfrac{\mathrm{d}}{\mathrm{d}t}+b\right)x_1(t)+\bar{\varepsilon}X_1(t)+\dfrac{1}{2}\delta[X_0(t)+X_2(t)]=0,\\ \qquad\qquad\cdots\cdots\cdots,\\ \left(\dfrac{\mathrm{d}^2}{\mathrm{d}t^2}+a\dfrac{\mathrm{d}}{\mathrm{d}t}+b\right)x_N(t)+\bar{\varepsilon}X_N(t)+\dfrac{1}{2}\delta[X_{N-1}(t)]=0\end{cases} \tag{6}$$

where $X_i(t)$, $i = 0, 1,\ldots, N$, are polynomial functions of $x_i(t)$, which can be derived symbolically by MAPLE. If the ergodic random excitation $F_2(t)$ in Eq. (6) is replaced by its representative sample process, usually a pseudo-random one in practice, then, Eq. (6) is fully an equivalent deterministic nonlinear system for the stochastic system (1). When $N \to \infty$ in Eq. (3), $\sum_{i=0}^{\infty}x_i(t)U_i(u)$ is strictly equivalent to $x(t,u)$ of the stochastic dynamical system. Otherwise, if N is finite, Eq. (3) is just approximately valid with a minimal residual mean square error. For example, we can take $N = 4$ and obtain the numerical solution $x_i(t), (i=0,1,\cdots,4)$ of Eq. (6) by available effective numerical methods. Then, the approximate stochastic response and its ensemble mean for system (1) can be expressed respectively as

$$x(t,u)\approx\sum_{i=0}^{4}x_i(t)U_i(u)\quad\text{and}\quad E[x(t,u)]\approx\sum_{i=0}^{4}x_i(t)E[U_i(u)]=x_0(t). \tag{7}$$

It is worth noting if any $x_i(t)$ is found to be chaos, then $x(t, u)$ must be stochastic chaos. Hence, Eq. (6) is of vital importance to studying stochastic chaos.

3 Typical Phenomena of Stochastic Chaos

3.1 Example 1 - Stochastic Chaos in Duffing Oscillator

Consider a Duffing oscillator with a bounded random parameter and driven by a harmonic excitation

$$\ddot{x} + a\dot{x} + b(x + cx^3) = f\sin(\omega t),\ a > 0$$

where a, c, and f are constants, b is the random parameter, expressed as $b = \bar{b} + \alpha u$, with u as a bounded random variable with an arch-like PDF.

In calculation, system parameters are taken as a = 0.3, $\bar{b} = -1.0$, $c = -1.0$, $\omega = 1.2$, and $f = 0.46$. Two sample chaotic attractors are shown in Figs. 1(a) and 1(b) with TLE ≈ 0.3 and 0.2, for α = 0.1 and 0.0, respectively, [4]. One can see that the stochastic chaos attractor in Fig. 1(a) is much more diffusive than the deterministic one in Fig. 1(b).

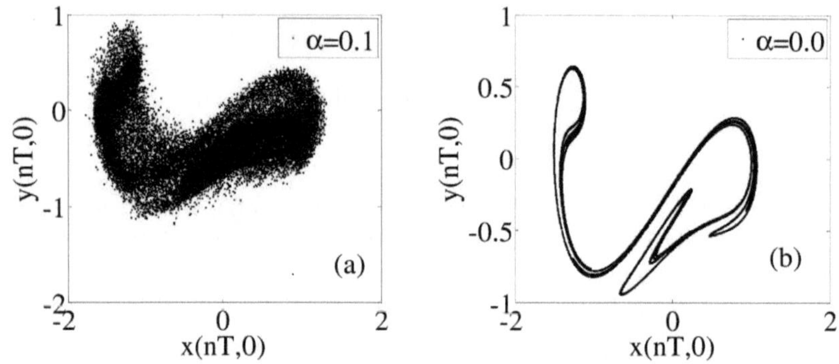

Fig. 1 Chaotic attractors. (The ordinate **y** in Poincaré maps stands for dx/dt.)

3.2 Example 2 - Control of Chaos by Random Noise or Chaotic Driving

The differential equation of a Φ^6-DVP system is described as follows [5]:

$$\ddot{x} - \mu(1 - x^2)\dot{x} + \alpha x + \gamma x^3 + \beta x^5 = f\cos\Omega_1 t + DU(t)$$

where μ, α, β, and γ are system parameters, f and Ω_1 are amplitude and frequency of the harmonic excitation, $U(t)$ is an ergodic bounded noise with intensity D. Using the ergodic assumption, the random noise can be replaced by its representative sample, and the stochastic problem can be treated as a deterministic one.

At first, letting $D = 0$, we obtain the TLE diagram for the system responses via the bifurcation parameter f, shown in Fig. 2. In calculation, the system parameters are taken as $\mu = 0.4$, $\alpha = 0.212$, $\beta = 0.1$, $\gamma = 1.0$, and $\Omega_1 = 0.86$.

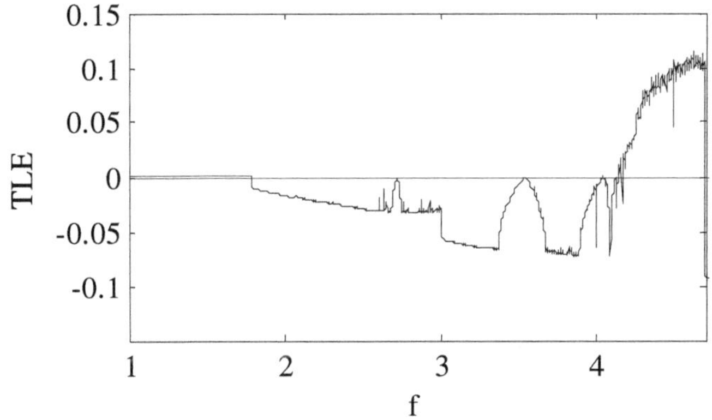

Fig. 2 Variation of TLE against f

Now consider the effect of bounded noise on chaos control. Bounded noise $U(t)$ may be defined as follows:

$$U(t) = \cos(\Omega_2 t + \psi), \qquad \psi = \sigma B(t) + \Gamma$$

where, Ω_2 stands for center frequency of noise, $B(t)$ is a standard Wiener process, σ is the noise intensity, and Γ is a uniform random variable defined on $[0, 2\pi]$. In calculation, we take $\sigma = 0.2$, $\Omega_2 = 0.43$, and $f = 4.5$ for chaos suppression. The variations of TLE against D for system responses are shown in Fig. 3(a), while the results for Rössler chaotic driving in Fig. 3(b). The effect of chaos suppression is clearly seen after the TLE becomes negative.

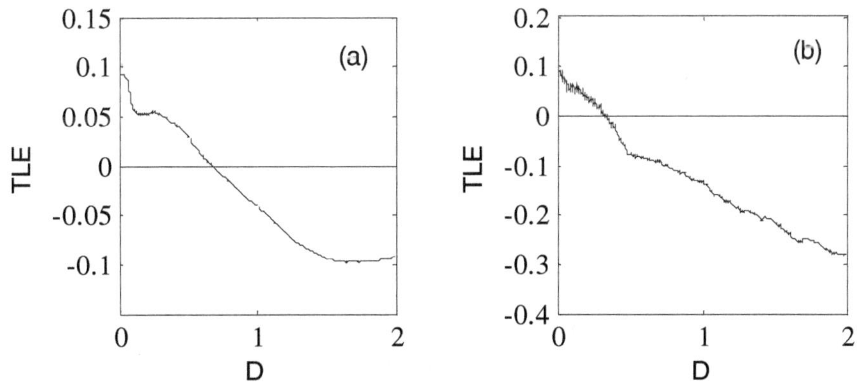

Fig. 3 Variation of TLE vs. D for suppressing chaos, by (a) bounded noise, (b) chaotic driving

For the case of inducing chaos, we take $\sigma = 0.2$, $\Omega_2 = 0.43$, and $f = 3.5$. The variations of TLE against D for system responses under both the harmonic and bounded noise excitations are shown in Fig. 4(a), while the results for Rössler chaotic driving are shown in Fig. 4(b). The inducing of chaos happens in the region in Fig. 4 where there TLE becomes positive. Figure 3(b) and 4(b) show that chaotic driving can play the similar role as bounded random noise in chaos control. Fig. 5 shows the two different kinds of attractors of DVP system before and after the happening of inducing chaos. However, the induced chaos is actually a stochastic chaos.

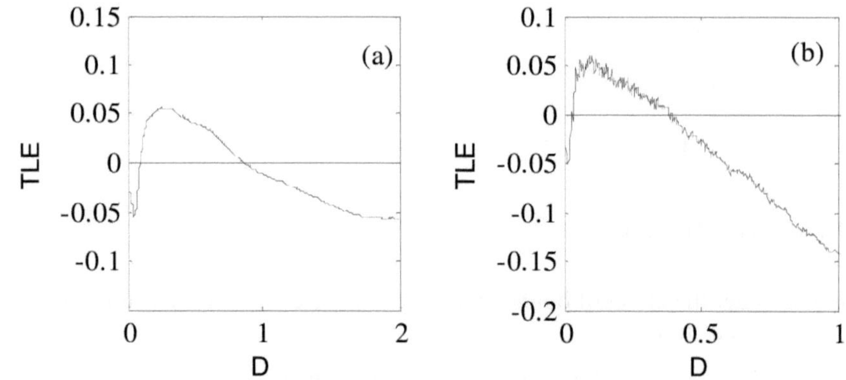

Fig. 4 Variation of TLE vs. D for inducing chaos, by (a) bounded noise, (b) chaotic driving

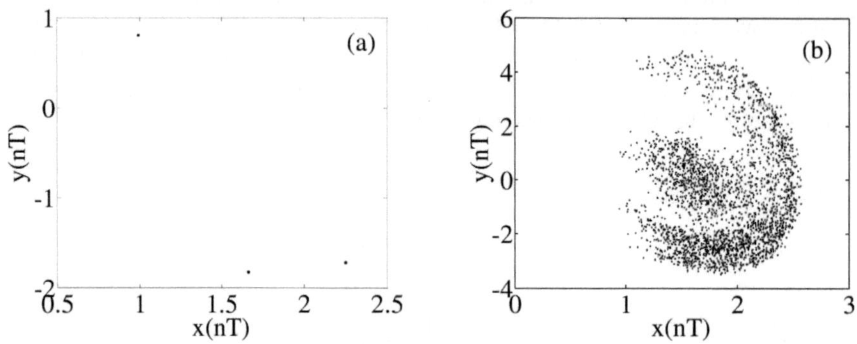

Fig. 5 Attractors of DVP system for different values of D, (a) for $D = 0$, (b) for $D = 0.3$

3.3 Example 3- Synchronization Induced by Common Noise

Noise induced synchronization originally refers to the phenomenon where two uncoupled identical nonlinear oscillators achieve complete synchronization under the action of a common noise. Consider the complete synchronization induced by bounded noise in a pair of Φ^6-Duffing systems via numerical simulations. The Φ^6-Duffing system implies its potential is of a sixth-order polynomial. The system is governed by the following equation

$$\ddot{x} + a\dot{x} + bx + cx^3 + dx^5 = f\cos\Omega t + DU(t)$$

where $U(t)$ stands for a bounded noise with strength D. Fig. 6 shows how the TLE of the system is evolving with varying D [6]. As the strength D gradually increases, the general trend of evolution of TLE is descending from a positive value, until the TLE across abscissa for the first time at a critical value D_1 (≈ 0.24). Then, the TLE fluctuates about abscissa in a small interval of D, and the TLE across abscissa finally at another critical value D_2 (≈ 0.42). From then on, as D further increases from D_2, the general trend of TLE repeats its descending way, and the TLE remains negative. We did not find any synchronization during $0 < D < D_1$, but we did find some kind of synchronization when $D > D_2$. Meanwhile we obtained three kinds of attractors in the Poincaré maps, as shown in Figs. 7(a) – (c). Fig. 7(a) shows the one for $D = 0$, Fig. 7(b) is for $D = 0.1$ ($0 < D < D_1$), and Fig. 7(c) is for $D = 1.0$ ($D > D_2$). The first one indicates surely deterministic chaos, since $D = 0$, and TLE > 0. The second one, thicker than the first one and smeared with some random diffusions, indicates stochastic chaos, since $D_1 > D > 0$ and TLE > 0. However, the third kind of attractor cannot be chaotic, since TLE < 0; yet must indicate steady state random responses due to the excitation of bounded noise, since $D > D_2 > 0$.

Synchronization of stochastic chaos does not happen when the TLE is positive, but it does happen when the TLE becomes negative. Hence, the steady state random responses are no longer chaotic, though the random attractor may look even

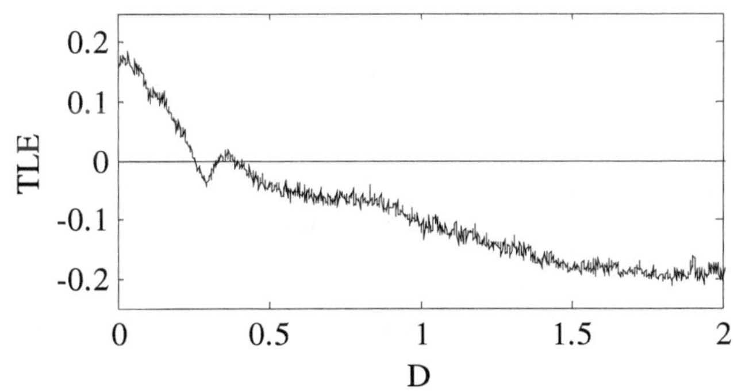

Fig. 6 TLE vs. strength of bounded noise D

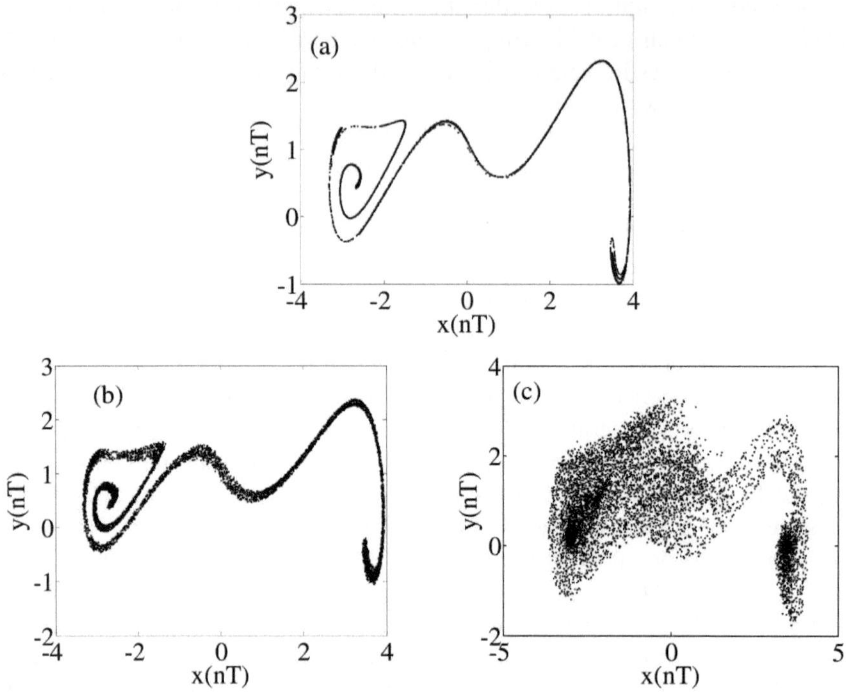

Fig. 7 Three kinds of different attractors in a Φ^6-Duffing system with different noise strength, (a) for D = 0.0; (b) for D = 0.1; (c) for D = 1.0

much more diffusive. In fact, not all motion that looks in disorder is chaos. Besides, it is found that synchronization of two identical nonlinear systems can also be induced by a common chaotic driving in a similar way as by the bounded noise [6].

3.4 Example 4 - Forced Synchronization in Master-Slave Systems

Chaos synchronization is a specific chaos control aiming at making the responses of two nonlinear systems to move in some synchronized way. Since Pecora LM and Carroll TL [7] discovered the phenomenon, chaos synchronization has been attracting great attention. However, one may want to know what kind of motion the synchronized one is. The following example may reveal some clues.

Consider a harmonically driven stochastic Duffing system, described as

$$\ddot{x}+a\dot{x}+b(x+cx^3)=f\sin\omega t, \qquad a>0 \tag{8}$$

where $b=\bar{b}+\sigma\mu$, and μ is a bounded random variable with a given λ-PDF. Treating system (8) as a master system, its equivalent deterministic system can be obtained using the Gegenbauer polynomial approximation. It can be rewritten in the form of first order differential equations as follows, named as a master system as well,

$$\dot{X} = AX + G(X) + F(t) \tag{9}$$

where $X = [x_0 \cdots x_4 \, \dot{x}_0 \cdots \dot{x}_4]^T$ and $F(t) = [0,0,0,0,0, f \sin \omega t, 0,0,0,0]^T$. Then choose an identical Duffing system with state feedback as a slave system

$$\ddot{z} + a\dot{z} + b(z + cz^3) + k'(z - x) + k''(\dot{z} - \dot{x}) = f \sin \omega t, \; a > 0 \tag{10}$$

The equivalent deterministic system of the slave Duffing system (10) can be obtained also by the Gegenbauer polynomial approximation, and further expressed in terms of the following first order differential equations, called as a slave system as well,

$$\dot{Z} = AZ + G(Z) + F(t) + K(X - Z) \tag{11}$$

where $Z = [z_0 \cdots z_4 \, \dot{z}_0 \cdots \dot{z}_4]^T$, and $K = \mathrm{diag}[k', \cdots, k', k'', \cdots, k'']$ as the feedback gain matrix .

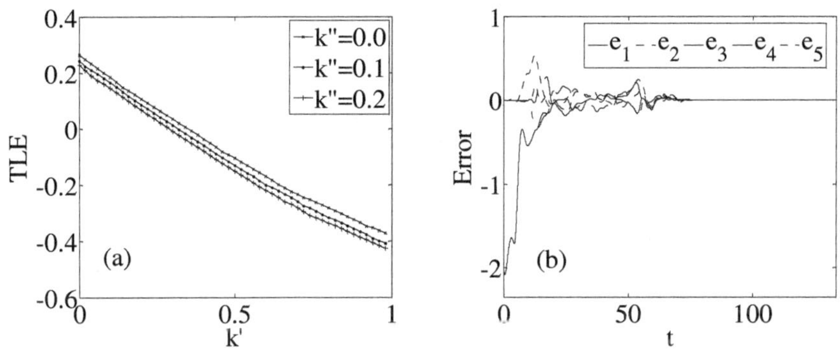

Fig. 8 (a) TLE for the slave system , (b) Errors between z_i and x_i , for k'=0.34, k"=0.2

Numerical simulations were conducted on synchronization phenomena between systems (9) and (11) under initial conditions

$$X(0) = [-1, 0, 0, 0, 0, 0, 0, 0, 0, 0]^T \text{ and } Z(0) = [1, 0, 0, 0, 0, 1, 0, 0, 0, 0]^T$$

The master system is treated as an independent system, whose ensemble mean sample response with given initial conditions $X(0)$ is taken as a specific reference time function for the slave system. Then, due to the pre-designed feedback control, an ensemble mean sample response of the slave system under the initial condition $Z(0)$, different from $X(0)$, finally synchronizes with the given reference motion within satisfactory accuracy. It seems that the synchronization is nearly perfect in this numerical experiment. However, one may ask what kind of motion the synchronized response could be. In Fig. 8(a), we can see that when k′ = k″= 0, the master system and the slave system are identical, and they all have a positive TLE. Hence, the ensemble mean response $E[x(t,\mu)]$, obtained from the master system, is a deterministic chaos, and $x(t,\mu)$ is a stochastic chaos. On the other hand, we know that reference motion is a deterministic time function, no matter how irregular or strange it looks like. In addition, for k′ = 0.34, k″ = 0.2, the slave system just has a negative TLE, so that the only way out for the negative feedback slave system is to duplicate the reference motion. Fig. 8(b) shows the errors between the master system and the slave system, defined as $e_i = z_i - x_i$, $i = 0, \ldots, 4$. It is seen that then errors approach zero after certain period of time, indicating that the given sample of random chaos can be duplicated within a prescribed accuracy. However, at this moment the slave system is essentially no longer a chaotic one.

4 Conclusions

A practical strategy for studying stochastic chaos for two kind of stochastic nonlinear systems is introduced. Stochastic chaos is a specific kind of random process with both the features of chaos and random process. Stochastic chaos holds an infinite numbers of deterministic samples, each having at least one positive TLE as its identification. Hence, the study and the control of stochastic chaos can be effectively made through its deterministic sample motion. The orthogonal polynomial approximation method is shown to be practical for studying stochastic chaos in dynamic systems with random parameters and polynomial nonlinearities, especially for control of stochastic chaos. The results of our analysis show that synchronization can be realized even in two stochastic nonlinear systems. However, the essence of “synchronization in chaotic systems” in master-slave mode is simply the duplication of the control reference signal, a deterministic realization of a given chaos. The synchronized motion of two identical nonlinear systems induced by a common noise is found merely a kind of non-chaotic motion.

Acknowledgments. This research was supported by NSFC under Grant Nos. 10902062, 10872156, and 10932009. The authors deeply appreciate Dr. H. Q. Zhang’s timely help in computer graphics and document preparing.

References

[1] Fang, T.: Understanding of chaos together with its control and synchronization. Chinese J. Vib. Eng. 20(5), 447–453 (2007)
[2] Ott, E.: Chaos in dynamical systems, 2nd edn. Cambridge University Press, Cambridge (2002)
[3] Tel, T., Gruiz, M.: Chaotic dynamics. Cambridge University Press, Cambridge (2006)

[4] Wu, C.L.: Random responses, bifurcation and chaos with their control in random structure and stochastic dynamic systems. Ph.D. Thesis, Northwestern Polytechnical University, Xian, China (2007)
[5] Yang, X.L., Xu, W.: Two-fold influence of non-periodic force on chaos control. Acta Physica Sinica 58(6), 3722–3728 (2009)
[6] Yang, X.L.: Study on random chaos, chaos control and synchronization of typical nonlinear dynamical systems. Ph.D. Thesis, Northwestern Polytechnical University, Xian, China (2007)
[7] Pecora, L.M., Carroll, T.L.: Synchronization in chaotic systems. Phys. Rev. Letts. 64(8), 821–824 (1990)
[8] Wu, C.L., Fang, T., Rong, H.W.: Chaos synchronization of two stochastic Duffing oscillators by feedback control. Chaos, Solitons and Fractals 32(3), 1201–1207 (2007)

Fractal Basin Boundaries and Chaotic Dynamics in the Randomly-Driven Henon-Heiles Oscillator

C.B. Gan

Department of Mechanical Engineering, Zhejiang University,
Hangzhou 310027, P.R. China

Abstract. Chaotic scattering is usually associated with the transient chaotic dynamics in open Hamiltonian systems. Our goal is twofold. First, the Henon-Heiles oscillator with the bounded noisy excitation and/or weak dissipation is chosen as a chaotic scattering paradigmatic example to observe the noisy basin boundaries and compute their corresponding fractal characteristics. Second, we investigate several sample invariant sets, and employ some previous methods to identify the noisy dynamics. It is shown that, fractal structure of basins is robust in the presence of the bounded noisy excitation for the Henon-Heiles oscillator, which is also verified by the evaluation of quantitive fractal dimension. The stable and unstable manifolds of sample chaotic invariant sets in some typical two-dimensional section, as well as the noisy chaotic time series, are presented and discussed.

Keywords: Henon-Heiles oscillator, bounded noisy excitation, fractal basin boundary, chaotic dynamics.

1 Introduction

Chaotic scattering has been studied for more than two decades because it is relevant to many areas such as nonlinear dynamics, fluid mechanics, astrophysics, optics, nanophysics, etc., see [1] and other references therein. It is very important to investigate which dynamical phenomena are robust and may persist in the presence of various kinds of perturbation, since perturbations such as noisy excitation and/or weak dissipation can be present in most realistic situations.

Originally, the Henon-Heiles (H-H) system was proposed in 1964 to address the question whether there exist more than two constants of motion in the dynamics of a galaxy model ([2]). It is described by the Hamiltonian

$$H = \frac{1}{2}(\dot{x}^2 + \dot{y}^2) + \frac{1}{2}(x^2 + y^2) + x^2 y - \frac{1}{3} y^3 \tag{1}$$

which defines the motion of a particle with unit mass in the two-dimensional potential

$$U(x, y) = \frac{1}{2}(x^2 + y^2) + x^2 y - \frac{1}{3} y^3 \tag{2}$$

W.Q. Zhu et al. (Eds.): Nonlinear Stochastic Dynamics & Control, IUTAM BOOK SERIES 29, pp. 183–190.
springerlink.com

Since then it has become a paradigmatic model for studying nonlinear and chaotic dynamics in continuous-time Hamiltonian systems ([3-7]). In the H-H system, for energies below a certain threshold value, the motions are bounded and the particles cannot leave the scattering region, but for energies above this threshold value, three exits appear and particles can escape towards infinity through any one of the exits.

In this study, the H-H oscillator is further used to investigate the effects of the bounded noise on the exit basin boundaries and the scattering dynamics. In Section 2, we briefly introduce the model, and explore the noisy fractal basin boundaries in the nondissipative and dissipative H-H oscillator respectively. Section 3 discusses the fractal characterizations, the stable and unstable manifolds of sample chaotic invariant sets, as well as the noisy chaotic dynamics. In the final section, the results of this work are summarized.

2 Noisy Exit Basins in the H-H Oscillator

Here, the H-H oscillator subjected to the bounded noisy excitation ([8]) is written by

$$\begin{aligned} &\ddot{x}+x+2xy+\alpha\dot{x}=0 \\ &\ddot{y}+y+x^2-y^2+\beta\dot{y}=f\cos[\Omega t+\sigma B(t)+\gamma] \end{aligned} \tag{3}$$

where α and β are the damping coefficients, f and Ω are the strength and central circular frequency of the bounded noisy excitation respectively, $B(t)$ is a standard Wiener process with strength σ, and γ is a random variable uniformly distributed in the interval $[0,2\pi)$. When $\alpha=\beta=f=0$, the H-H oscillator (3) has C_{3v} symmetry: it is invariant under rotations around the origin by $2\pi/3$ and $4\pi/3$, and under reflections at three symmetry lines with the angles $\pm\pi/6$ and $\pi/2$ with respect to the x axis.

Fig. 1 shows the contours of the potential $U(x,y)$ (see Eq. (2)). It exhibits three saddles at energy $E_{sad}=1/6$ lying at $(x,y)=(0,1)$, $(-\sqrt{3}/2,-1/2)$ and $(\sqrt{3}/2,-1/2)$. The equipotential lines at $E=E_{sad}$ form an equilateral triangle with side length $\sqrt{3}$. For the sake of clarity, we call Exit 1 the upper exit ($y\to+\infty$), Exit 2 the left exit ($x\to-\infty, y\to-\infty$), and Exit 3 the right exit ($x\to+\infty, y\to-\infty$). In this H-H system, the phase space is mixed with the KAM islands and chaotic seas, and a small amount of dissipation can convert the islands into sinks, or attractors.

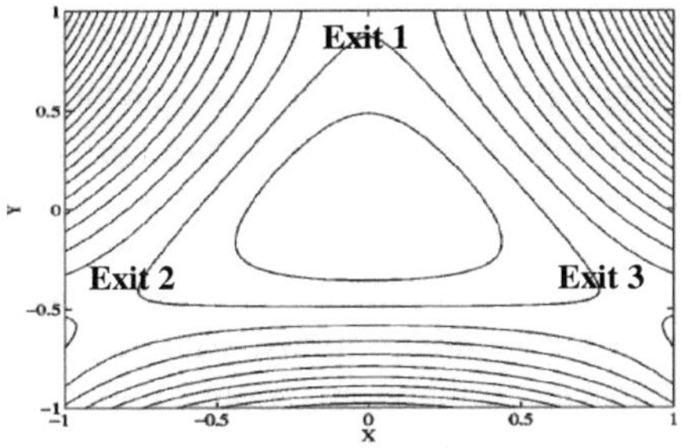

Fig. 1 Equipotential lines of the H-H potential function $U(x,y)$

To plot the exit basin diagrams for the H-H oscillator (3), we must calculate each trajectory by solving the differential equations of motion from a fine grid of initial conditions. The phase space of (3) depends on $(x, y, \dot{x}, \dot{y})$ and the energy, so three variables must be fixed to define a trajectory. Throughout this paper, we define some specific Poincare surfaces of section to display the results. For each particle, its initial velocity is expressed by

$$v = \sqrt{\dot{x}^2 + \dot{y}^2} = \sqrt{2E - x^2 - y^2 - 2x^2 y + \frac{2}{3} y^3} \tag{4}$$

In [4], two different choices on initial conditions are discussed in detail. Here, we also fix the initial conditions as $x = 0, y \in (y_{\min}, y_{LO})$, $\theta \in (0,2\pi)$, and the initial energies of all the particle are the same as E=0.19 . The value y_{LO} is the distance between the origin of coordinates and the position of each Lyapunov orbit ([3]), and is required to be calculated numerically. θ is the angle that v forms with the positive y axis in the counterclockwise sense. The Poincare map is defined by the plane $x = 0$ and $\dot{x} > 0$, and for this choice of initial conditions, Eq. (4) becomes $v = \sqrt{2E - y^2 + \frac{2}{3} y^3}$. As the radicand must be positive, y must be bigger than $y_{\min}$, where $y_{\min}$ is the real solution of $2E - y^2 + \frac{2}{3} y^3 = 0$. For each initial angle θ, the initial vertical velocity $\dot{y}_i$ is given by $\dot{y} = v\cos\theta$, so we use $(y, \dot{y})$ as initial conditions and plot the exit diagram by generating 2000×2000 grid points uniformly in the region $(y_{\min}, y_{LO}) \times (0,2\pi)$.

From [9], each physical realization $\eta(t)$ of the bounded noisy process $\cos[\Omega t + \sigma B(t) + \gamma]$ can be approximated by

$$\eta(t) \approx \sum_{k=1}^{N} A\cos(\omega_k t + \psi_k), \qquad N \to \infty \tag{5}$$

where $A = \sqrt{2S_{\xi}\Delta\omega}$, $\{\omega_k \mid k = 1,2,\cdots,N\}$ are independent and nonnegative random variables over the interval $[\omega_l, \omega_r]$, $\Delta\omega$ $(= (\omega_r - \omega_l)/N)$ is a frequency increment, $\{\psi_k \mid k = 1,2,\cdots,N\}$ are identically uniformly distributed over the interval $[0,2\pi)$ and N is a fixed positive integer. For crucial aspects on this approximate description, e.g., the minimal number N in Eq. (5), see [9]. For large positive integer N, the physical realization generated by Eq. (5) is almost ergodic, and numerical results show that the influence of this large integer can be neglected.

In many applications, systems are affected by the same external perturbations, so it is reasonable to use the identical almost-ergodic realization to simulate the motions of all chosen particles in the H-H system. If an orbit escapes through Exit 1, its initial condition belongs to the Exit 1 basin, and the same applies for Exit 2 and Exit 3. In order to visualize it, we plot the initial conditions with different color codes, according to the exit through which they escape. The color code we have chosen is black for Exit 1, dark gray for Exit 2, and light gray for Exit 3.

Typical plots of the exit basin structure of the randomly-driven H-H oscillator (3) are shown in Figs. 2 (a) and 2(b), the basin boundaries are clearly fractal for E=0.19. The complex basin topology associated with chaotic scattering turns out to be robust in the presence of the bounded noisy excitation. Here, the phase space has four different regions: three of them correspond to the three different exits, and the white region inside the plotted structure represents the particles that can not escape from the three exits. More numerical results show that similar fractal structures can be obtained from different physical realizations due to their ergodicity, which are not presented here.

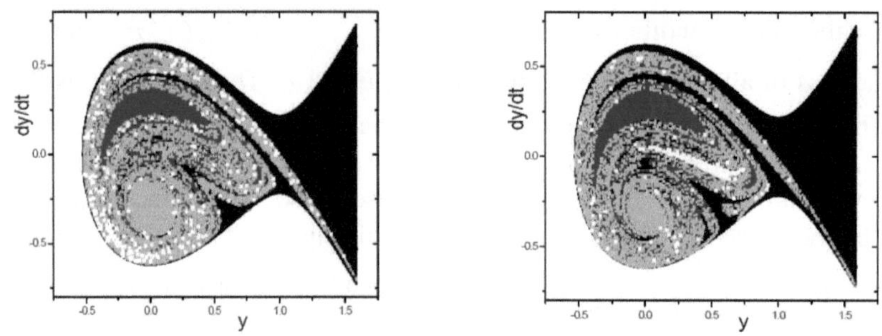

(a) Nondissipative ($\alpha = \beta = 0$), (b) Dissipative ($\alpha = \beta = 0.001$).

Fig. 2 Fractal exit basins in the randomly-driven H-H oscillator, where $f = 0.025$

3 Noisy Chaotic Dynamics

A chaotic attractor is usually characterized by the fractal dimension. The fractal dimension of the set of singularities in a scattering function can be calculated by employing the uncertainty algorithm ([9]). For a fixed value of the "uncertainty" ε, we should randomly choose an initial angle θ_0 and compute the quantity $\Delta \equiv |T(\theta_0) - T(\theta_0 + \varepsilon)|$, where T is the delay time (see Fig. 3 (a)). If $\Delta > s$ (s is a small positive number), then the angle θ_0 is uncertain with respect to ε. Otherwise, θ_0 is certain. A large number of initial conditions should be chosen randomly, which yields $f(\varepsilon)$, i.e., the fraction of the uncertain initial conditions. The quantity $f(\varepsilon)/\varepsilon$ typically scales with ε as $f(\varepsilon)/\varepsilon \sim \varepsilon^{-D}$, where D is the uncertainty dimension related to the dimension $\overline{D}$ of the chaotic attractor as $\overline{D} = M - D$, and M is the dimension of the phase space (M=3 in our case).

Fig. 3(b) shows $f(\varepsilon)/\varepsilon$ versus ε on a logarithmic scale for the dissipative H-H oscillator subjected to the bounded noisy excitation, where the same physical realization is used as in Fig. 2 and the constant s is arbitrarily chosen to be 0.01. The estimated slope is $D = 0.8 \pm 0.003 < 1$, and the fractal dimension of the sample chaotic attractor $\overline{D} = 3 - D$. Here, the terminology "sample" means that the result is obtained from a deterministic physical realization generated by Eq. (5). Thus, the fractal dimension of a sample chaotic attractor may also be estimated from the uncertainty algorithm even though the bounded noisy excitation is imposed on the H-H oscillator.

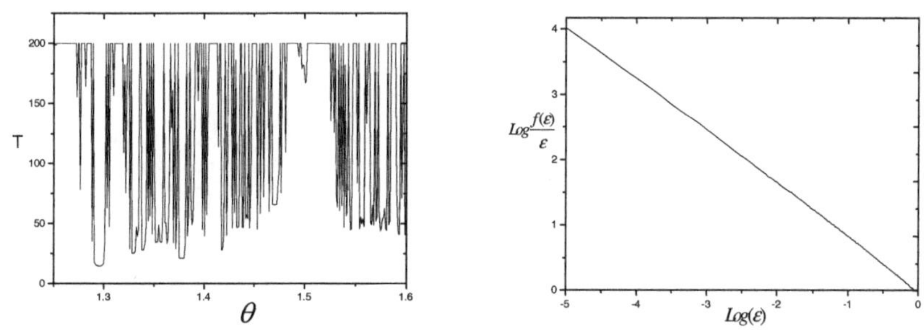

(a) Typical delay-time function, (b) Uncertainty dimension.

Fig. 3 Fractal characteristics of exit basin boundaries in the randomly-driven H-H oscillator (3), where $\alpha = \beta = 0.001$, $f = 0.025$

As shown in [11], each stable manifold of a sample chaotic invariant set can be approximated by the "sprinkler algorithm". To find the unstable manifold of the sample chaotic invariant set, it is better to change the sign of every differential equation and draw the stable manifold of the dynamical system. The result is exactly the unstable manifold of the original system. As the dimension of these fractal sets is between two and three, we can only plot its intersection with a Poincare map. Fig. 4 shows two sample sets of the stable and unstable manifolds in the Poincare surface of section for E=0.19. As compared with the results presented by Aguirre et al [4], we can see that there are no significant changes when the bounded noisy excitation is imposed on the H-H oscillator. However, the stable and unstable manifolds of each sample chaotic set seem not symmetric to each other, since we have used the time-reversal physical realization to solve the equations of motion. Moreover, the traces of the sample manifolds are thicker than those in the deterministic case due to the dispersive role of the bounded noisy excitation.

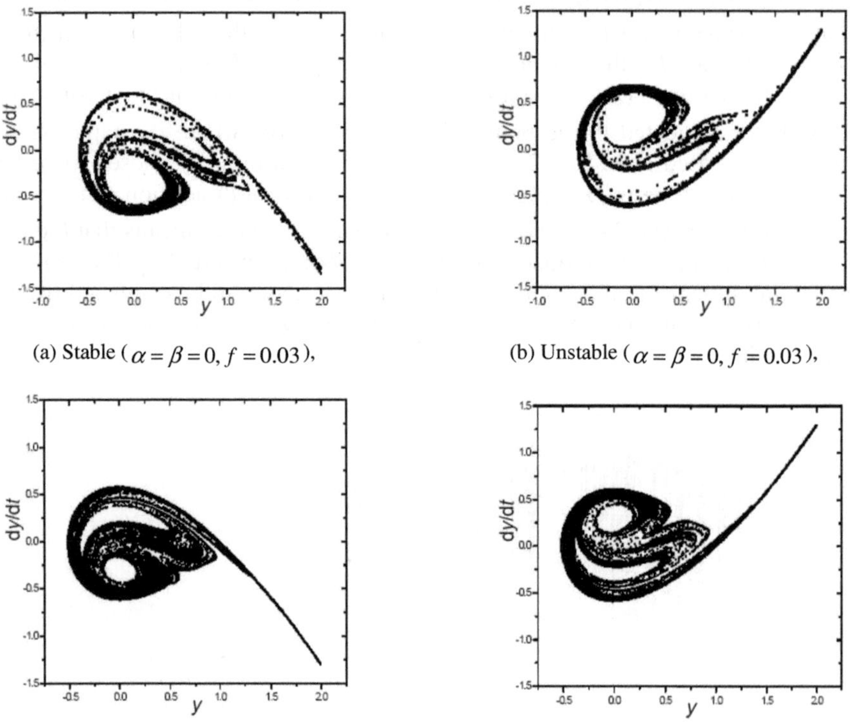

(a) Stable ($\alpha=\beta=0, f=0.03$), (b) Unstable ($\alpha=\beta=0, f=0.03$),

(c) Stable ($\alpha=\beta=0.0001, f=0.01$), (d) Unstable ($\alpha=\beta=0.0001, f=0.01$).

Fig. 4 Typical sample invariant manifolds in the randomly-driven H-H oscillator (3)

In the end of this section, we explore chaotic dynamics from the noisy signals in the randomly-driven H-H oscillator (3) by some previous methods ([12-16]). Here, the leading Lyapunov exponent of each sample time series can be evaluated by the least-square fit from the mean divergence <Ln(Divergence)> vs t curve in Fig. 5 (b) ([14]). For the noisy time series shown in Fig. 5 (a), the original signal and its corresponding surrogates are clearly distinct when the viewing scale ε_0 tends to zero, and the null hypothesis of a periodic orbit with correlated noise should be rejected ([15]), see Fig. 5 (c). The correlation dimensions are estimated according to the algorithm by Judd ([17]) for each original time series and 30 PPS data sets. The algorithm described in [17] estimates correlation dimension d_c as a function of viewing scale ε_0, so d_c for each time series is not a single number, but a curve. According to the initial obvious linear slope shown in Fig. 5 (b), as well as the rejected null hypothesis from Fig. 5 (c), the time series is chaotic-dominant.

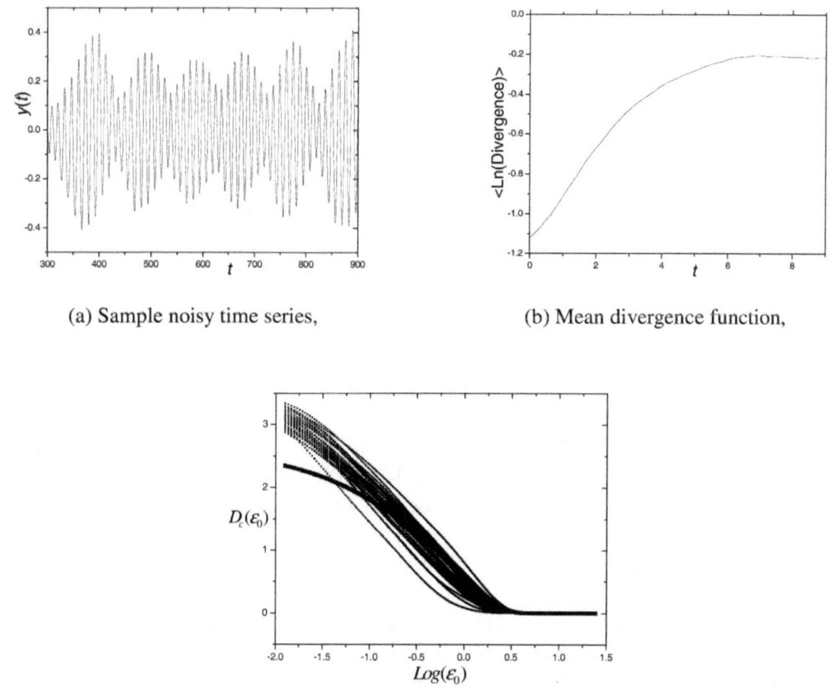

(a) Sample noisy time series, (b) Mean divergence function,

(c) Correlation dimensions.

Fig. 5 Typical chaotic-dominant response in the randomly-driven H-H oscillator, in which $\alpha = \beta = 0.003, f = 0.025$. The top panel shows the original data from system (3) for (a) sample noisy time series, (b) the mean divergence from the data in (a). Comparisons of the correlation dimensions for the original time series and its surrogates are presented in (c), where the thick scatter line and the thin scatter lines represent the results from the original data and its surrogates respectively. The embedding dimension m=7, the reconstruction delay $J = 31$, and the noise radius $\rho = 0.0003$.

4 Conclusions

In this paper, we have studied the noisy scattering dynamics in the randomly-driven H-H oscillator when the energy is enough large to permit particles to escape from the exits. We paid special attention to the computation of the exit basins, which show a rich pattern of noisy fractal structures and the uncertainty dimensions of the fractal sets. Computations of the stable and unstable manifolds of two sample chaotic invariant sets have also been carried out, from which we can see no significant differences as compared with those previously presented for the deterministic H-H oscillator. In addition, an arbitrarily chosen noisy time series is identified as the chaotic-dominant one by previous methods.

References

[1] Seoane, J.M., Aguirre, J., Sanjuan, M.A.F., Lai, Y.C.: Basin topology in dissipative chaotic scattering. Chaos 16, 023101-1-8 (2006)
[2] Henon, M., Heiles, C.: The applicability of the third integral of motion: Some numerical experiments. Astronomical Journal 69, 73–79 (1964)
[3] Contopoulos, G.: Asymptotic curves and escapes in Hamiltonian systems. Astronomy and Astrophysics 231, 41–55 (1990)
[4] Aguirre, J., Vallejo, J.C., Sanjuan, M.A.F.: Wada basins and chaotic invariant sets in the Henon-Heiles system. Physical Review E 64, 066208-1-11 (2001)
[5] Seoane, J.M., Sanjuan, M.A.F., Lai, Y.C.: Fractal dimension in dissipative chaotic scattering. Physical Review E 76, 016208-1-6 (2007)
[6] Zhao, H.J., Du, M.L.: Threshold law for escaping from the Henon-Heiles system. Physical Review E 76, 027201-1-4 (2007)
[7] Seoane, J.M., Sanjuan, M.A.F.: Exponential decay and scaling laws in noisy chaotic scattering. Physics Letters A 372(2), 110–116 (2008)
[8] Lin, Y.K., Cai, G.Q.: Probabilistic Structural Dynamics-Advanced Theory and Applications. McGraw-Hill, Singapore (1995)
[9] Shinozuka, M.: Digital simulation of random processes and its applications. Journal of Sound and Vibration 25, 111–128 (1972)
[10] McDonald, S.W., Grebogi, C., Ott, E., Yorke, J.A.: Fractal basin boundaries. Physical D 17(2), 125–153 (1985)
[11] Kantz, H., Grassberger, P.: Repellers, semi-attractors, and long-lived chaotic transients. Physica D 17(1), 75–86 (1985)
[12] Gan, C.: Noise-induced chaos in Duffing oscillator with double wells. Nonlinear Dynamics 45(3-4), 305–317 (2006)
[13] Gan, C.: Pseudo-periodic surrogate test to sample time series in stochastic softening Duffing oscillator. Physics Letters A 357(3), 204–208 (2006)
[14] Rosenstein, M.T., Collins, J.J., Luca, C.J.: A practical method for calculating leading Lyapunov exponents from small data sets. Physica D 65, 117–134 (1993)
[15] Small, M., Xu, D., Harrison, R.G.: Surrogate test for pseudo-periodic time series data. Physical Review Letters 87(18), 188101-1-4 (2001)
[16] Zhu, W.Q., Liu, Z.H.: Homoclinic bifurcation and chaos in coupled simple pendulum and harmonic oscillator under bounded noise excitation. International Journal of Bifurcations and Chaos 15(1), 233–243 (2005)
[17] Judd, K.: An improved estimator of dimension and some comments on providing confidence intervals. Physica D 56, 216–228 (1992)

Moment Lyapunov Exponent for a Three Dimensional Stochastic System

Shenghong Li [1,2] and Xianbin Liu[1]

[1] College of Aerospace Engineering, Nanjing University of Aeronautics and Astronautics, Nanjing, China
[2] Department of Mathematics and Physics, Jiangsu university of Science and Technology, Zhenjiang, Jiangsu Province, China

Abstract. In the present paper, for an arbitrary finite real number p, the pth moment Lyapunov exponent for a codimension two bifurcation system that is on a three-dimensional center manifold and is subjected to a parametric excitation by a small intensity white noise is investigated. Via a perturbation method and a linear stochastic transformation introduced by Wedig, an eigenvalue problem associated with the moment Lyapunov exponent is obtained. The eigenvalue problem is then solved approximately via a Fourier cosine series, and for whom the convergence rate is illustrated numerically. Furthermore, the stability regions of pth moment are also obtained.

Keywords: Moment Lyapunov exponent, Ito equation, perturbation method, Stability region.

1 Introduction

For a stochastic system that is driven by a white noise or an ergodic and real noise excitation, there are many results refer to the asymptotic expansions of maximal Lyapunov exponents. However, for an almost sure stable system, there exists a probability that the mean square response for the system may still exceed some threshold and may grow exponently, which implies that the mean square response is unstable.

Let $x(t,x_0)$ be a solution to a random dynamical system, in order to describe the exponential growth rate of its pth$(p>0)$ moment, the moment Lyapunov exponent is defined as

$$\Lambda(p,x_0) = \lim_{t\to\infty}\frac{1}{t}\log E\left\|x(t,x_0)\right\|^p, \quad p\in R \tag{1}$$

which implies that if $\Lambda(p,x_0)<0$, then $E\|x(t,x_0)\|^p\to 0$ as $t\to\infty$ and if $\Lambda(p,x_0)>0$, $E\|x(t,x_0)\|^p\to\infty$ as $t\to\infty$. It has been proven [1,2] that the limit is independent of the initial value x_0, and $\Lambda(p)$ is a convex analytic function of $p\in\mathbf{R}$, and

$$\lambda = \frac{\partial\Lambda}{\partial p}\Big|_{p=0} = \lim_{t\to\infty}\frac{1}{t}\log\left\|x(t,x_0)\right\|^p, \quad p\in R \tag{2}$$

is in fact the maximal Lyapunov exponent.

W.Q. Zhu et al. (Eds.): Nonlinear Stochastic Dynamics & Control, IUTAM BOOK SERIES 29, pp. 191–200.
springerlink.com

Although the moment Lyapunov exponents are important in the study of the stability of stochastic dynamical systems, it is very difficult to determine the expressions of the moment Lyapunov exponent in actual cases, and there are very few results concerning this aspect.

For a white and real noises excited two dimensional system and a system of two coupled oscillators that is driven by a real noise, Arnold et al. [3,4] and Namachchivaya et al. [5] obtain the asymptotic expansions of the pth moment Lyapunov exponents in the case of small noise intensity and small p. Khasminskii and Moshchuk [6] consider a two dimensional system with small white noise excitations. They prove that for a system in which the system matrix has two purely imaginary eigenvalues, the pth moment Lyapunov exponent possesses an asymptotic expansion in terms of the small noise intensity for a finite value of p. For a system of two coupled oscillators that is driven by real noises, Namachchivaya and Roessel [7] obtained an asymptotic expansion of the finite pth moment Lyapunov exponent. In Ref. [7], the extension of the perturbation method that is first introduced by Arnold et al. [8] is applied. For a two dimensional system that are under real noise excitation and bounded noise excitation, via applying a procedure similar to that employed in Khasminskii and Moshchuk [6], Xie [9] obtain weak-noise expansions of moment Lyapunov exponent, the Lyapunov exponent, and the stability index.

In this paper, we consider a codimension two bifurcation system, that is on a three dimensional center manifold and is excited parametrically by a white noise. the maximal Lyapunov exponent for this system has been investigated by Liu and Liew[10]. Applying the procedure exposed by in Wedig [11] and the perturbation method, an eigenvalue problem for the first order term of the expansion of the moment Lyapunov exponent is obtained. The eigenvalue problem is then solved by Fourier cosine series, through which an infinite matrix is yielded and the leading eigenvalue of this infinite matrix is the first order approximation of the pth moment Lyapunov exponent. The numerical results of the pth moment Lyapunov exponent are illustrated and in addition, the stability regions of pth moment are also obtained numerically.

2 Equations

Consider a typical deterministic codimension two bifurcation system which is on a three-dimension central manifold and possesses one zero-eigenvalue and a pair of pure imaginary eigenvalues [12]

$$\begin{aligned} \dot{r} &= \mu_1 r + a_1 rz + (a_2 r^3 + a_3 r^2 z) + O(|r,z|^4) \\ \dot{z} &= \mu_2 z + (b_1 r^2 + b_2 z^2) + (b_3 r^2 z + b_4 z^3) + O(|r,z|^4) \\ \dot{\Theta} &= \omega + O(|r,z|^2) \end{aligned} \tag{3}$$

Where μ_1 and μ_2 are unfolding parameters, a_1, a_2, a_3, b_1, b_2, b_3 and ω are real constants. This normal form arises in the classic fluid dynamic stability study of coquette flow. Via the transformation of $r=(x_1^2+x_2^2)^{1/2}$, $z=x_3$, $\Theta=\arctan(x_1/x_2)$, and in the vicinity of equilibrium point $(x_1, x_2, x_3)=(0,0,0)$, the linearization of the original system (3), that is subjected to a white noise excitation is then obtained, i.e.

$$\mathrm{dx} = (\mathrm{A}_0\mathrm{x} - \varepsilon \mathrm{A}_1\mathrm{x})\mathrm{d}t + \sqrt{\varepsilon}\mathrm{Bx} \circ \mathrm{d}W(t) \tag{4}$$

where, $W(t)$ is a unit Wiener process, "$\circ$" means that Eq.(2) is a Stratonovitch stochastic differential equation, and

$$\mathrm{A}_0 = \begin{bmatrix} 0 & \omega & 0 \\ -\omega & 0 & 0 \\ 0 & 0 & 0 \end{bmatrix}, \quad \mathrm{A}_1 = \begin{bmatrix} \delta_1 & 0 & 0 \\ 0 & \delta_1 & 0 \\ 0 & 0 & \delta_2 \end{bmatrix}, \quad \mathrm{B} = \begin{bmatrix} b_{11} & b_{12} & b_{13} \\ b_{21} & b_{22} & b_{23} \\ b_{31} & b_{32} & b_{33} \end{bmatrix}$$

In Eq.(4), the parameters μ_1, μ_2 have been rescaled such that

$$\mu_1 = -\varepsilon\delta_1, \ \mu_2 = -\varepsilon\delta_2 \tag{5}$$

Via appling a spherical polar transformation

$$\begin{aligned} &x_1 = a\cos\theta\sin\phi,\ x_2 = a\cos\theta\cos\phi,\ x_3 = a\sin\theta,\ a = \|x\| \\ &\phi(t) = \omega t + \varphi(t),\ \theta \in [-\pi/2, \pi/2],\ \phi, \varphi \in [0, 2\pi] \end{aligned} \tag{6}$$

and introducing a pth norm $P=a^p$, then the Stratonovitch equations about the norm process P and the phase processes θ and ϕ is then obtained as

$$\begin{aligned} &\mathrm{d}P = \varepsilon pP\rho_1\mathrm{d}t + \sqrt{\varepsilon}\, pP\rho_2 \circ \mathrm{d}W \\ &\mathrm{d}\theta = \varepsilon\theta_1\mathrm{d}t + \sqrt{\varepsilon}\theta_2 \circ \mathrm{d}W, \quad \mathrm{d}\phi = \omega\mathrm{d}t + \sqrt{\varepsilon}\phi_2 \circ \mathrm{d}W \end{aligned} \tag{7}$$

where

$$\begin{aligned} &\rho_1 = -\delta_1\cos^2\theta - \delta_2\sin^2\theta \\ &\rho_2 = \frac{1}{4}(f_{11} - 2f_{13})\cos(2\theta) + \frac{1}{2}f_{12}\sin 2\theta + \frac{1}{4}(f_{11} + 2f_{13}) \\ &\theta_1 = \frac{1}{2}(\delta_1 - \delta_2)\sin 2\theta \\ &\theta_2 = \frac{1}{2}f_{21}\cos(2\theta) - \frac{1}{4}(f_{22} - 2f_{23})\sin 2\theta + \frac{1}{2}f_{24} \\ &\phi_2 = f_{31}\tan\theta + f_{32} \\ &f_{11} = f_{22} = k_1 + k_2\cos 2\phi + k_3\sin 2\phi,\ f_{12} = f_{21} = k_5\sin\phi + k_7\cos\phi \\ &f_{13} = f_{23} = b_{33},\ f_{24} = k_6\sin\phi + k_8\cos\phi \\ &f_{31} = b_{13}\cos\phi - b_{23}\sin\phi,\ f_{32} = -\frac{1}{2}(k_2\sin 2\phi - k_3\cos 2\phi + k_4) \\ &k_1 = b_{22} + b_{11},\ k_2 = b_{22} - b_{11},\ k_3 = b_{12} + b_{21},\ k_4 = b_{21} - b_{12} \\ &k_5 = b_{13} + b_{31},\ k_6 = b_{31} - b_{13},\ k_7 = b_{23} + b_{32},\ k_8 = b_{32} - b_{23} \end{aligned} \tag{8}$$

By introducing the Wong-Zakai correction terms, Eq.(7) can then be converted to Ito stochastic differential equations, i.e.

$$\begin{aligned} &\mathrm{d}P = \varepsilon pP\tilde{\rho}_1\mathrm{d}t + \sqrt{\varepsilon}\, pP\rho_2\mathrm{d}W \\ &\mathrm{d}\theta = \varepsilon\tilde{\theta}_1\mathrm{d}t + \sqrt{\varepsilon}\theta_2\mathrm{d}W, \quad \mathrm{d}\phi = (\omega + \varepsilon\tilde{\phi}_1)\mathrm{d}t + \sqrt{\varepsilon}\phi_2\mathrm{d}W \end{aligned} \tag{9}$$

where

$$\tilde{\theta}_1 = \theta_1 + \frac{1}{2}[\theta_2\partial_\theta + \phi_2\partial_\phi]\theta_2, \quad \tilde{\phi}_1 = \frac{1}{2}[\theta_2\partial_\theta + \phi_2\partial_\phi]\phi_2$$
$$\tilde{\rho}_1 = \rho_1 + \frac{1}{2}\{p\rho_2^2 + [\theta_2\partial_\theta + \phi_2\partial_\phi]\rho_2\}, \quad \partial_\theta = \frac{\partial}{\partial\theta}, \quad \partial_\phi = \frac{\partial}{\partial\phi} \tag{10}$$

3 Asymptotic Analysis

In this section, a linear stochastic transformation, which was firstly introduced by Wedig [11] and is used to obtain the eigenvalue problem for the pth moment Lyapunov exponents of a two dimensional linear Ito stochastic system, is applied, i.e.

$$S = T(\theta,\phi)P, \quad P = T^{-1}(\theta,\phi)S, \quad -\frac{\pi}{2} \le \theta \le \frac{\pi}{2}, 0 \le \phi \le 2\pi \tag{11}$$

where, $T(\theta, \phi)$ is defined as a scalar function with respect to the stationary phase processes θ and ϕ, then the Ito equation for the new scalar diffusion process S is obtained via Ito lemma, i.e.

$$\mathrm{d}S = P\{\omega T'_\phi + \varepsilon[p\tilde{\rho}_1 T + (\tilde{\theta}_1 + p\rho_2\theta_2)T'_\theta + (\tilde{\phi}_1 + p\rho_2\phi_2)T'_\phi + \theta_2\phi_2 T''_{\theta\phi}$$
$$+\frac{1}{2}\theta_2^2 T''_{\theta\theta} + \frac{1}{2}\phi_2^2 T''_{\phi\phi}]\}\mathrm{d}t + \sqrt{\varepsilon}P(p\rho_2 T + \theta_2 T'_\theta + \phi_2 T'_\phi)\mathrm{d}W \tag{12}$$

where

$$T'_\phi = \partial_\phi T, \; T'_\theta = \partial_\theta T, \; T''_{\theta\phi} = \partial^2_{\theta\phi}T, \; T''_{\theta\theta} = \partial^2_\theta T, \; T''_{\phi\phi} = \partial^2_\phi T$$
$$\partial^2_{\theta\phi} = \frac{\partial^2}{\partial\theta\partial\phi}, \quad \partial^2_\theta = \frac{\partial^2}{\partial\theta^2}, \quad \partial^2_\phi = \frac{\partial^2}{\partial\phi^2} \tag{13}$$

If the transformation function $T(\theta, \phi)$ is bounded and non-singular, both of the processes P and S possess the same stochastic stability. Then, according to Wedig [11], the function $T(\theta, \phi)$ is chosen such that the drift term of Eq.(12) is independent of the phase processes θ and ϕ, i.e.

$$\mathrm{d}S = \Lambda S\mathrm{d}t + \sqrt{\varepsilon}ST^{-1}(\theta,\phi)(p\rho_2 T + \theta_2 T'_\theta + \phi_2 T'_\phi)\mathrm{d}W \tag{14}$$

where $\Lambda=\Lambda(p)$, which is an unknown function of p. By comparing Eq.(8) and Eq.(9), we see that the transformation function $T(\theta, \phi)$ is govern by the following equation

$$(L_0 + \varepsilon L_1)T(\theta,\phi) = \Lambda(p)T(\theta,\phi) \tag{15}$$

where

$$L_0 = \omega\partial_\phi$$
$$L_1 = p\tilde{\rho}_1 + (\tilde{\theta}_1 + p\rho_2\theta_2)\partial_\theta + (\tilde{\phi}_1 + p\rho_2\phi_2)\partial_\phi + \theta_2\phi_2\partial^2_{\theta\phi} + \frac{1}{2}\theta_2^2\partial^2_\theta + \frac{1}{2}\phi_2^2\partial^2_\phi \tag{16}$$

Eq.(15) defines an eigenvalue problem for a second-order differential operator in which $\Lambda(p)$ is the eigenvalue and $T(\theta,\phi)$ is the ralevant eigenfunction. In Eq.(14), the eigenvalue $\Lambda(p)$ is the pth moment Lyapunov exponent of system (4).

Through a perturbation method, both $\Lambda(p)$ and $T(\theta,\phi)$ are expanded in power series of ε, i.e.

$$\begin{aligned} \Lambda(p) &= \Lambda_0(p) + \varepsilon\Lambda_1(p) + \cdots + \varepsilon^n \Lambda_n(p) + \cdots \\ T(\theta,\phi) &= T_0(\theta,\phi) + \varepsilon T_1(\theta,\phi) + \cdots + \varepsilon^n T_n(\theta,\phi) + \cdots \end{aligned} \tag{17}$$

Substituting Eq.(17) into Eq.(15) and equating terms of the equal powers of ε, we then obtain the following recurrent equations

$$\begin{aligned} &\varepsilon^0 \quad L_0T_0 = \Lambda_0T_0 \\ &\varepsilon^1 \quad L_0T_1 + L_1T_0 = \Lambda_0T_1 + \Lambda_1T_0 \\ &\varepsilon^2 \quad L_0T_2 + L_1T_1 = \Lambda_0T_2 + \Lambda_1T_1 + \Lambda_2T_0, \cdots, \\ &\varepsilon^n \quad L_0T_n + L_1T_{n-1} = \Lambda_0T_n + \Lambda_1T_{n-1} + \cdots + \Lambda_nT_0, \cdots \end{aligned} \tag{18}$$

3.1 The ε^0 Order Equation

The first equation in Eq.(18) is

$$\omega[\partial T_0/\partial\phi] = \Lambda_0T_0 \tag{19}$$

The moment Lyapunov exponent $\Lambda(p)$ has the property that $\Lambda(0)=0$, which means that $\Lambda_0(0)=0$. Since the left hand of Eq.(19) does not contain the parameter p, then the eigenvalue$\Lambda_0(p)$ is independent of p, which leads to a fact that $\Lambda_0(p)=0$. Eq.(19) is then equivalent to

$$\omega[\partial T_0/\partial\phi] = 0 \tag{20}$$

to which, the solution is

$$T_0(\theta,\phi) = \frac{1}{2\pi}F_0(\theta), \quad \theta\in[-\frac{\pi}{2},\frac{\pi}{2}], \phi\in[0,2\pi] \tag{21}$$

Since $\Lambda_0(p)=0$, then the associated adjoint differential equation of Eq.(20) is

$$-\omega\left[\partial T_0^*/\partial\phi\right] = 0 \tag{22}$$

Via a direct integral, one can easily obtain the solution, i.e.

$$T_0^*(\theta,\phi) = \frac{1}{2\pi}F_0^*(\theta), \quad \theta\in[-\frac{\pi}{2},\frac{\pi}{2}], \phi\in[0,2\pi] \tag{23}$$

where both $F_0(\theta)$ and $F_0^*(\theta)$ are arbitrary bounded functions with respect to θ, and $F_0(\theta)$ is needed to be determined through the solvability condition of the second equation in Eq.(18).

3.2 The ε^1 Order Equation

The second equation in Eq.(18) is

$$L_0 T_1 = \Lambda_1 T_0 - L_1 T_0 \tag{24}$$

The result in Eq.(21) leads to

$$L_1 T_0 = \frac{1}{2\pi}[p\tilde{\rho}_1 F_0(\theta) + (\tilde{\theta}_1 + p\rho_2\theta_2)F_0'(\theta) + \frac{1}{2}\theta_2^2 F_0''(\theta)] \tag{25}$$

The solvability condition of Eq.(24) is found to be

$$\int_{-\pi/2}^{\pi/2}\left[\Lambda_1 F_0(\theta) - \frac{1}{2}\sigma^2(\theta)F_0''(\theta) - \mu(\theta)F_0'(\theta) - pq(\theta)F_0(\theta)\right]F_0^*(\theta)\mathrm{d}\theta = 0 \tag{26}$$

where

$$\begin{aligned}
\mu(\theta) &= (p\beta_1 - \beta_1 + \beta_9)\sin(4\theta) - \beta_3\cos(2\theta) + (\beta_2 - \frac{p}{2}\beta_5 - \beta_{10})\sin(2\theta) \\
&\quad + [\alpha_8 - \alpha_7 - \beta_3\sin(2\theta) + 2\beta_4\cos^2(\theta)]\tan(\theta) + \beta_3 \\
\sigma^2(\theta) &= \beta_1\cos(4\theta) + \beta_5\cos(2\theta) - \beta_1 + \beta_{11} \\
q(\theta) &= (1 - \frac{1}{2}p)\beta_1\cos(4\theta) + (\beta_6 - \beta_2)\cos(2\theta) \\
&\quad + [2\beta_3\tan(\theta) - 4\beta_8]\cos^2(\theta) - [\beta_3 - \beta_4\tan(\theta)]\sin(2\theta) \\
&\quad + \beta_8\sin^2(2\theta) - \beta_1 - \pi\Delta^+ - \beta_4 + \beta_7 + \alpha_7 + \beta_6
\end{aligned} \tag{27}$$

and

$$\begin{aligned}
\beta_1 &= \frac{\pi}{32}(8k_1b_{33} - 8b_{33}^2 - 2k_1^2 - k_2^2 - k_3^2 + 4k_5^2 + 4k_7^2) \\
\beta_2 &= \frac{\pi}{16}p(8b_{33}^2 - 2k_1^2 - k_2^2 - k_3^2) - \frac{1}{2}\beta_5 + \alpha_7 - \beta_4 + \pi\Delta^- \\
\beta_3 &= \frac{1}{12}(k_3b_{23} - 2k_2b_{13}), \quad \beta_4 = -\frac{1}{8}(k_5b_{23} + k_7b_{13}) \\
\beta_5 &= \frac{\pi}{2}(k_5k_6 + k_7k_8), \quad \beta_6 = \frac{\pi}{8}(k_2^2 + 3k_3^2) \\
\beta_7 &= \frac{\pi}{64}p(8k_1b_{33} + 24b_{33}^2 + 6k_1^2 + 3k_2^2 + 3k_3^2 + 4k_5^2 + 4k_7^2) \\
\beta_8 &= \frac{\pi}{16}pk_1k_2, \quad \beta_9 = \frac{\pi}{8}k_7^2 \\
\beta_{10} &= -\frac{\pi}{16}(k_2^2 - 4k_3^2), \quad \beta_{11} = \frac{\pi}{4}(k_5^2 + k_6^2 + k_7^2 + k_8^2) \\
\Delta^{\pm} &= \delta_1 \pm \delta_2, \quad \alpha_i = \frac{\pi}{4}(k_ib_{23} + k_{i-2}b_{13}), \quad i = 7, 8,
\end{aligned} \tag{28}$$

4 Solution of the Eigenvalue Problem

Since the equality in Eq.(26) is satisfied for arbitrary $F_0^*(\theta)$, then Eq.(26) can be simplified as an ordinary differential equation, i.e.

$$L(p)F_0(\theta) = \Lambda_1(p)F_0(\theta), \quad L(p) = \frac{1}{2}\sigma^2(\theta)\frac{\mathrm{d}^2}{\mathrm{d}\theta^2} + \mu(\theta)\frac{\mathrm{d}}{\mathrm{d}\theta} + pq(\theta) \tag{29}$$

To solve Eq.(29), appropriate boundary conditions for the process θ are needed. As in [7], the boundary conditions are determined by considering the adjoint equation with p=0, i.e.

$$L^*F^*(\theta) = 0, \quad L^* = \frac{1}{2}\frac{\mathrm{d}^2}{\mathrm{d}\theta^2}[\sigma^2(\theta)] - \frac{\mathrm{d}}{\mathrm{d}\theta}[\mu(\theta)] \tag{30}$$

is the relevant Fokker-Planck operator. The solution to Eq.(30) had been studied in [10], and only the case, in whch, both of the boundaries $\theta=\pm\pi/2$ are entrances, are considered in the present paper. In this case, it is easy to verify that for Eq.(29), at both of $\theta=\pm\pi/2$, zero Neumann boundary conditions are obtained.

According to Wedig [8], the solution to Eq.(29) can be obtained by an orthogonal expansion. As in Namachchivaya[7], $F_0(\theta)$ may be expanded as a Fourier cosine series, i.e.

$$F_0(\theta) = \sum_{n=0}^{\infty} z_n \cos(2n\theta) \tag{31}$$

By inserting Eq.(31) into Eq.(29), multiplying by $\cos(2n\theta)$ in both sides and integrating for θ, we obtain

$$\sum_{m=0}^{\infty} a_{mn} z_m = \Lambda_1 z_n, \quad n = 0,1,2,\cdots \tag{32}$$

Where

$$a_{mn} = \frac{1}{\pi^2}\int_{-\frac{\pi}{2}}^{\frac{\pi}{2}} \{L(p)[\cos(2m\theta)]\}\cos(2n\theta)d\theta \tag{33}$$

In Eq.(32), for each z_n, to guarantee the existence of the nontrivial solution, the determinant of the coefficients must vanishes. Therefore, the problem for evaluating $\Lambda_1(p)$ is converted into evaluating the leading eigenvalue of an infinite sequence of sub-matrices of the matrix **A** that is defined in Eq.(34), which construct an approximate sequence of $\Lambda_1(p)$.

$$\mathrm{A} = \begin{bmatrix} a_{00} & a_{01} & a_{02} & \cdots \\ a_{10} & a_{11} & a_{12} & \cdots \\ a_{20} & a_{21} & a_{22} & \cdots \\ \vdots & \vdots & \vdots & \ddots \end{bmatrix} \tag{34}$$

The set of approximate eigenvalues obtained by this method converges to the exact eigenvalues as $n\to\infty$. However, the mass of calculation increases drastically with the increase of n, so we obtain the approximate eigenvalues by the truncation of n. For example, in the case that n=0, we obtain $\Lambda_1(p)=a_{00}$. For n=1, we obtain the second order approximation of $\Lambda_1(p)$ by solving the eigenvalue of the second order sub-matrix, and for n=2, by solving the eigenvalue of the third order sub-matrix, the third order approximation of $\Lambda_1(p)$ is obtained. Because of the complexity of expressions, here we only present the elements of the second order sub-matrix

$$
\begin{aligned}
a_{00} &= \frac{p}{8}[(3p+2)(\frac{k_1^2}{4}+b_{33}^2)+(p-2)(\frac{k_5^2}{2}+\frac{k_7^2}{2}+k_1b_{33})]+\frac{p}{64}[(3p+10)k_2^2 \\
&\quad +(3p+26)k_3^2]-p(\delta_1+\delta_2)+\frac{p}{4}(k_7b_{23}+k_5b_{13})-\frac{3}{16}p^2k_1k_2 \\
a_{01} &= \frac{p}{32}[(p+2)k_2^2+(p+6)k_3^2]+\frac{p}{8}[k_5(k_6-b_{13})+k_7(k_8-b_{23})] \\
&\quad +\frac{p^2}{16}[k_1^2-2k_1k_2-4b_{33}^2]-\frac{p}{2}(\delta_1-\delta_2) \\
a_{10} &= \frac{p+2}{16}[p(k_1^2-4b_{33}^2)+2(k_5k_6+k_7k_8)-8(\delta_1-\delta_2)]+\frac{1}{32}[(p+2)^2k_2^2 \\
&\quad +(p^2+8p-8)k_3^2]-\frac{1}{8}(p-2)(k_7b_{23}+k_5b_{13})-\frac{1}{2}(k_6b_{13}+k_8b_{23})-\frac{1}{8}p^2k_1k_2 \\
a_{11} &= \frac{1}{128}(7p^2+6p-8)(k_1^2+4b_{33}^2)+\frac{1}{256}[(7p^2-8)(k_2^2+k_3^2)+2p(11k_2^2+27k_3^2)] \\
&\quad +\frac{1}{64}(p^2-6p-8)(k_5^2+k_7^2)+\frac{1}{32}(p-2)[(p-4)k_1b_{33}+4(k_5b_{13}+k_7b_{23})] \\
&\quad -\frac{1}{4}(k_6^2+k_8^2-k_6b_{13}+k_8b_{23})-\frac{1}{2}p(\delta_1+\delta_2)-\frac{7}{64}p^2k_1k_2
\end{aligned}
$$

5 Numerical Results and Conclusions

Because of the expressions of the elements of the matrix **A**, it is difficult to solve the eigenvalues of matrix defined in Eq.(34), especially in the case that n>2. We now investigate the numerical approximations of the eigenvalues to show the convergence rate.

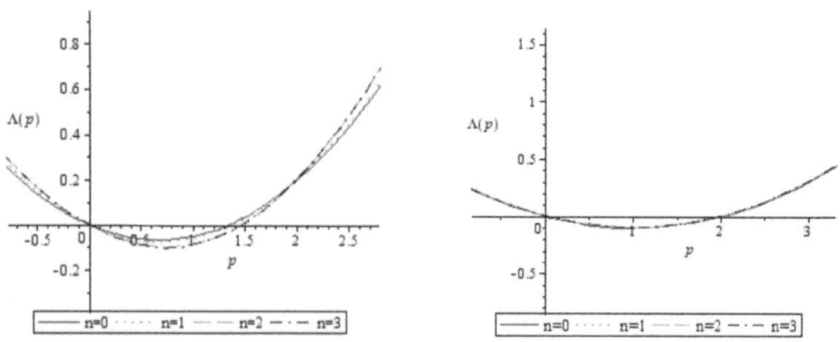

(a) (b)

Fig. 1 Variation of moment Lyapunov exponent with n and p for the case: (a), ε=0.1, $k_1=k_5=k_7=2$, $k_4=-2$, $k_2=k_3=k_6=k_8=0$, $\delta_2=\delta_2=1$; (b), ε=0.1, $k_1=k_4=k_6=k_8=2$, $k_2=k_3=k_5=k_7=0$, $\delta_2=\delta_2=1$

In Fig.1(a), it appears that the deviation of approximate eigenvalues $g(p)$ becomes less and less with the increase of n, and in the cases of n=2 and n=3, the values of $g(p)$ almost coincide. In Fig.1(b), it is clear that the first four order approximations completely coincide, which means that this method is efficient. The two figures also indicate that the system is almost-sure stable because the value of $g'(0)$, which is in fact the maximal Lyapunov exponent for the stochastic system, is negative, but for sufficient large p, the system is unstable in pth moment sense.

In addition, the stability boundaries shown in Fig.2(a) and (b) are obtained by setting the moment Lyapunov exponent to be equal zero. By comparing these regions, it is seen that these stability regions are strongly dependent on the value of p. In Fig.2(a), the differences among the stability boundaries in the cases of the first three order approximations are neglectable, and all the lines virtually coincide at p=2, and in Fig.2(b), all the three-order approximations of the stability boundary coincide.

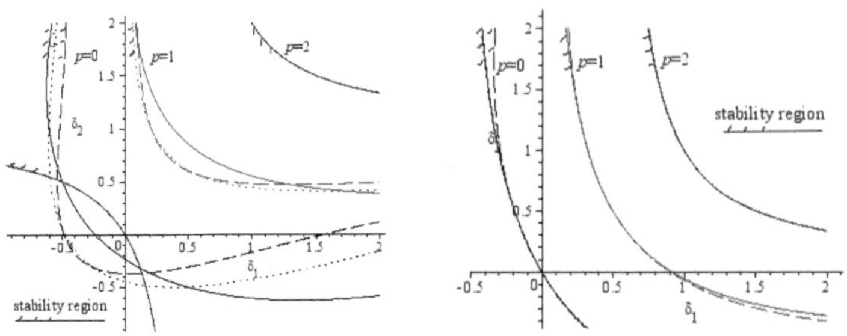

(a) (b)

Fig. 2 Variation of stability region with n and p for the case: (a), ε=0.1, $k_1=k_5=k_7=2$, $k_4=-2$, $k_2=k_3=k_6=k_8=0$, n=1—Solidline, n=2—Dotline, n=3—Dashline; (b), ε=0.1, $k_1=k_4=k_6=k_8=2$, $k_2=k_3=k_5=k_7=0$, n=1—Solidline, n=2—Dotline, n=3—Dashline

In this paper, the pth moment Lyapunov exponent of a codimension two bifurcation system, that is on a three dimensional center manifold and is excited parametrically by a white noise, is investigated. By applying the procedure exposed by in Wedig [11] and the perturbation method, an eigenvalue problem of the first expansion of the moment Lyapunov exponent is obtained. The eigenvalue problem is then solved by Fourier cosine series. The numerical results of the pth moment Lyapunov exponent are illustrated and in addition, the stability regions of pth moment are also obtained numerically.

Acknowledgments. This research was supported by the National Natural Science Foundation of China (Grant No. 10672074).

References

1. Arnold, L.A.: A formula connecting sample and moment stability of linear stochastic systems. SIAM Journal of Applied Mathematics 44, 793–802 (1984)
2. Arnold, L., Kliemann, W., Oejeklaus, E.: Lyapunov exponents of liear stochastic systems. In: Lyapunov exponents. Lecture Notes in Math., vol. 1186, pp. 85–125. Springer, New York (1986)
3. Arnold, L., Oejeklaus, E.: Almost sure and moment stability for linear Ito equations. In: Lyapunov exponents. Lecture Notes in Math., vol. 1186, pp. 129–159. Springer, New York (1986)
4. Arnold, L., Doyle, M.M., Namachchivaya, N.S.: Small noise expansion of moment Lyapunov Exponents for two dimensional systems. Dynamics an Stability of Systems 12(3), 187–211 (1997)
5. Namachchivaya, N.S., Roessel, H.J., Van Doyle, M.M.: Moment Lyapunov exponent for two coupled oscillators driven by real noise. SIAM Journal of Applied Mathematics 56, 1400–1423 (1996)
6. Khasminskii, R., Moshchuk, Z.N.: Moment Lyapunov exponent and stability index for linear conservative system with small random perturbation. SIAM Journal of Applied Mathematics 58(1), 245–256 (1998)
7. Namachchivaya, N.S.: Moment Lyapunov exponent and stochastic stability of two coupled oscillators driven by real noise. Journal of Applied Mechanics 68, 903–914 (2001)
8. Arnold, L., Papanicolaou, G., Wihstutz, V.: Asymptotic analysis of the Lyapunov exponents and rotation numbers of the random oscillator and applications. SIAM Journal of Applied Mathematics 46, 427–450 (1986)
9. Xie, W.C.: Moment Lyapunov exponents of a two dimensional system under real noise excitation. Journal of Sound and Vibration 239(1), 139–155 (2001)
10. Liu, X.B., Liew, K.M.: Maximal Lyapunov exponent of a codimension two bifurcation system excited by a white noise. International Journal of Non-linear Mechanics 40, 653–668 (2005)
11. Wedig, W.: Lyapunov exponent of stochastic systems and related bifurcation problems. In: Stochastic Structural Dynamics—Progress in Theory and Applications, pp. 315–327. Elsevier Applied Science, New York (1988)
12. Guckenheimer, G., Holmes, P.: Nonlinear Oscillations, Dynamical Systems, and Bifurcations of Vector Fields. Springer, New York (1983)

Bifurcation Analysis of Stochastic Non-smooth Systems

Nicole Gaus and Carsten Proppe

Institute of Engineering Mechanics Karlsruhe Institute of Technology,
Kaiserstr. 10, 76131 Karlsruhe, Germany

Abstract. Non-smooth systems with stochastic parameters are important models e.g. for brake and cam follower systems. They show special bifurcation phenomena, such as grazing bifurcations. This contribution studies the influence of stochastic processes on bifurcations in non-smooth systems. As an example, the classical mass on a belt system is considered, where stick-slip vibrations occur. Measurements indicate that the friction coefficient which plays a large role in the system behavior is not deterministic but can be described as a friction characteristic with added white noise. Therefore, a stochastic process is introduced into the non-smooth model and its influence on the bifurcation behavior is studied. It is shown that the stochastic process may alter the bifurcation behavior of the deterministic system.

Keywords: Bifurcation, Non-smooth system, Lyapunov exponent.

1 Introduction

The system consists of a mass (m) on a belt, see Fig. 1. The belt moves with constant velocity v_0. The mass is attached to the surrounding by a spring (spring constant c) and a damper (damping constant d). It is externally excited by the periodically driven force $Q_0\cos(\Omega t)$. The friction force $F_R=\mu(v_{rel})F_N \mathrm{sign}(v_{rel})$ changes its sign due to the direction of the relative velocity $v_{rel} = v_0 - \dot{x}$. As normal force F_N merely the dead weight is considered.

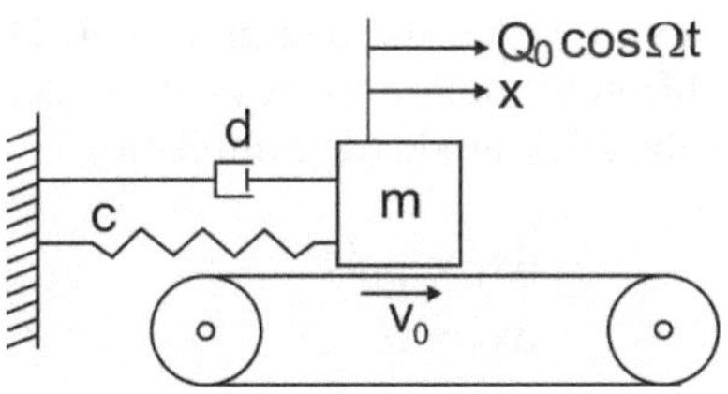

Fig. 1 The model

W.Q. Zhu et al. (Eds.): Nonlinear Stochastic Dynamics & Control, IUTAM BOOK SERIES 29, pp. 201–209.
springerlink.com

Two different motion modes may occur, the stick mode where the mass sticks to the belt and the slip mode where the mass slips due to the restoring force. The conditions for the stick mode are $\dot{x} = v_0$ and $|F_R| > |cx + d\dot{x} - Q_0 \cos(\Omega t)|$.

The stochastic friction coefficient $\mu(v_{rel})$ is modeled as a sum of Gaussian white noise $\xi\tau$ with intensity μ_S and a deterministic friction coefficient $\mu_0(v_{rel})$ which follows a characteristic like Coulombs or Stribecks law

$$\mu(v_{rel}) = \mu_0(v_{rel}) + \mu_S \, \xi_\tau. \tag{1}$$

The deterministic friction coefficient following Stribecks law has been obtained by Hinrichs [1] from experimental data:

$$\mu_0(v_{rel}) = \frac{0.3}{1 + 1.42\,|v_{rel}|} + 0.1 + 0.01 v_{rel}^{\,2} \tag{2}$$

A realisation of the stochastic friction characteristic is shown in Fig. 2. For the stick phase $v_{rel}=0$ the friction characteristic $\mu(v_{rel})=0.4$ is modeled to be deterministic.

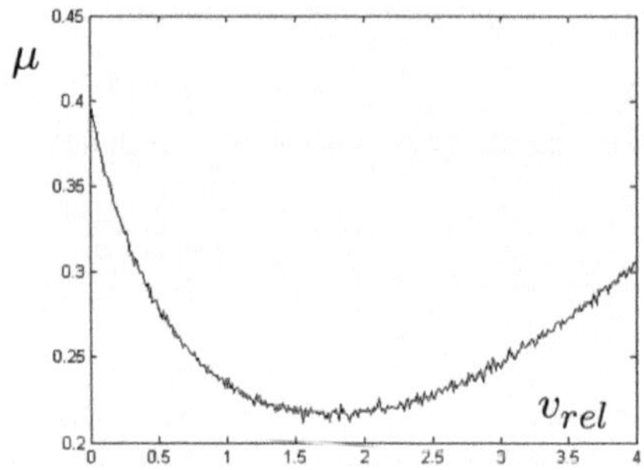

Fig. 2 Stochastic friction characteristic

Introducing the stochastic friction characteristic $\mu(v_{rel})$, the normalized time $\tau=\omega_0 t$, the eigenfrequency $\omega_0^2=c/m$, the damping ratio $D=d/(2m\omega_0)$ and frequency ratio $\eta=\omega_0/\Omega$, the two different motion modes can be described by first order differential equations. For the stick mode (deterministic)

$$\begin{aligned} dx_1 &= \gamma_0 d\tau \\ dx_2 &= 0 \end{aligned} \tag{3}$$

where γ_0 is the ratio of belt velocity to eigenfrequency $\gamma_0=v_0/\omega_0$. For the slip mode (stochastic)

$$\begin{aligned} dX_1 &= X_2 d\tau \\ dX_2 &= [-X_1 - 2DX_2 + F\mu_0 \text{sign}(v_0 - \omega_0 X_2) + Q\cos(X_3)]d\tau \\ &\quad + [F\mu_S \text{sign}(v_0 - \omega_0 X_2)]dW_\tau \\ dX_3 &= \eta d\tau \end{aligned} \tag{4}$$

where $F=F_N/c$, $Q=Q_0/c$ and dW_τ is the increment of a Wiener process. The numerical integration is done by the Euler- Maruyama method, see [2].

In the phase plane shown in Fig. 3, we can see that the mass has the constant velocity of the belt $\dot{x} = \gamma_0$ during the stick phase. Up to the point where the stick condition no longer holds and the slip phase starts. Once the velocity is the same as the belt velocity and the stick condition is fulfilled the mass reattaches to the belt and moves with constant velocity $\dot{x} = \gamma_0$ again. The motion is described by a limit cycle.

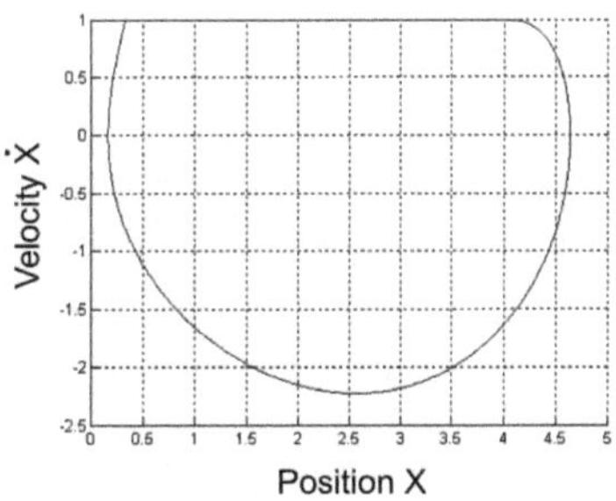

Fig. 3 Phase plane (γ_0=1, ω_0=1, Q=0, D=0, μ_S=0)

2 Damping Induced Bifurcation

First the external excitation is not considered. The system parameters are ω_0=1 and D=0. The initial conditions are $X(0) = 0$, $\dot{X}(0) = \gamma_0$. If thc damping coefficient is increased, as can be seen in Fig. 4, the limit cycle becomes narrower and the difference between the position $X_{sl\text{-}st}$ (changing point from slip to stick) and $X_{st\text{-}sl}$ (changing point from stick to slip) is reduced, up until the point where the limit cycle no longer exists. In this case, here for D=0.4 there is a fixed point with the velocity $\dot{X} = 0$. At a damping coefficient D=0.376 the distance is almost zero. The limit cycle still exists, see Fig. 5.

In the stochastic case for μ_S=0.005 however, it can be seen in the phase plane in Fig. 6, that only some trajectories still reach the limit cycle but most of them are moving to the fixed point. The mean of all trajectories therefore also approaches to the fixed point thus in the mean the limit cycle no longer exists.

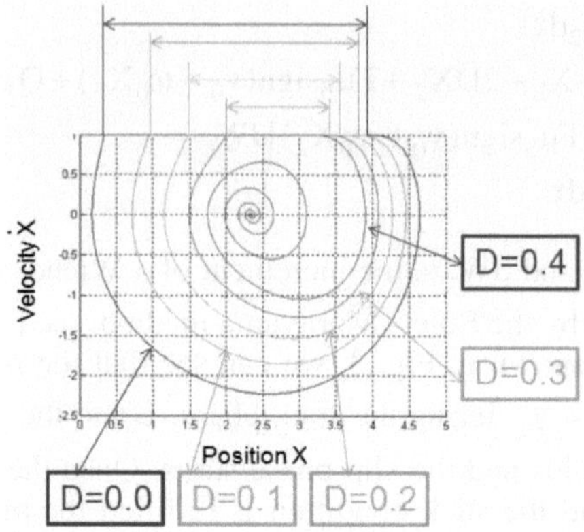

Fig. 4 Limit cycle to fixed point

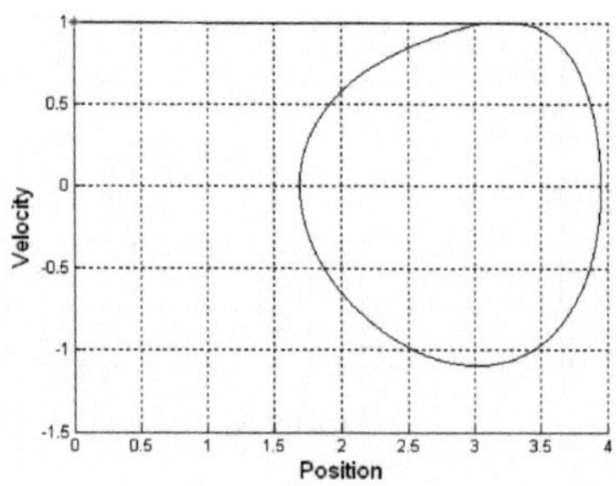

Fig. 5 Phase plane D=0.376 (γ_0=1, $\boldsymbol{\omega}_0$=1, Q=0, μ_s=0)

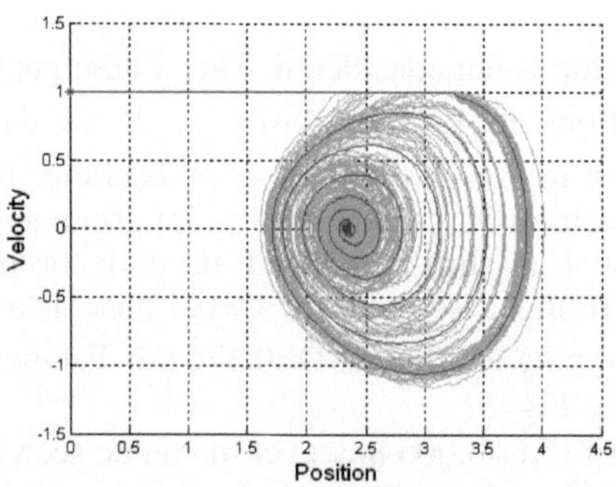

Fig. 6 Phase plane for D=0.376: all trajectories (grey), mean value (black)

3 Excitation Induced Bifurcation and Sensitivity to Initial Conditions

Another bifurcation can be examined if the external excitation is considered. Now higher periodic solutions can occur in the deterministic case as well as in the stochastic one, see Fig. 7. Damping is neglected, the eigenfrequency $\omega_0=1$, the initial conditions are $X(0)=0$, $\dot{X}(0)=\gamma_0$ and the amplitude of the external excitation is Q=1m. In grey the stochastic trajectories are shown ($\mu_s=0.005$), in black the deterministic ones.

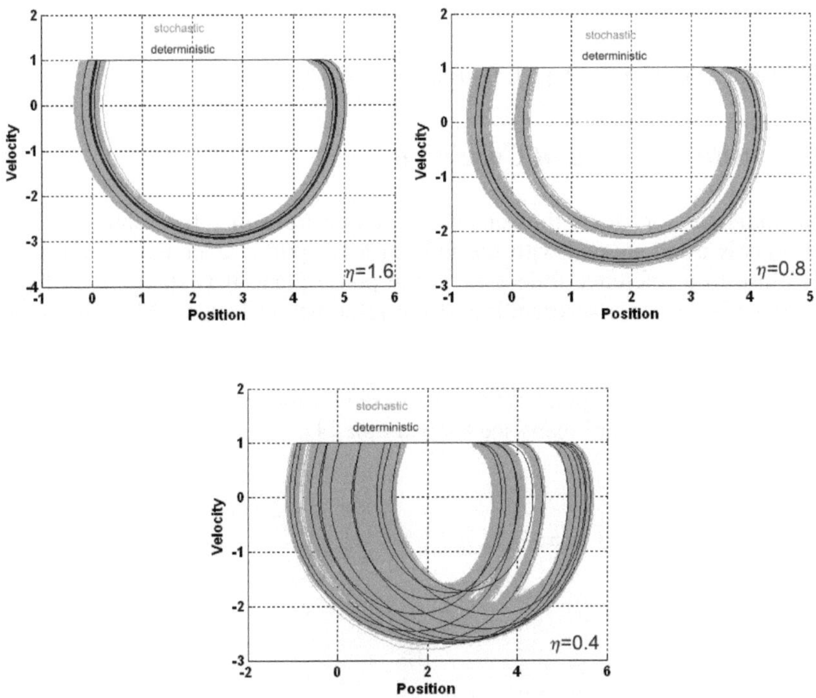

Fig. 7 Phase plane for η=1.6, η=0.8, η=0.4 (D=0, $\gamma_0=1$, Q=1m)

For the frequency ratio η=0.4 the trajectory is chaotic, for η=0.8 it has two periods and for η=1.6 one period. In Fig. 8 the Poincaré diagram of the coordinate $X_{sl\text{-}st}$ (changing point from slip to stick) is shown. The different behavior (one-periodic, higher periodic, chaos) is visible in the deterministic case (black) as well as in the stochastic case (grey).

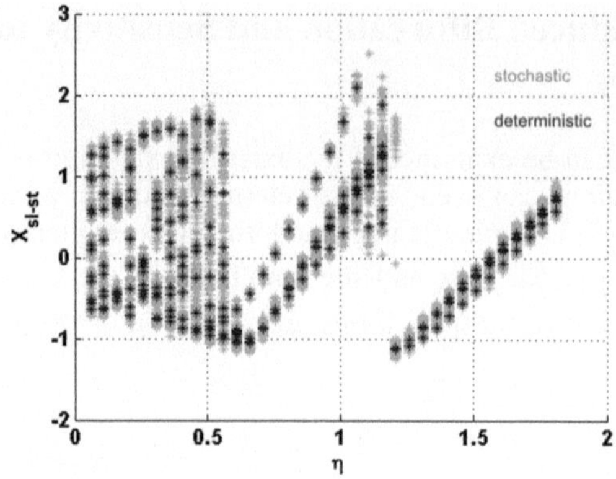

Fig. 8 Poincaré diagram (D=0, γ_0=1, Q=1m, μ_s=0.005)

3.1 Sensitivity to Initial Condition

As an indicator for chaos the sensitivity of one trajectory to a change in initial conditions is often studied with the help of a Lyapunov exponent. If a trajectory starts close to a reference trajectory in the phase plane it is stable, if it does not move outside a certain surrounding of the reference trajectory. More information can be found for example in [3]. Here the distance is measured not for the entire trajectory but for the position of the changing point slip to stick $X_{sl\text{-}st}$. The sensitivity is examined by a Lyapunov exponent based on the Poincaré map. For more information about Poincaré maps see for example [4].

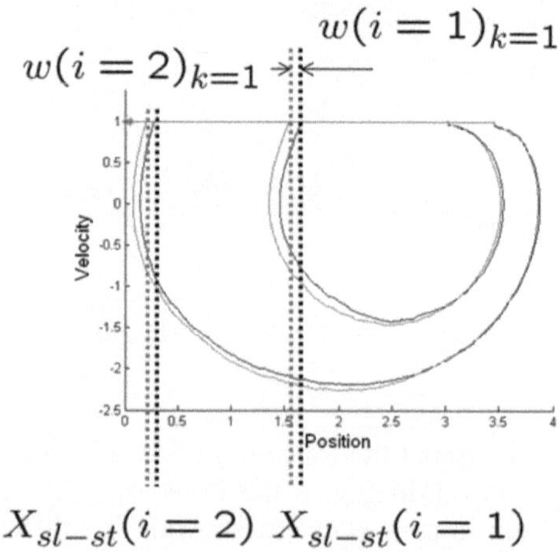

Fig. 9 Distance of $X_{sl\text{-}st}$ of disturbed trajectory to reference trajectory

First a reference trajectory is calculated. Then a small disturbance is put on the initial conditions and the trajectory is calculated once more. As shown in Fig. 9

the difference w of the position at the slip to stick changing point of the disturbed trajectory to the one of the reference trajectory is calculated. This distance will either decrease or increase during the motion which is described by the Lyapunov exponent

$$\lambda = \lim_{i\to\infty}\frac{1}{i}\ln\frac{w(i)}{w(1)} \tag{5}$$

where for a positive exponent ($\lambda>0$), the trajectory is unstable, and for a negative exponent ($\lambda<0$), the trajectory is stable.

In the stochastic case the reference and the disturbed trajectory are calculated *M* times and the mean value of each difference is calculated with

$$E(W(i)) = \frac{1}{M}\sum_{i=1}^{M} w(i,k). \tag{6}$$

The Lyapunov exponent can be calculated from the mean value of the differences by

$$\lambda = \lim_{i\to\infty}\frac{1}{i}\ln\frac{E(W(i))}{E(W(1))}. \tag{7}$$

For longer time spans and thus a large number of changes the Lyapunov exponent should approach asymptotically a constant value.

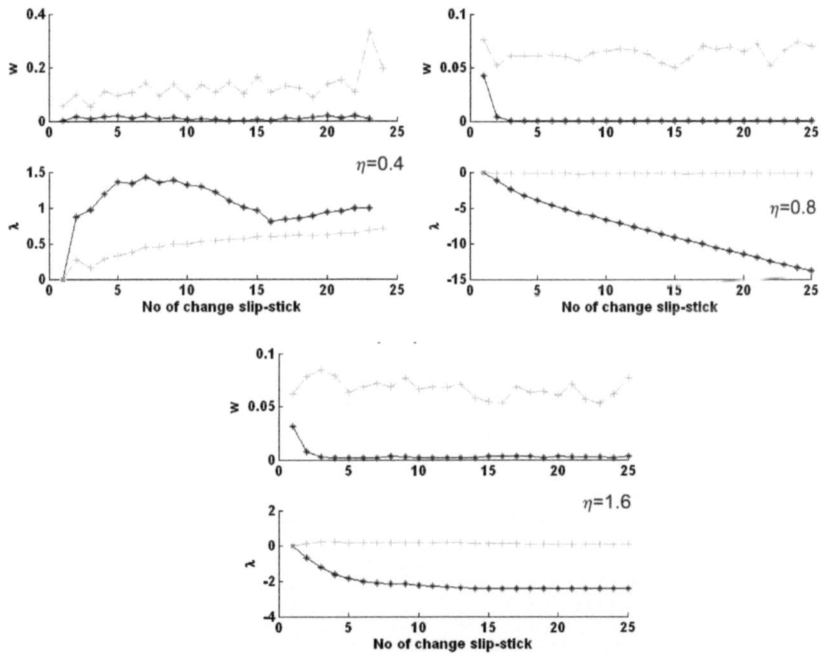

Fig. 10 Distance w and Lyapunov exponent **λ** for $\eta=0.4$, $\eta=0.8$, $\eta=1.6$ ($D=0$, $\gamma_0=1$, $Q=1m$)

The comparison of distance w and the development of λ are shown in Fig.10 for the deterministic and the stochastic system with three different frequency ratios and a simulation time of 200 seconds. Again the stochastic case is colored grey and the deterministic one black. The stochastic and the deterministic Lyapunov exponent have the same sign for these frequency ratios. For η=0.6 the stochastic case is unstable while the deterministic one is stable, see Figs. 11 and 12. For η=1.2 we can see in the Poincare diagram in Fig. 8 that the deterministic trajectory is one periodic but the stochastic one is two-periodic. However in Fig. 13 it can be seen that the deterministic case is unstable. If a small disturbance is added the deterministic trajectory is also two periodic.

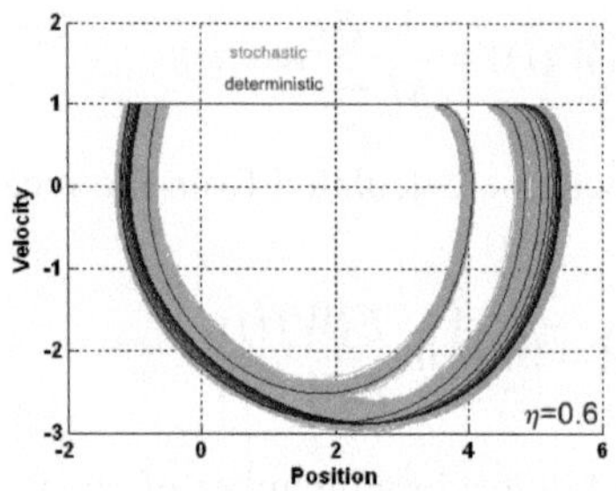

Fig. 11 Phase plane for η=0.6

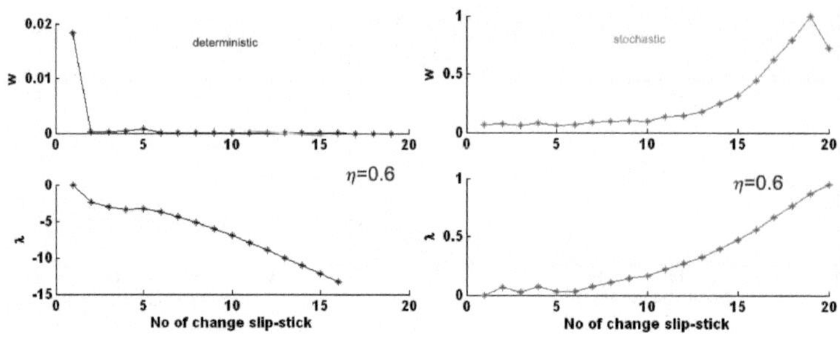

Fig. 12 Distance w and Lyapunov exponent **λ** for η=0.6 (D=0, γ_0=1, Q=1m)

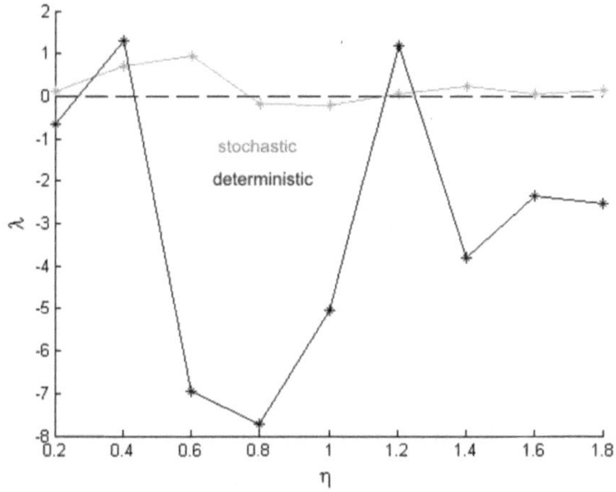

Fig. 13 Lyapunov exponent for η=0.2-1.8 (D=0, γ_0=1, Q=1m)

4 Conclusions

The non-smooth mass on a belt system shows several bifurcation behaviors:

- Considering damping we can see that the stable limit cycle in the deterministic case does not exist anymore if the stochastic friction coefficient is introduced. In the mean the trajectories meet in a stable fixed point.
- Considering the external excitation the bifurcation from one periodic to two periodic solutions to chaos can be detected with a Poincare diagram. The behavior is similar for the stochastic case and the deterministic one. However looking at the sensitivity, in this case a Lyapunov exponent of the switch point from slip to stick, we can see that for some excitation parameters the trajectories are stable in the deterministic case but unstable in the mean of the stochastic case or vice versa.

References

[1] Hinrichs, N.: Reibschwingungen mit selbst- und Fremderregung: Experiment, Modellierun und Berechnung. VDI Verlag GmbH, Düsseldorf (1997)
[2] Arnold, L.: Stochastische Differentialgleichungen. Oldenbourg, München (1973)
[3] Moon, F.C.: Chaotic Vibrations. Wiley, New York (2004)
[4] Nayfeh, A.H., Balachandran, B.: Applied Nonlinear Dynamics. Wiley, New York (1995)

4 Conclusions

Part 3
Resonance and Synchronization of Nonlinear Stochastic Systems

Synchronization Behavior of a Clustered Neuronal Network in a Noisy Environment

X.J. Sun [1] and Q.S. Lu[2,*]

[1] Zhou Pei-Yuan Center for Applied Mathematics, Tsinghua Univeristy, Beijing 100084, China
[2] Department of Dynamics and Control, Beihang University, Beijing 100191, China
e-mail: qishaolu@hotmail.com

Abstract. Synchronization behavior is discussed in a clustered neuronal network, in which Rulkov maps with additive noise are applied as building blocks. Additive noise applied here is used to make the Rulkov map generate spiking activities. It is revealed that clustered structure of networks in noisy environments is able to make the subnetwork more synchronous but suppress the synchrony of the entire network meanwhile. The effects of the cluster number, the coupling strength and the noise intensity on spatiotemporal synchronization in the network are discussed. The obtained results are helpful for understanding the clustered structure in cortical systems of the brain from a new viewpoint of synchronization.

Keywords: Neuronal network, Synchronization, Noise, Pattern.

1 Introduction

Synchronization is ubiquitous in nature and plays an important role in several different situations such as predator-prey process, birds migratory and heartbeats. Synchronization has also been experimentally observed in nervous systems, such as hippocampal CA1 area [1], retina [2], and striate cortex [3]. Synchronization involves at least two elements in interaction. Several works addressing synchronization of large populations, which constitutes a complex network, have been intensively studied [4-7] and comprehensively reviewed [8]. In nervous systems, neurons are connected via synapses-forming complex neuronal networks. Synchronization of neuronal complex networks has been discussed in many articles. The obtained results indicate that various factors, such as coupling connections and external stochastic forces, may have important influences on synchronization behavior of neuronal complex networks. For example, Hasegawa found that with introducing the weak heterogeneity to regular networks, the synchronization may be slightly increased for diffusively couplings, while it is decreased for sigmoid couplings [9]. Shi et. Al. showed that noise not only can induce complete synchronization in uncoupled identical neurons but also can enhance the synchronization of weakly coupled neurons [10].

---

* Corresponding author.

W.Q. Zhu et al. (Eds.): Nonlinear Stochastic Dynamics & Control, IUTAM BOOK SERIES 29, pp. 213–220.
springerlink.com

In the human brain, neurons in some communities usually behave synchronously to realize nervous information functions. Nevertheless, as we know, many of disorder diseases, such as Parkinson and epilepsy, are bound up with ideas of widespread rhythmic synchronization of neuronal elements. Thus, synchronization is essential in carrying out nervous functions but is not always a fine phenomenon in nervous systems. A question emerges from our mind, i.e. how our brain can sustain these two paradoxical states simultaneously. In this paper, we devote to giving some explanations on this problem from the perspective of nonlinear dynamics.

2 Mathematical Model

It is commonly accepted that cortical architecture and connections are organized in a hierarchical and clustered (modular) way. Integration of many clusters, including large numbers of neurons forms a particular cortical area. Generally, the connectivity within each individual cluster is dense, while connections between neurons which are in different clusters are sparse. This kind of network is called complex clustered network.

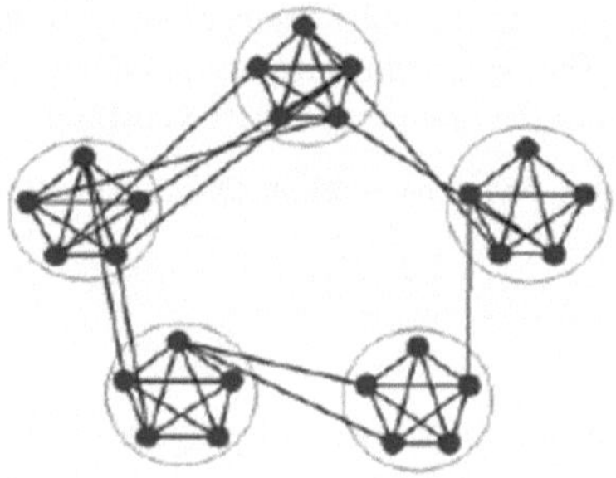

Fig. 1 Illustration of a clustered network with regular subnetworks. Here $N = 25$, $M = 5$, $m = 2$, $p = 0.1$.

In this paper, we consider a clustered network [11], in which N nodes are grouped into M clusters and each cluster contains $n = N/M$ nodes. Nodes are distributed on a ring so that the subnetwork is a regular one. Each node connects to m nearest neighbors in the same cluster, and each pair of nodes in the nearest clusters is connected with probability p . A clustered network model with regular subnetworks is illustrated in Fig.1. In this paper, we take the size of the clustered network N as 200 . Let each node inside a cluster connects to its $m = 8$ nearest neighbors.

The local neuronal model for each node in the clustered network is taken as Rulkov map [12, 13]. In nervous systems, neurons are usually subject to random fluctuations. These random fluctuations arise from many different sources, such as the quasi-random release of neurotransmitter by the synapses, the random switching of ion channels, and most importantly random synaptic input from other neurons. Thus, the existence of noise in dynamics of nervous systems should be concerned. In this paper, we introduce additive noise into the neuronal network. The mathematical equations of the studied network with additive noise terms are presented as follows:

$$\begin{cases} x_i(n+1) = \dfrac{\alpha}{1+x_i^2(n)} + y_i(n) + \eta_i(n) + \varepsilon \sum_{j=1}^{N} a(i,j)(x_j(n) - x_i(n)), \\ y_i(n+1) = y_i(n) - \beta x_i(n) - \gamma, \end{cases} \tag{1}$$

where $x_i(n)$ is the membrane potential of the neuron and $y_i(n)$ is the variation of the ion concentration. n is the iterated time index, ε is the coupling strength, and $a(i,j)=1$ if nodes i and j are connected, otherwise $a(i,j)=0$. $\eta_i(n)$ is the additive noise with

$$\begin{cases} \eta_i(n) = 0, \\ \eta_i(n)\eta_j(m) = 2\sigma\delta_{i,j}\delta_{n,m}, \end{cases} \tag{2}$$

where σ denotes the noise intensity of $\eta_i(n)$. α, β and γ are system parameters. Here β, γ are both taken as 0.001. In this case, the map is governed by a single excitable steady state when $\alpha < 2.0$, and the excitable steady state loses its stability via a Hopf bifurcation when $\alpha > 2.0$. In this paper, we take $\alpha = 1.99$. It means that the studied system will not generate any spiking activities in the absence of noise $\eta_i(n)$. Here, the noise intensity σ is taken as 0.001. Under this noise intensity, the neuronal network can stay in the firing state.

3 Main Results

Firstly, we fix the network topology, namely we fix the probability p as 0.004 and the cluster number M as 2. We take the coupling strength ε as the control parameter to study the effect of ε on the firing activities of the studied neuronal network. Fig.2 shows four spatiotemporal patterns observed in the clustered network for four different coupling strengths. When the neurons are weakly coupled with each other, the clustered neuronal network exhibits chaotic spatiotemporal pattern, as shown by Fig.2(a) with $\varepsilon = 0.0005$. When ε increases a bit to 0.0008, although the clustered neuronal network remains still chaotic (see Fig.2(b)), some black stripes (none in Fig.2(a)) can be observed from the right side of this figure. This implies that there are some ordered parts shown by Fig.2(b), even though the pattern is disordered as a whole. With the coupling strength ε increasing to 0.001 (see Fig.2(c)), we find that the black stripes become more clear and the spatiotemporal chaotic pattern becomes more ordered. When the coupling strength ε increases further to 0.005, the clustered neuronal network exhibits synchronized pattern (see Fig.2(d)). The obtained results show that the clustered neuronal network can stay in synchronous state when the coupling strength ε is large enough.

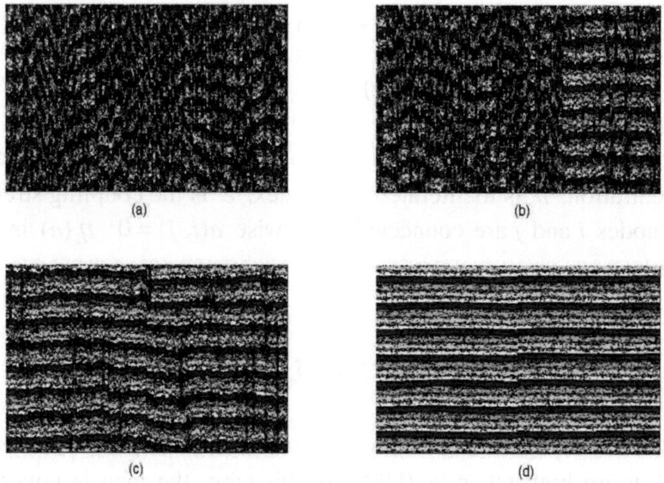

Fig. 2 Space-time plots illustrate the transition from chaotic spatiotemporal patterns to spiking synchronized ones with the coupling strength ε increasing. $\varepsilon = 0.0005$, 0.0008, 0.001 and 0.005, respectively. The color profile is linear, black depicting $x_i(n) = 0$ and white $x_i(n) = -1.7$.

Next, we study the effect of the cluster number M on the firing activities of the studied neuronal network. Namely, we take M as the control parameter. As mentioned before, the size of the studied neuronal network N is taken as 200 and it is required that each neuron connects to its eight nearest neighbors. Thus, 20 is the maximum value of M. In the following, we take the coupling strength $\varepsilon = 0.005$, where the entire clustered neuronal network with two clusters ($M = 2$) is synchronized (see Figs.2(d) and 3(a)). We increase the cluster number M from 2 to 8, the spatiotemporal pattern for $M = 8$ is presented by Fig.3(b). From this figure, we can see that synchronization of the entire clustered neuronal network has been broken down, but neurons which belong to the same cluster are still synchronized with each other. Here, we call this dynamical behavior as cluster-synchronization. In order to understand the meaning of cluster-synchronization better, we plot the time series of four neurons as shown in Fig.4. In this figure, neurons $\#1$ and $\#2$ are randomly chosen from the first cluster, while neurons $\#3$ and $\#4$ are randomly chosen from the second cluster. From this figure, we can see that neurons #1 and #2 (or #3 and #4) inside the first (or second) cluster are synchronized with each other. While neurons #1 and #3, belonging to different clusters, are not synchronized. Figs.3(c) and (d) are the spatiotemporal patterns corresponding to $M = 10$ and $M = 20$, respectively. Similar to Fig.3(b), cluster-synchronization can also be observed from these two figures. Thus, it is conclude from the results shown in Figs.3 and 4 that the synchronized patterns transit to cluster-synchronized ones with M increasing. Namely, the synchronization of the whole clustered neuronal network is weakened, while the synchronization of each cluster still sustain.

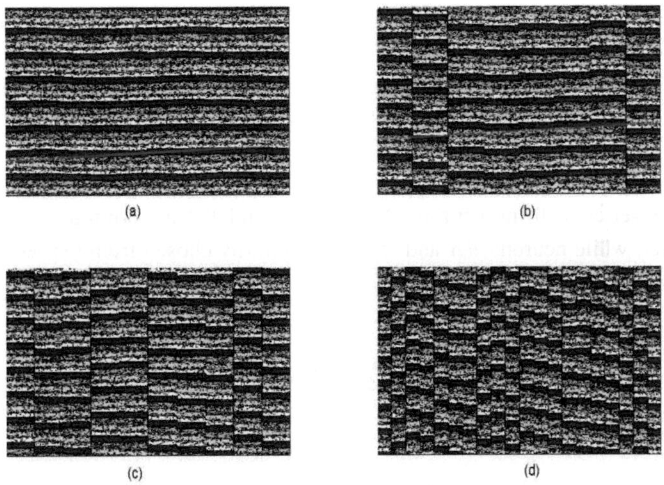

Fig. 3 Space-time plots illustrate transition from spiking synchronized patterns to clustered synchronized ones with the cluster number M increasing. $M = 2, 8, 10$ and 20, respectively. The profile is linear, where black is for $x_i(n) = 0$ and white for $x_i(n) = -1.7$. Here $\varepsilon = 0.005$.

In order to study the degree of spatiotemporal synchronization in the network quantitatively and confirm the above assessments of wave front propagation via space-time plots, we introduce a synchronization parameter S by means of the standard deviation which can be calculated effectively according to [14]

$$S = \frac{1}{T}\sum_{n=1}^{T} S(n), S(n) = \frac{1}{N}\sum_{i=1}^{N}[x_i(n)]^2 - [\frac{1}{N}\sum_{i=1}^{N} x_i(n)]^2 \tag{3}$$

Similarly, we can define an synchronization index S_i to quantify synchrony of the i-th cluster ($i = 1, \cdots, M$) as

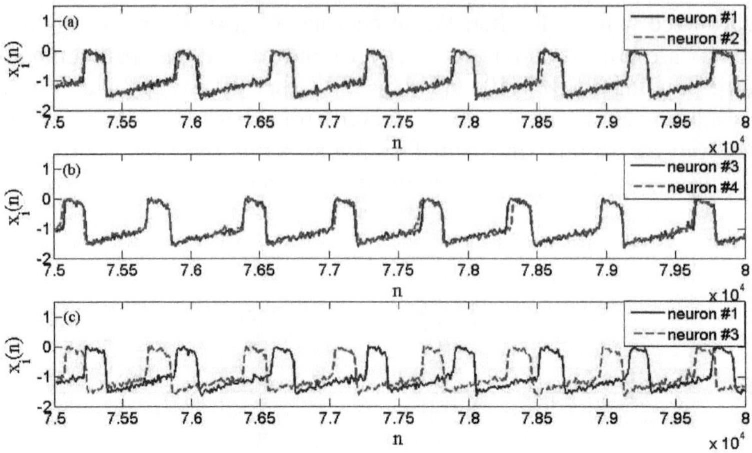

Fig. 4 Time series of firing neurons. Neurons #1 and #2 are arbitrarily chosen from the first cluster, while neurons #3 and #4 are arbitrarily chosen from the second cluster. In this figure, $M = 8$ and $\varepsilon = 0.005$.

$$S_i = \frac{1}{T}\sum_{n=1}^{T} S_i(n), \tag{4}$$

where

$$S_i(n) = \frac{1}{n}\sum_{j=1}^{n} x_{m(i-1)+j}^2 - [\frac{1}{n}\sum_{i=1}^{n} x_{m(i-1)+j}]^2 \tag{5}$$

The indices S and S_i are used to denote the synchronization levels of the entire network and the i-th cluster, respectively.

Variations of S and S_i with respect to the coupling strength ε are plotted in Figs.5(a) and 5(b). These two figures are corresponding to the spatiotemporal patterns shown in Fig.2. From Figs 5(a) and 5(b), we find that both of S and S_i decreases with increasing ε. This is an intuitive result that coupling strength can enhance the synchrony of the network. The influence of the cluster number M on synchronization of the network exhibited in Fig.3 is quantitatively exhibited by Figs.5(c) and 5(d). It is clearly seen that S increases while S_i decreases with M increasing. In other words, synchronization of subnetwork can be enhanced by clustered structures at the expense of desynchronization of the entire network.

Finally, we discuss the effect of the noise intensity on the observed spatiotemporal synchronization phenomena briefly. Fig.6 exhibits the dependence of S (see Fig.6(a)) and S_i (see Fig.6(b)) on the coupling strength ε and the noise intensity σ, respectively. As shown in Fig.6, we can see that S and S_i increases when the noise intensity σ becomes larger for fixed coupling strength ε. Stronger coupling is required to make the entire clustered neuronal network (or subnetworks) synchronized for larger noise intensities. This indicates that additive noise may reduce the synchrony of both the entire clustered neuronal network and its clusters.

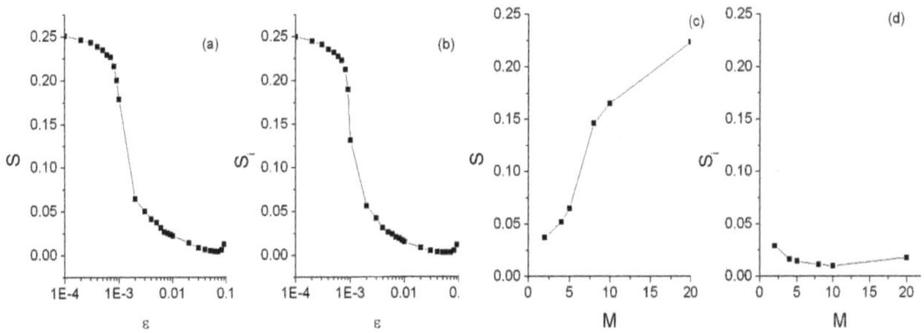

Fig. 5 Variations of synchronization index S (a) and S_i (b) with respect to ε and variations of synchronization index S (c) and S_i (d) with respect to M

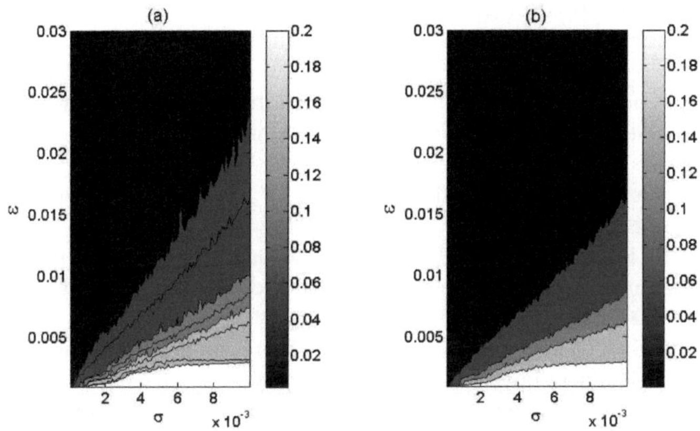

Fig. 6 Contour plots of S (a) and S_i (b) on the coupling strength ε and the noise intensity σ. The cluster number $M = 2$ and the probability $p = 0.004$. The profile is linear, where black is for $S = 0$ (a) or $S_i = 0$ (b) and white for $S = 0.2$ (a) or $S_i = 0.2$ (b).

4 Summary

Synchronization behavior in a clustered neuronal network with additive noise is discussed to reveal the role played by the clustered structure of networks. The effects of the coupling strength and the cluster number are considered. Clustered structures have been found in several cortical systems and may be ideal for achieving high functional complexity [4]. These results can help us to understand the existence of clustered structure in cortical systems from the viewpoint of synchronization.

Acknowledgements. This work was supported by the National Natural Science Foundation of China Nos. 10872014 and 10972018. X.J. Sun is also thankful for support from the China Postdoctoral Science Foundation Project (Fund No. 20090460337).

References

[1] Fujiwara-Tsukamoto, Y., Isomura, Y., Nambu, A., Takada, M.: Excitatory GABA input directly drives seizure-like rhythmic synchronization in mature hippocampal CA1 pyramidal cells. Neurosci. 119, 265–275 (2003)

[2] Neuenschwander, S., Castelo-Branco, M., Singer, W.: Synchronous oscillations in cat retina. Vision Research 39, 2485–2497 (1999)

[3] Gray, C.M., Engel, A.K., Konig, P., Singer, W.: Synchronization of oscillatory neuronal responses in cat striate cortex: temporal properties. Visual Neurosci. 8, 337–347 (1992)

[4] Barahona, M., Pecora, L.M.: Synchronization in small-world systems. Phys. Rev. E 89, 54101 (2002)

[5] Zhou, C.S., Motter, A.E., Kurths, J.: Universality in the synchronization of weighted random networks. Phys. Rev. Lett. 96, 34101 (2006)

[6] Synchronization in complex networks with age ordering. Phys. Rev. Lett. 94, 138701 (2005)

[7] Wang, X.F., Chen, G.R.: Synchronization in scale-free dynamical networks: robustness and fragility. IEEE Transactions on Circuits and systems I: Fundation Theory and applications 49, 54–62 (2002)

[8] Arenas, A., Diaz-Guilera, A., Kurths, J., Moreno, Y., Zhou, C.S.: Synchronization in complex networks. Phys. Rep. 469, 93–153 (2008)

[9] Zhigulin, V.P., Robinovich, M.I., Huerta, R., Abardanel, H.D.I.: Robustness and enhancement of neural synchronization by activity-dependent coupling. Phys. Rev. E 67, 21901 (2003)

[10] Shi, X., Lu, Q.S.: Coherence resonance and synchronization of Hindmarsh-Rose neurons with noise. Chinese Physics 14, 1088–1094 (2006)

[11] Huang, L., Lai, Y.C., Gatenby, R.A.: Alternating synchronizability of complex clustered networks with regular local structure. Phys. Rev. E 77, 16103 (2008)

[12] Rulkov, N.F.: Regularization of synchronized chaotic bursts. Phys. Rev. Lett. 86, 183–186 (2001)

[13] Rulkov, N.F.: Modeling of spiking-bursting neural behavior using two-dimensional map. Phys. Rev. E 65, 41922 (2002)

[14] Gao, Z., Hu, B., Hu, G.: Stochastic resonance of small-world networks. Phys. Rev. E 65, 16209 (2001)

Stochastic Parameter Resonance of Road-Vehicle Systems and Related Bifurcation Problems

Walter V. Wedig

Institut für Technische Mechanik, Universität Karlsruhe
76128 Karlsruhe, Germany
e-mail: wwedig@t-online.de

Abstract. The paper investigates stochastic dynamics of road-vehicle systems and related bifurcation problems. The ride on rough roads generates vertical car vibrations whose root-mean-squares are resonant for critical car speeds and vanish when the car velocity is increasing, infinitely. These investigations are extended to wheel suspensions with progressive spring characteristics. For weak but still positive damping, the car vibrations become unstable when the velocity reaches the parameter resonance near twice the critical speed bifurcating into stochastic chaos of larger non-stationary car vibrations.

Keywords: Nonlinear road-vehicle systems, random road models, covariance equations, critical car speeds, parameter resonances, stability, chaos.

1 Introduction to the Problem

To introduce road-vehicle systems of interest (see e.g. [1]), consider the quarter car model with one degree of freedom, shown in Fig. (1). The car is riding on randomly shaped road surfaces, that generates vertical car vibrations described by the following system equations [2]:

$$\ddot{X}_t + 2D\omega_1\dot{R}_t + \omega_1^2(1+\gamma R_t^2)R_t = 0, \qquad \gamma \geq 0, \tag{1}$$

$$dZ_s = -\Omega Z_s ds + \sigma dW_s, \qquad S_z(w) = \frac{\sigma^2}{\Omega^2 + w^2}\,. \tag{2}$$

In Eq. (1), ω_1 is the natural frequency of the car and D is its damping measure. The parameters γ determine the cubic-progressive spring characteristic of the wheel suspension. X_t describes the absolute vertical car vibration in dependence on time t and R_t denotes the associated relative motion, given by $R_t = X_t - Z_t$. The base excitation Z_s of the car is defined by Eq. (2) in dependence on the way coordinate s where Ω is a fixed road or way frequency and σ denotes the noise intensity of the Wiener increments dW_s. In the stationary case, Z_s possesses the variance $\sigma_z^2 = \sigma^2/(2\Omega)$ and the way spectrum $S_z(w)$, noted in Eq. (2).

W.Q. Zhu et al. (Eds.): Nonlinear Stochastic Dynamics & Control, IUTAM BOOK SERIES 29, pp. 221–228.
springerlink.com

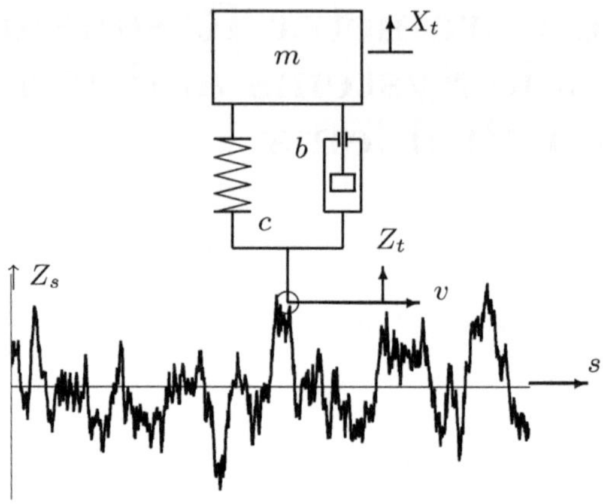

Fig. 1 Scheme of road-vehicle systems

The transformation from way to time is performed by means of the car speed v applying the increments $ds = vdt$ and $dW_s = \sqrt{v}\,dW_t$ with the expectations $E(dW_s^2) = ds$ and $E(dW_t^2) = dt$. Therewith, Eq. (2) yields

$$dZ_t = -\Omega v Z_t dt + \sigma\sqrt{v}dW_t, \qquad S_z(\omega) = \frac{\sigma^2 v}{\omega^2 + (\Omega v)^2}\,. \tag{3}$$

The integration of the time spectrum $S_z(\omega)$ over all frequencies $|\omega| < \infty$ leads to the same variance $\sigma_z^2 = \sigma^2/(2\Omega)$, already noted above. Obviously, this road spectrum (see also [3]) is vanishing for sufficiently slow and fast car speeds, respectively. Its variance, however, is independent on speed.

2 Spectral Analysis of Road-Vehicle Systems

In the linear case that $\gamma, \delta = 0$, the complex transfer functions are applied to Eq. (1) and (3) in order to obtain the complete response spectrum

$$S_x(\omega) = \frac{\sigma^2 v[\omega_1^4 + (2D\omega_1\omega)^2]}{[(\omega_1^2 - \omega^2)^2 + (2D\omega_1\omega)^2](\omega^2 + \Omega^2 v^2)}\,. \tag{4}$$

It is valid for all frequencies $|\omega| < \infty$. The square root of the response spectrum related to the excitation spectrum is plotted in Fig. (2) versus the frequency ratio ω/ω_1 for three damping values. According to [4], the applied frequency scaling is linear with $\omega/\omega_1 = \mu$ in the under-critical range and rational with $\omega/\omega_1 = 1/(2-\mu)$ for over-critical frequencies. Note, that the spectral diagram in Fig. (2) coincides exactly with the resonance diagram of cars under harmonic base excitations showing no amplification in the low frequency range, high resonances near the critical frequency ratio $\omega/\omega_1 = 1$ and complete absorption when the spectral frequencies tend to infinity.

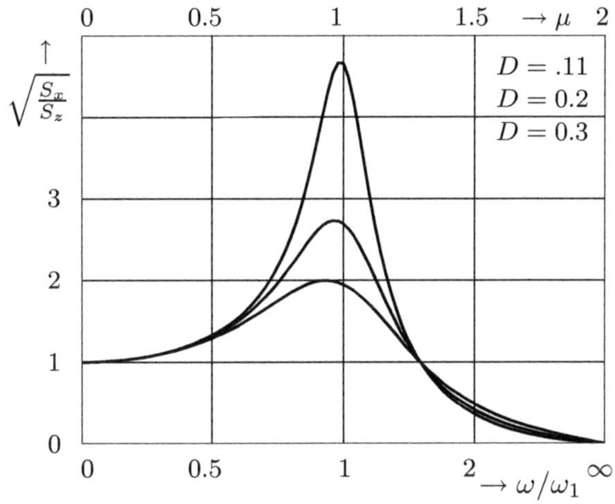

Fig. 2 Resonance diagram of related spectra

The integration of the power spectrum (4) over all frequencies leads to the following rms-ratio of the stationary car response and its base excitation:

$$\frac{\sigma_x}{\sigma_z} = \sqrt{\frac{2D + (1+4D^2)\nu}{2D[1+(\nu+2D)\nu]}}\,, \qquad \text{with } \nu = v\frac{\Omega}{\omega_1}\,. \tag{5}$$

Herein, the dimensionless parameter ν is proportional to the car speed v times the ratio of road and car frequency. The calculated rms ratio in Eq. (5) is plotted in Fig. (3) for three different dampings showing similar resonance properties, as above; i.e. the rms-ratio is one and zero for sufficiently low or high car speeds, respectively. In between both limiting speeds, there is a critical car speed near $\nu = 1$. However, due to the broad-banded base excitation applied, the resonance range around the critical speed is lower but much broader in comparison with that one of harmonic base excitations.

3 Covariance Analysis of Road-Vehicle Systems

To derive the same rms-ratio, as already noted in Eq. (5), by means of the Lyapunov matrix equation, the equation (1) of motion is rewritten into a first order increment equation system. For these purposes, the velocity $Y_t = \dot{X}_t/\omega_1$ is introduced as well as dimensionless time and noise by $d\tau = \omega_1 dt$ and $dW_\tau = \sqrt{\omega_1}dW_t$. Eliminating the displacement X_τ, the state variables Z_τ, R_τ, Y_τ are related to σ_z to obtain the state equations, as follows:

$$\begin{aligned} d\bar{Z}_\tau &= -\nu\bar{Z}_\tau d\tau + \sqrt{2v}\,dW_\tau\,, \qquad & Z_\tau = \sigma_z\bar{Z}_\tau, \quad \bar{\gamma} = \gamma\sigma^2, \\ d\bar{R}_\tau &= (\bar{Y}_\tau + \nu\bar{Z}_\tau)d\tau - \sqrt{2v}\,dW_\tau\,, \qquad & (R_\tau, Y_\tau) = \sigma_z(\bar{R}_\tau, \bar{Y}_\tau), \\ d\bar{Y}_\tau &= -(1+\bar{\gamma}\bar{R}_\tau^2)\bar{R}_\tau d\tau - 2D(\bar{Y}_\tau + \nu\bar{Z}_\tau)d\tau + 2D\sqrt{2v}\,dW_\tau\,. \end{aligned} \tag{6}$$

For the linear case that $\bar{\gamma} = 0$, these equations are written into the vector form (7), which gives the Lyapunov covariance equation, noted in (8).

$$d\mathbf{V}_\tau = A\mathbf{V}_\tau\, d\tau + \mathbf{g}\sqrt{2\nu}\, dW_\tau\,, \qquad \mathbf{V}_\tau = (\bar{Z}_\tau, \bar{R}_\tau, \bar{Y}_\tau)^T, \tag{7}$$

$$dE(\mathbf{V}_\tau \mathbf{V}_\tau^T)/d\tau = AE(\mathbf{V}_\tau \mathbf{V}_\tau^T) + E(\mathbf{V}_\tau \mathbf{V}_\tau^T)A^T + 2\nu\, \mathbf{g}\,\mathbf{g}^T. \tag{8}$$

The matrix equation (8) possesses the stationary solutions $E(\bar{Z}_\tau^2) = 1$ and

$$E(\bar{R}_\tau^2) = \frac{\nu}{2D}\frac{1+2D\nu}{1+\nu^2+2D\nu}\,, \quad E(\bar{X}_\tau^2) = \frac{1}{2D}\frac{2D+\nu+4D^2\nu}{1+\nu^2+2D\nu}\,. \tag{9}$$

The square root of these expectations yields the rms-values of interest.

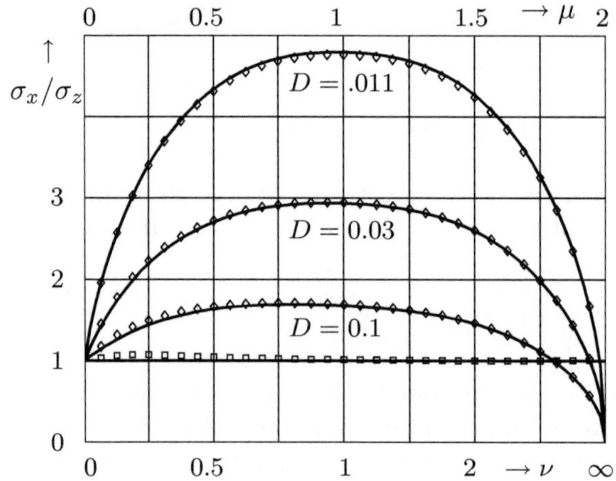

Fig. 3 RMS-ratios of car vibrations vs car speed

Fig. (3) shows analytical results and numerical approximations of the rms-values plotted versus the related car speed $\nu = v\Omega/\omega_1$ in the range $0 \leq \nu < \infty$. As already explained, the applied ν- scaling is linear in the under-critical range and rational for $1 \leq \nu < \infty$. Solid lines represent analytical results calculated by Eq. (5) or (9). Simulation results are obtained applying Euler schemes to the linear system (6) for the scan rate $\Delta\tau = 0.001$ and $N_s = 10^6$ sample points. Hereby, the Wiener increment dW_τ is approximated by $\Delta W_n = \sqrt{\Delta\tau}\, R_n$, where R_n is a sequence of normally distributed numbers with zero mean and $E(R_n^2) = 1$. The simulation results are marked by squares for the excitation values and by diamonds for the response. Note, that the applied Euler scheme may be improved by some higher order schemes [5].

4 Extensions to Nonlinear Car Vibrations

The investigations above are extended to progressive spring characteristics by means of the normalized nonlinear state equations

$$d\tilde{R}_\tau = [(\sigma_{\bar{y}}\tilde{Y}_\tau + \bar{Z}_\tau)d\tau - \sigma_{\bar{r}}\, dW_\tau]/\sigma_{\bar{r}}, \qquad d\bar{Z}_\tau = -\nu\bar{Z}_\tau d\tau + \sqrt{2\nu}\, dW_\tau\,, \quad (10)$$

$$d\tilde{Y}_\tau = [-\sigma_{\bar{r}}(1+\bar{\gamma}\sigma_{\bar{r}}^2\tilde{R}_\tau^2)\tilde{R}_\tau d\tau - 2D(\sigma_{\bar{y}}\tilde{Y}_\tau + \nu\bar{Z}_\tau)d\tau + 2D\sqrt{2\nu}dW_\tau]/\sigma_{\bar{y}}\,. \quad (11)$$

These equations are derived by introducing the normalized state processes $\bar{R}_\tau = \sigma_{\bar{r}}\tilde{R}_\tau$ and $\bar{Y}_\tau = \sigma_{\bar{y}}\tilde{Y}_\tau$ into Eq. (6) utilizing the rms-values

$$\sigma_{\bar{r}} = \sqrt{\frac{\nu}{2D}\frac{1+2D\nu}{1+\nu^2+2D\nu}}\,, \qquad \sigma_{\bar{y}} = \sqrt{\frac{\nu}{2D}\frac{1+4D^2+2D^3\nu}{1+\nu^2+2D\nu}}\,. \quad (12)$$

For extensions to nonlinear damping in stochastic systems see [6].

The stationary solutions of Eq. (10) and (11) are investigated by means of the Hermite moments $P_{k\ell m} = E[H_k(\tilde{R}_\tau)H_\ell(\tilde{Y}_\tau)H_m(\bar{Z}_\tau)]$. Applying the multiplication and derivation rules, the Hermite moments are derived to

$$\begin{aligned} &dP_{k\ell m}/d\tau = -\nu m(P_{k\ell m} + P_{k\ell m-2}) - 2D\ell(P_{k\ell m} + P_{k\ell-2m}) \\ &+k[\sigma_{\bar{y}}(P_{k-1\ell+1m} + P_{k-1\ell-1m}) + \nu(P_{k-1\ell m+1} + P_{k-1\ell m-1})]/\sigma_{\bar{r}} \\ &-\ell[2D\nu(P_{k\ell-1m+1} + P_{k\ell-1m-1}) + \sigma_{\bar{r}}(P_{k+1\ell-1m} + P_{k-1\ell-1m})]/\sigma_{\bar{y}} \\ &-\ell\bar{\gamma}\sigma_{\bar{r}}^3(P_{k+3\ell-1m} + 3P_{k+1\ell-1m} + 3P_{k-1\ell-1m} + P_{k-3\ell-1m})/\sigma_{\bar{y}} \\ &+\nu[k(k-1)P_{k-2\ell m}/\sigma_{\bar{r}}^2 + 4D^2\ell(\ell-1)P_{k\ell-2m}/\sigma_{\bar{y}}^2 + m(m-1)P_{k\ell m-2} \\ &+4D\ell m P_{k\ell-1m-1}/\sigma_{\bar{y}} - 2kmP_{k-1\ell m-1}/\sigma_{\bar{r}} - 4Dk\ell P_{k-1\ell-1m}/\sigma_{\bar{r}}/\sigma_{\bar{y}}]. \end{aligned} \quad (13)$$

Eq. (13) is an infinite equation system for all indices $k, \ell, m = 0, 1, 2, \ldots$ and $P_{000} = 1$. Decoupling higher order moments with indices $k + \ell + m > 2$ by $P_{k+3\ell-1m} \approx 0$, a second order approximation is calculated to

$$E(\bar{R}_\tau^2) = \frac{\nu}{2D}\,\frac{1+2D\nu+3\bar{\gamma}\sigma_{\bar{r}}^2}{1+\nu^2+2D\nu+3\bar{\gamma}\sigma_{\bar{r}}^2}\,. \quad (14)$$

Obviously, Eq. (14) represents an asymptotic result that leads to the correct rms value for the limiting procedure $\bar{\gamma} \to 0$ of linear road-vehicle systems.

Solid lines in Fig. (4) show evaluations of the approximated result (14) in comparison with the interrupted lines of corresponding linear road-vehicle systems. Numerical results of Monte-Carlo simulations are marked by circles, valid for the rms-values of the relative car vibration. They are obtained for three different damping values and the nonlinearity parameter $\bar{\gamma} = 0.2$ applying the Euler scheme to Eq. (6) for $N = 10^7$ samples and the step size $\Delta\tau = 0.001$. For strong damping, there is a good coincidence between the numerical circle results and the solid lines calculated by the approximation (14). For decreasing damping, however, deviations between both become larger particularly in the parameter resonance range near $\nu = 2$ where the stationary car vibrations are unstable as shown in the next section.

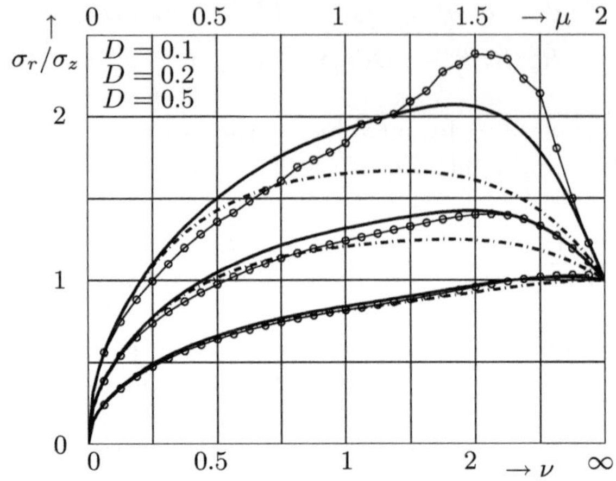

Fig. 4 Relative rms-values vs car speed for $\bar{\gamma} = 0.2$

5 Asymptotic Stability of Stationary Car Vibrations

To investigate the asymptotic stability of nonlinear car vibrations (see e.g. [7]), the perturbation equations associated to Eq. (6) are derived for both, the relative displacement R_τ and the absolute velocity Y_τ by means of $\bar{R}_\tau = R_{st} + \Delta R_\tau$ and $\bar{Y}_\tau = Y_{st} + \Delta Y_\tau$. The insertion of both perturbations into Eq. (6) gives the linearized perturbation equations

$$d(\Delta R_\tau) = \Delta Y_\tau \, d\tau \,, \qquad \Delta R_\tau, \Delta Y_\tau << 1, \tag{15}$$
$$d(\Delta Y_\tau) = -(2D\Delta Y_\tau + \Delta R_\tau + 3\bar{\gamma} R_{st}^2 \Delta R_\tau) \, d\tau. \tag{16}$$

The introduction of the polar coordinates [8] by $\Delta R_\tau = A_\tau \cos \Phi_\tau$ and $\Delta Y_\tau = A_\tau \sin \Phi_\tau$ into Eq. (15) and (16) leads to the transformed equations

$$d\Phi_\tau = -d\tau - (3\bar{\gamma} R_{st}^2 \cos \Phi_\tau + 2D \sin \Phi_\tau) \cos \Phi_\tau d\tau, \tag{17}$$
$$d(\ln A_\tau) = -(3\bar{\gamma} R_{st}^2 \cos \Phi_\tau + 2D \sin \Phi_\tau) \sin \Phi_\tau \, d\tau. \tag{18}$$

According to the multiplicative ergodic theorem [9], the ln-amplitude equation is integrated in the time domain yielding the top Lyapunov exponent

$$\lambda_{top} = \lim_{T\to\infty} \frac{1}{T} \int_0^T [-3\bar{\gamma} R_{st}^2 \cos \Phi_\tau - 2D \sin \Phi_\tau] \sin \Phi_\tau d\tau. \tag{19}$$

Fig. (5) shows evaluations of the top Lyapunov exponent for the damping measures $D = 0.07, 0.05$ and the parameter values $\bar{\gamma} = 0.1, 0.12$. The scan rate selected was $\Delta\tau = 0.001$ applied for $N = 10^7$ sample points.

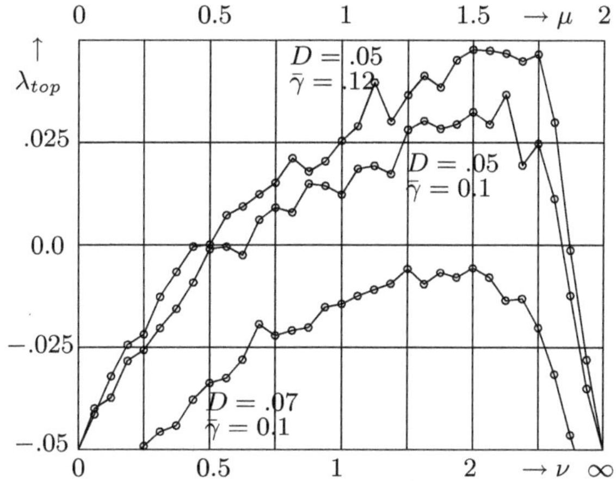

Fig. 5 Top Lyapunov exponents vs car speed

Note, that the results obtained above show Kramer's effect [10]; i.e. for weak positive damping stationary system vibrations can be destabilized even if they are excited by white noise. In our case of road noise, destabilization is effected by car speeds, as well. In Fig. (5), the calculated Lyapunov exponents are plotted for three data sets. The lowest curve shows that the damping is so strong, that the top Lyapunov exponents are negative in the whole speed range. For same nonlinearity, but smaller damping, the top Lyapunov exponents of the middle curve are negative only for slow or high car speeds. In between these two speed ranges, the top Lyapunov exponents are positive indicating that the stationary car vibrations become non-stationary bifurcating into stochastic chaos. The highest curve in Fig. (5) is obtained for the same weak damping but higher nonlinear parameter. Again, the largest top Lyapunov exponents are situated in the parameter resonance near twice the critical car speed of the linear vehicle system.

6 Conclusions

The present paper investigates road-vehicle systems and related bifurcation problems. Riding on rough roads, vertical car vibrations are generated in dependence on the velocity of the car. For increasing car speeds, the vertical vibrations become resonant, when the related car speed is critical near $\nu = 1$. However, the car vibrations are absorbed and vanishing completely, when the car is running with infinite velocity.

In the second part, the vehicle modelling is extended to nonlinear wheel suspensions. Introducing relative displacement coordinates, stationary solutions are approximated by means of Hermite polynomials. The result obtained show a good coincidence between both, the analytical and numerical approximations only for small nonlinearities and strong damping.

Larger deviations from the stationary situation are due to instabilities of the nonlinear car vibrations. For sufficiently weak damping they become unstable and bifurcate into stochastic chaos when the car speeds reach the parameter resonance near twice the critical speed of the linear system.

References

1. Ammon, D.: Modellbildung und Systementwicklung in der Fahrzeugdynamik. B.G. Teubner, Stuttgart (1997)
2. Wedig, W.: Dynamics of Cars Driving on Stochastic Roads. In: Spanos, P.D., Deodatis, G. (eds.) Proceedings of Computational Stochastic Mechanics, pp. 647–654. Millpress, Rotterdam (2003)
3. Popp, K., Schiehlen, W.: Fahrdynamik. B.G. Teubner, Stuttgart (1993)
4. Klotter, K.: Schwingungen, Hütte, Theoretische Grundlagen, vol. 28, pp. 565–639. Aufl., Ernst & Sohn, Berlin (1955)
5. Talay, D.: Simulation and Numerical Analysis of Stochastic Differential Systems: A Review. In: Krée, P., Wedig, W. (eds.) Probabilistic Methods in Applied Physics. Lecture Notes in Physics, vol. 451, pp. 54–96. Springer, Heidelberg (1995)
6. Wedig, W.: Vertical Dynamics of Riding Cars under Stochastic and Harmonic Base Excitations. In: Rega, Vestroni (eds.) IUTAM Symposium on Chaotic Dynamics and Control of Systems and Processes in Mechanics, pp. 371–381. Springer, Dordrecht (2005)
7. Wedig, W.: Stochastic Dynamics of Road-Vehicle Systems and Related Bifurcation Problems. In: Macucci, M., Basso, G. (eds.) AIP Conference Proceedings 1129 of the 20th International Conference on Fluctuations and Noise, ICFN 2009, pp. 33–36 (2009)
8. Khasminskii, R.Z.: Necessary and sufficient conditions for asymptotic stability of linear stochastic systems. Theory Prob. Appl. 12, 144–147 (1974)
9. Oseledec, V.I.: A multiplicative ergodic theorem, Lyapunov characteristic numbers for dynamical systems. Trans. Moscow Math. Soc. 19, 197–231 (1968)
10. Arnold, L., Imkeller, P.: The Kramers Oscillator Revisited. Lecture Notes in Physics, pp. 280–291 (2000)

A Review of Parameter-Induced Stochastic Resonance and Current Applications in Two-Dimensional Image Processing

Yibing Yang and Bohou Xu

Department of Mechanics, State Key Laboratory of Fluid Power Transmission and Control, Zhejiang University, Hangzhou 310027, P.R. China

Abstract. Stochastic resonance (SR) is a phenomenon that nonlinear system synchronizes with noise to boost a resonant-like behavior. Parameter-induced stochastic resonance (PSR) proposed in this paper is realized by optimally tuning system parameters without adding any noise. It has been proved the performance of PSR is better than traditional SR technique which is in fact a particular case in PSR region. The applications of PSR to signal processing and target detection in shallow water reverberation have been reviewed. A new concept of two-dimensional parameter-induced stochastic resonance (2D-PSR) and its applications in the restoration of degraded image and pattern recognition of remote sensing image are developed.

Keywords: Stochastic resonance; Nonlinear system; Image processing.

1 Review of Parameter-Induced Stochastic Resonance

Stochastic Resonance (SR) is a phenomenon manifested in nonlinear systems in which generally feeble input information, such as a weak signal, can be amplified and optimized by the presence of noise. This concept was first proposed by Benzi in 1981 [1], addressing the problem of the periodically recurrent ice ages. The effect requires three basic ingredients:

1. An energetic activation barrier or, more generally, a form of threshold;
2. A weak coherent input, such as a periodic signal;
3. A source of noise that is inherent in the system or that adds to the coherent input.

Given these features, the response of the system undergoes resonant-like behavior as a function of the noise level, hence the name Stochastic Resonance. Over the last few decades, Stochastic Resonance has continuously attracted considerable attention, and has been applied to large variety of fields, including physics, chemistry, biomedical sciences and engineering systems.

W.Q. Zhu et al. (Eds.): Nonlinear Stochastic Dynamics & Control, IUTAM BOOK SERIES 29, pp. 229–238.
springerlink.com

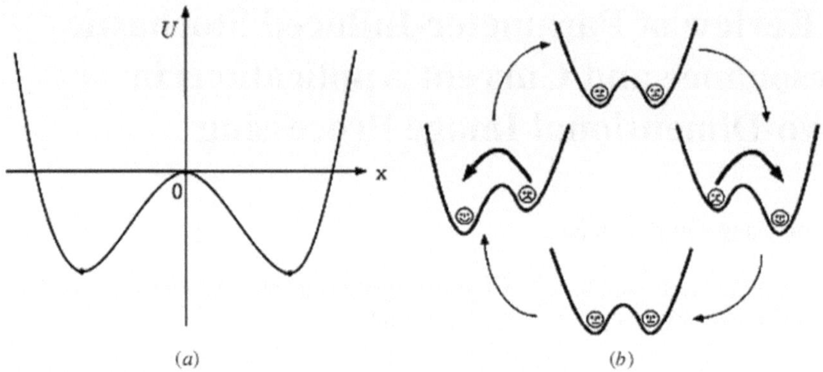

Fig. 1 (a) Symmetric double-well potential. (b) Double-well potential under periodic weak forcing.

In order to introduce the mechanism of stochastic resonance, let us consider a heavily damped particle of mass m and viscous friction γ moving in a symmetric double-well potential $U(x)$ (see Fig. 1(a)). If we apply a weak periodic forcing to the particle, the double-well potential would be titled asymmetrically up and down, periodically raising and lowering the potential barrier, as shown in Fig 1(b). Generally, the weak force itself cannot cause the particle to shift between the two wells. However, if the system is interfered by noise with certain intensity which is synchronized with the periodic force, the particle will roll between the two wells in accordance with the periodic force. This synchronization is called the stochastic resonance.

Conventional SR is realized by adding an optimal amount of noise into the system. However, this method has some shortcomings. For example, in most situations, the input signal is corrupted by noise with a given quantity. So it is easy to increase the input noise but impossible to decrease it. If the input signal is already suprathreshold, Conventional SR can do nothing with it. Recently, a new approach called Parameter-induced Stochastic Resonance (PSR) is proposed to realize the SR effect by optimally tuning the system parameters. Unlike Conventional SR, PSR which in fact tunes the height of the barrier to affect the system's response speed can occur both in subthreshold case and in suprothreshold case. Moreover, it has been proved that SR realized by conventional method (adding noise) is an optimization problem in a subregion of the parameter space [2]. Parameter-induced Stochastic Resonance has been successfully applied to many fields including one-dimensional (1D) digital signal and analog signal processing, target detection in shallow-water reverberation, two-dimensional (2D) image processing and etc.

In one-dimensional case, the theory of Parameter-induced Stochastic Resonance and the principle of parameter optimization have been founded. It is proved that the low SNR baseband binary pulse amplitude modulated (PAM) signals can get a high SNR gain after being processed by PSR technique with optimally tuned parameters. Besides, a marked enhancement of the channel capacity for binary PAM signal is achieved [3]. Since the PSR system is double-well and nonlinear, if we process analog signal directly by PSR technique, strong distortion may occur in outputs. To overcome this problem, a simple but effective nonlinear inversion method is developed to remove the distortion, which makes it possible to tackle with analog signal using PSR technique [4]. The target detection with active sonar is often limited by the presence of reverberation which is highly correlated with the transmitted signal. Using PSR-based technique, the detection performance can be optimized by tuning system parameters. Theoretical and experimental results have shown that PSR technique performs well under the condition of weak signal-to-reverberation-noise ratio [5]. Image processing is another important field to which SR technique can be applied [6]. Based on PSR theory, we propose a two-dimensional parameter-induced stochastic resonance (2D-PSR) system with parameters a, b to be adjustable. It will be proved in this paper that the SNR gain of 2D-PSR system can surpass one which is impossible for linear filtering [7].

This paper is organized as follows. In Section 2, we describe the 2D-PSR system and its output probability density function which is derived by solving corresponding Fokker-Planck Equation. In Section 3, we will introduce how to use theories discussed in Section 2 to improve SNR gain. Experimental results of degraded image processing and an application to pattern recognition of remote sensing image by our proposed technique are presented in Section 4 and Section 5. Finally, we draw the conclusions of 2D-PSR technique and its prospect in Section 6.

2 2D-PSR System and the Corresponding Fokker-Planck Equation

Similar to one-dimensional case, we propose the following two-dimensional parameter-induced stochastic resonance system

$$\frac{\partial w}{\partial x}+\frac{\partial w}{\partial y}=a\cdot w-b\cdot w^3+h+\Gamma(x,y) \tag{1}$$

where $w=w(x,y)$ is the state variable (also taken as the system output), a, b are system parameters to be adjustable, h is the original signal and $\Gamma(x,y)$ is the Gaussian white noise with $\langle\Gamma(x,y),\Gamma(x_1,y_1)\rangle=2D\cdot\delta(x-x_1,y-y_1)$. Here D is the noise intensity, which is related to noise variance σ^2 by $\sigma^2=\dfrac{2\cdot D}{\Delta x\cdot\Delta y}$ in two-dimensional case, where $\Delta x, \Delta y$ are the sampling intervals along horizontal and vertical directions. According to the characteristic method in Partial Differential Equation (PDE) theory [8], Equation (1) is equivalent to a set of Ordinary Differential Equations (ODE)

$$\frac{dx}{1}=\frac{dy}{1}=\frac{dw}{aw-bw^3+h+\Gamma} \tag{2}$$

The characteristic line is $\frac{dy}{dx}=1$ or $y=x+C$ with C being the constant, which indicates that in any arbitrarily small region, the solution of Eq. (1) is symmetric along the diagonal direction. Thus we can solve Eq. (1) by independently dealing with Eq. (2)

$$\frac{dw}{dx}=aw-bw^3+h(x)+\Gamma(x) \tag{3-1}$$

$$\frac{dw}{dy}=aw-bw^3+h(y)+\Gamma(y) \tag{3-2}$$

The corresponding Fokker-Planck Equation (FPE) for Eq. (3-1) is

$$\frac{\partial\rho(w,x)}{\partial x}=-\frac{\partial}{\partial x}\left[f(w)\cdot\rho(w,x)\right]+\frac{D}{\Delta x}\cdot\frac{\partial^2\rho(w,x)}{\partial w^2} \tag{4}$$

where $\rho(w,x)$ is the output probability density function (PDF). When $x\to\infty$ we can obtain the static PDF of Eq. (4)

$$\rho(w,x\to\infty)=\rho_s(w)=N\cdot\exp[\varphi_0(w)]=N\cdot\exp\left[\int_{-\infty}^{+\infty}\Delta x\cdot\frac{aw-bw^3+h}{D}dw\right] \tag{5}$$

with N to be the normalized factor. The asymptotic dynamic PDF of Eq. (4) is [3]

$$\rho(w,x)=\sum_{i=0}^{n-1}N_i\cdot\Phi_i(w)\cdot\exp(-\lambda_i\cdot x)+\left[\rho_s(w)-\sum_{i=0}^{n-1}N_i\cdot\Phi_i(w)\right]\cdot\exp(-\lambda_n\cdot x) \tag{6}$$

where $0=\lambda_0<\lambda_1<\ldots<\lambda_n$ and $\Phi_0=\exp[\varphi_0(w)],\Phi_1,\ldots,\Phi_n$ are the n orders eigenvalues and eigenfunctions of Eq. (3-1), N_i is the constant to be determined by orthodonal condition of eigenfunctions [3]. λ_1 which is the dominant factor of system's settling down to steady state is regarded as the system response speed.

The static and dynamic PDF of Eq. (3-2) can be derived similarly.

3 Theory of 2D-PSR for Degraded Image Processing

Previously we set λ_1 to be around 3 so that the error between dynamic and static PDF is within $e^{-3} \approx 5\%$. However this constrained condition will restrict our choice for parameters a, b. Therefore we introduce the concept of dynamic signal-to-noise ratio (DSNR)

$$DSNR(a,b) = \frac{E[w]}{\sqrt{Var[w]}} \tag{7}$$

where $E[\bullet]$ is an expectation operator and $Var[\bullet] = E[\bullet^2] - E^2[\bullet]$. Without the restriction of $\lambda_1 \approx 3$, the system response speed can either be higher or lower. Consequently the valid sample points of the output will vary following the value of λ_1. Previously, when processing a noisy image through 2D-PSR system, we just directly pick up the last sample point of each sample period [6], which is assumed to have the best statistic performance in the means of probability density. According to the dynamic solution, when the system response speed grows faster, the output performance will decrease, however, there will be more sample points valid to be considered. If these available sample points are averaged, we will get a better result. The statistic characteristics of averaged outputs can be calculated by the theory of local average random field [9]. Assume $w(x)$ is a random field with expectation m and variance σ^2. $W_X(x)$ is length average of $w(x)$ over a period X. Here $W_X(x)$ is called the local average random field, which has expectation and variance

$$\begin{aligned} E[W_X(x)] &= m = \int w \cdot \rho(w,x)dw \\ Var[W_X(x)] &= \Omega(X) \cdot \sigma^2 \end{aligned} \tag{8}$$

where $\Omega(X)$ is called the variance function of $W_X(x)$. Let $\rho(\xi) = Cov(\xi)/\sigma^2$ be the normalized correlation function of $w(x)$. The relationship between $\Omega(X)$ and $\rho(\xi)$ can be described as follows

$$\Omega(X) = \frac{1}{X^2}\int_0^X \int_0^X \rho(x_1 - x_2)dx_1 dx_2 = \frac{2}{X}\int_0^X \left(1 - \frac{\xi}{X}\right)\rho(\xi)d\xi \tag{9}$$

In order to obtain variance function $\Omega(X)$ from Eq. (9), we have to figure out the second-order statistic characteristics of $w(x)$. Thus we rewrite Eq. (4) as

$$\frac{\partial \rho(w,x|w_0,x_0)}{\partial x} = -\frac{\partial}{\partial x}\left[f(w)\cdot \rho(w,x|w_0,x_0)\right] + \frac{D}{\Delta x}\cdot \frac{\partial^2 \rho(w,x|w_0,x_0)}{\partial w^2} \tag{10}$$

where $\rho(w,x|w_0,x_0)$ is the conditional PDF with $\rho(w,x_0|w_0,x_0)=\delta(w-w_0)$ and $\lim_{x\to\infty}\rho(w,x|w_0,x_0)=\rho(w,x)$. The first-order approximate solution of Eq. (10) is

$$\rho(w,x|w_0,x_0)=\rho(w,x)+[\delta(w-w_0)-\rho(w,x)]\cdot\exp[-\lambda_1\cdot(x-x_0)] \tag{11}$$

with λ_1 the system response speed. Thus

$$\Omega(X)=\frac{2}{X}\int_0^X\left(1-\frac{\xi}{X}\right)\exp(-\lambda_1\xi)d\xi \tag{12}$$

Replacing $E[w], Var[w]$ in Eq. (7) with Eqs. (8) and (12) we obtain

$$DSNR(a,b)=\frac{\int w\cdot\rho(w,x)dw}{\frac{2\cdot\sigma^2}{X}\int_0^X\left(1-\frac{\xi}{X}\right)\exp(-\lambda_1\xi)d\xi} \tag{13}$$

Here X is determined as follows. Define an allowance error: $\exp(-\lambda_1\cdot\xi)=err$ $\Rightarrow\xi=-\frac{1}{\lambda_1}\ln err$. If $\xi\geq X_b$ (here X_b means each length period), we directly use the last sample point to calculate the DSNR. If $\xi<X_b$, we take $X=X_b-\xi$. Our goal is to maximize DSNR by optimizing system parameters a,b, which can be solved by gradient descent algorithm very efficiently.

It has been proved that the SNR gain of linear systems cannot surpass one [7]. We will prove that the SNR gain of 2D-PSR system can outstrip one. The input SNR in our case is $SNR_{input}=h/\sqrt{2\cdot D}$ with h the original image and D the noise intensity. The SNR gain can be written as

$$SNR_{gain}=\frac{DSNR(a,b)}{SNR_{input}}=\frac{\sqrt{2\cdot D}\cdot\int w\cdot\rho(w,x)dw}{2h\cdot\sigma^2/X\int_0^X\left(1-\frac{\xi}{X}\right)\exp(-\lambda_1\xi)d\xi} \tag{14}$$

Fig. 2 shows the SNR gain as a function of parameter b with $a=1.5$. It indicates when system parameters a,b are properly set, the SNR gain will surpass one ,which is impossible for linear systems. In addition, the SNR gain remains high and descending slowly as b grows larger, which means 2D-PSR technique performs strong robustness to system parameters variations. If the parameters are not best optimized or even biased seriously, we can still obtain high SNR gain.

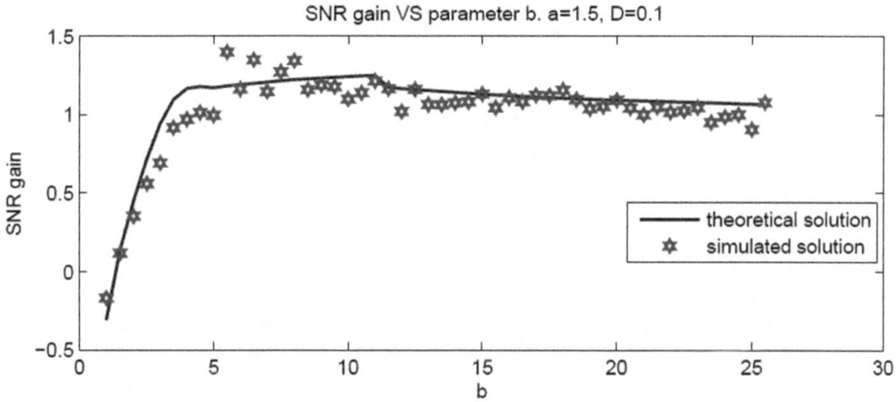

Fig. 2 SNR gain as a function of parameter b

4 Example of 2D-PSR Image Restoration

The simulation equation equivalent to Equation (3) is

$$\begin{cases} w_{m,n} = \left[aw_{m,n-1} - bw_{m,n-1}^3 + h_{m,n-1} + \Gamma_{m,n-1}\right] \cdot \Delta x + w_{m,n-1} \\ w_{m,n} = \left[aw_{m-1,n} - bw_{m-1,n}^3 + h_{m-1,n} + \Gamma_{m-1,n}\right] \cdot \Delta y + w_{m-1,n} \end{cases} \tag{15}$$

where the subscripts represent the locations of sample points, and $\Delta x, \Delta y$ are the sampling intervals along horizontal and vertical directions.

The intensity value of an image usually distributes in the region of [0, 255]. However, the double-well potential of 2D-PSR system is symmetric to zero. Thus we should first subtract the mean value of an image before processing with 2D-PSR technique and add it back later. Fig. 3(a) shows a computed tomography (CT) image corrupted by additive Gaussian white noise $N(0,57)$ (Fig. 3(b)). We sample the degraded image 5×5 times per pixel (five by row and five by column), and operate it on 2D-PSR system. After optimizing parameters a, b according to Eq. (13) we come up with the recovered image Fig. 3(c). Fig. 3(d) shows the result obtained by linear mean filtering, which takes the average value of every 5×5 blocks of Fig. 3(b). As a comparison, we have further processed Fig. 3(b) with total variation method [10], wavelet de-noising [11] and adaptive Wiener filtering [12]. The results are shown in Figs. 3(e), (f) and (g) respectively.

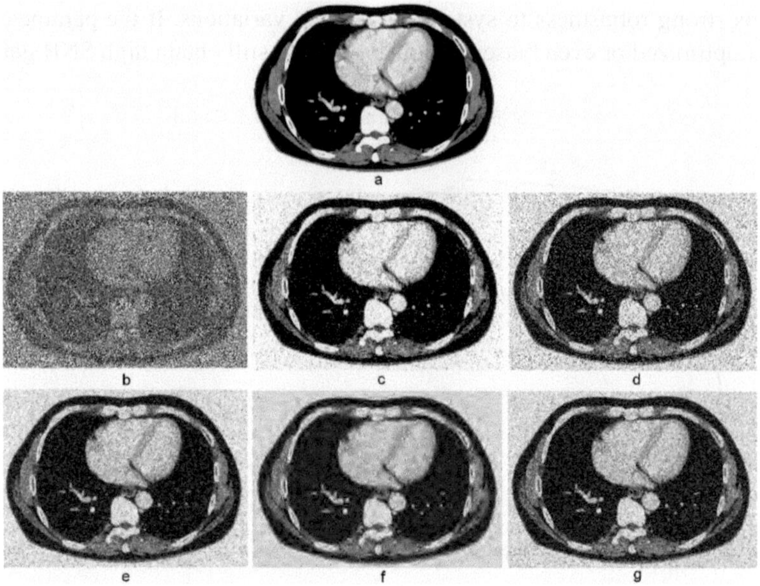

Fig. 3 (a) Original CT image. (b) Sampled CT image degraded by additive Gaussian white noise. (c) Image processed by 2D-PSR technique. (d)Image processed by mean filtering. (e) Image processed by total variation method. (f) Image processed by wavelet de-noising. (g) Image processed by adaptive Wiener filtering.

5 Example of 2D-PSR Pattern Recognition of Remote Sensing Image

The remote sensing multi-spectral imagc is a set of registered images with different bands of wavelength ranging from visible light to infrared. Fig. 4 shows a set of multi-spectral images of bands blue, green, red and infrared. Our goal is to classify rivers, buildings and forests from the multi-spectral images using 2D-PSR technique. To this end, we rewrite 2D-PSR equation in vector's form

$$\frac{\partial \mathbf{w}}{\partial \mathrm{x}}+\frac{\partial \mathbf{w}}{\partial y}=a\mathbf{w}-b\mathbf{w}^{3}+\mathrm{h}+\Gamma(x, y) \tag{16}$$

where $\mathtt{w}=[w_1, w_2,..., w_N]^T$, $\mathrm{h}=[h_1, h_2,..., h_N]^T$, $\Gamma=[\Gamma_1, \Gamma_2,..., \Gamma_N]^T$, N is the number of registered images. The static PDF of corresponding FPE in the horizontal direction is

$$\rho(\mathtt{w})=\mathtt{N}\cdot\exp\left(\frac{1/2\cdot a\mathtt{w}^2-1/4\cdot b\mathtt{w}^4+[h_1 w_1, h_2 w_2,..., h_N w_N]^T}{D}\right) \quad , \tag{17}$$

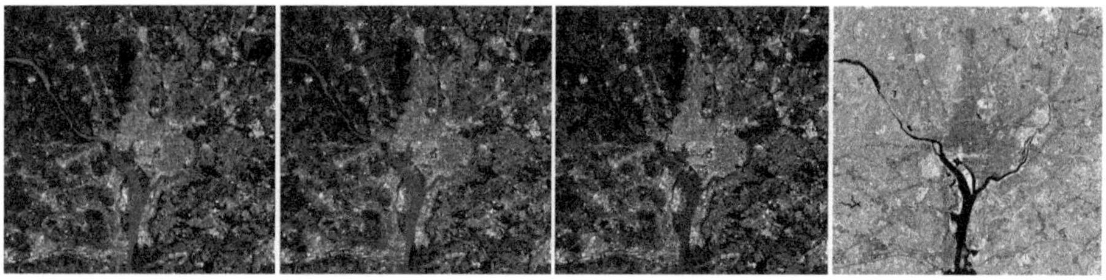

Fig. 4 Multi-spectral images with blue, green, red and infrared bands

Assume that the noise in the registered images is uncorrelated, thus $\rho(\mathrm{w})=\prod_{i=1}^{N}\rho(w_i)$ where $\rho(w_i)=N_i\cdot\exp\left[\left(1/2\cdot aw_i^2-1/4\cdot bw_i^4+h_iw_i\right)/D\right]$. The dynamic PDF in the horizontal direction is

$$\rho(\mathrm{w},x)=\prod_{i=1}^{N}\rho(w_i,x) \tag{18}$$

where $\rho(w_i,x)=\sum_{j=0}^{K-1}N_j\Phi_j(w)\exp(-\lambda_j x)+\left[\rho(w_i)-\sum_{j=0}^{K-1}N_j\Phi_j(w_i)\right]\exp(-\lambda_K x)$.

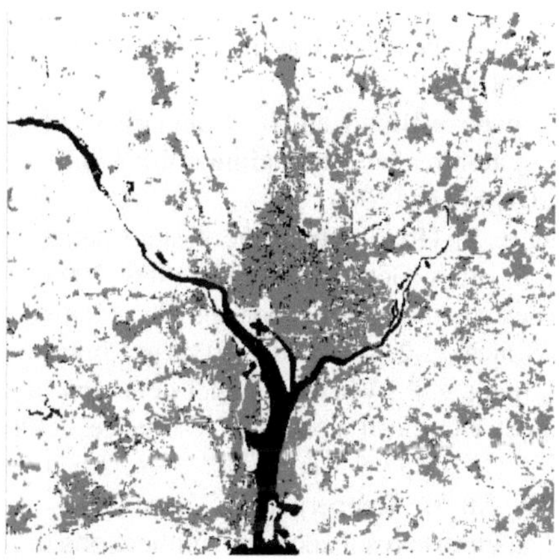

Fig. 5 Classified image by 2D-PSR technique with black, grey and white representing rivers buildings and forests respectively

The decision function of 2D-PSR based classifier is

$$D_k(\mathrm{w})=\ln(d_k(\mathrm{w}))=\ln\left[\rho(\mathrm{w},x|\omega_k)\cdot P(\omega_k)\right]=\ln\rho(\mathrm{w},x|\omega_k)+\ln P(\omega_k) \tag{19}$$

where ω_k is the number of classes (such as rivers, buildings and forests) and $P(\omega_k)$ is the probability of ω_k. In pattern recognition of remote sensing image, each class ω_k is determined by the input value h_k (for example the class of rivers buildings and forests can be classified by their grey levels). Thus we can replace ω_k in Eq. (19) by h_k. The final decision is

$$D(\mathtt{w}_{i,j}) = \{k | \max[D_k(\mathtt{w}_{i,j})], k = 1,2,...M\} \tag{20}$$

Fig 5 shows the classified result by 2D-PSR based pattern recognition.

6 Conclusions

In this paper, the researches on PSR technique and its applications in signal processing and target detection in shallow water reverberation have been reviewed. A current developed 2D-PSR technique is proposed. The corresponding static and dynamic PDF of the FPE is derived. A concept of dynamic signal-to-noise ratio (DSNR) is introduced to utilize more valid sample points to upgrade the output performance. It has been proved 2D-PSR technique can have SNR gain greater than one which is impossible in linear cases. Examples of 2D-PSR based image restoration and pattern recognition of remote sensing image have shown 2D-PSR technique to be promising in the field of image processing.

References

[1] Benzi, R., Sutera, A., vulpiani, A.: The mechanism of stochastic resonance. J. Phys. A 14, L453–L457 (1981)

[2] Xu, B., Duan, F., Chapeau-Blondeau, F.: Comparison of aperiodic stochastic resonance in a bistable system realized by adding noise and by tuning system parameters. Physical Review E 69, 61110 (2004)

[3] Xu, B., Li, J., Zheng, J.: How to tune the system parameters to realize stochastic resonance. J. Phys. A: Math. Gen. 36, 11969–11980 (2003)

[4] Li, H., Xu, B.: A new method to recover signals obtained by stochastic resonance. ACTA Mechanica Sinica 35-2, 194–198 (2003) (in Chinese)

[5] Zhang, H., Xu, B., Jiang, Z.P., Wu, X.: Target detection in shallow-water reverberation based on parameter-induced stochastic resonance. J. Phys. A: Math. Theor. 41, 105003 (2008)

[6] Yang, Y., Jiang, Z.P., Xu, B., Repperger, D.W.: An investigation of two-dimensional parameter-induced stochastic resonance and applications in nonlinear image processing. J. Phys. A: Math. Theor. 42, 145207 (2009)

[7] Reid, J.G.: Linear system fundamentals: continuous and discrete, classic and modern. McGraw-Hill, New York (1983); 0070518084

[8] Risken, H.: The Fokker-Planck Equation: Method of Solutions and Applications, Springer Series in Synergetics, 2nd edn. (1989)

[9] Vanmarcke, E.: Random fields: Analysis and synthesis. MIT Press, Cambridge (1983)

[10] Weickert, J.: Anisotropic diffusion in image processing. Vollzug der Promotion (1996)

[11] Gonzalez, R., Woods, R., Eddins, S.: Digital image processing using MATLAB. Publishing House of Electronics Industry, Beijing (2004)

[12] Lim, J.S.: Two-dimensional signal and image processing. Prentice-Hall Inc., Englewood Cliffs (1990)

Part 4
Control of Nonlinear Stochastic Systems

Design of Damper Viscous Properties for Semi-active Control of Asymmetric Structures

V. Gattulli, M. Lepidi, and F. Potenza

DISAT, University of L'Aquila 67040 Monteluco di Roio, L'Aquila, Italy

Abstract. A method to design semi-active control strategies of asymmetric structures is presented. The method is based on the optimal sizing of an equivalent Kelvin-Voight model describing the constitutive behavior of semi-active magneto-rheological devices, through the evaluation of the maximum achievable modal damping when they work in passive modality. The complex eigenvalue loci of the passively-controlled system versus the device mechanical characteristics are spanned for symmetric and asymmetric frame structures. A coherent representation of the reference effect ensured by an optimized linear active feedback on the eigenvalues loci is selected to drive the design of the adjustable properties of the semi-active device. A clipped-optimal control algorithm is used in a prototype experimental application whose performance are highlighted by the presented design method.

Keywords: Semi-active control, Earthquake engineering, Structural dynamics, Viscous devices.

1 Introduction

Great research effort has been focused over the last years on reducing the seismic response of engineering structures through dissipative systems [1]. Presently, an increasing attention is being paid to combine the reliable and cost-saving passive technology with the highly performing active strategies, by means of different hybrid and semi-active solutions. In this field, magnetorheological dampers are considered among the most promising devices to mitigate the structural vibrations, due to their mechanical simplicity, high dynamic range, low power requirements, large force capacity and mechanical robustness [2]. Experimental testing on large scale models show that the technology can be effectively implemented to control the structural dynamic response, and is suited for the seismic protection of civil structures [3,4]. Nonetheless, the full-scale applications are still circumscribed, owing probably to the relative youthfulness of the design guidelines currently available in the national and international codes.

The rich literature of theoretical and experimental studies existing on the topic [5-7] reveals that a number of challenging issues is calling for further research efforts. The accuracy level requested to the dynamical model describing the controlled structure, the adequacy of the rheological models used to reproduce the highly nonlinear hysteretic behaviour of the dampers, the stochastic versus deterministic approach to the optimization of the control strategies, the definition of

W.Q. Zhu et al. (Eds.): Nonlinear Stochastic Dynamics & Control, IUTAM BOOK SERIES 29, pp. 241–250.
springerlink.com

significant and synthetic performance indices, the representativeness of the reduced-scale specimens used in laboratory tests remain open investigation fields.

In this respect, the paper summarizes the authors' experience in the multifaceted task of semi-actively controlling the three-dimensional seismic response of a minimal building model, through interstorey chevron-type bracings embedding magnetorheological dampers. The work approaches the modellisation of the structure dynamics and the damper rheology, the model updating based on the experimental modal analysis, the design of a semi-active control strategy, and finally the experimental verification of the effectiveness of the adopted solutions in the seismic protection of the prototypal structure. Here, the issues related to overall semi-active control design method, for optimal sizing and placement of the dampers, the dynamic description of the relationship between the applied magnetic field and the damper rheological properties, the optimization of the semi-active-to-passive proportion in the control strategy for the energy dissipation are treated primarily.

2 Device Sizing for Seismic Excited Frame Structures

Consider a structural system equipped with control devices, described by a dynamic discrete model. Denoting **u** the displacement vector related to the N degrees-of-freedom, the forced response of the structure to a seismic action, represented by the monodirectional ground acceleration $\ddot{u}_g(t)$, is governed by

$$\mathbf{M}\ddot{\mathbf{u}} + \mathbf{C}_s\dot{\mathbf{u}} + \mathbf{K}\mathbf{u} + \mathbf{f}_d(\dot{\mathbf{u}}, \mathbf{u}) = -\mathbf{M}\mathbf{r}\,\ddot{u}_g(t) \quad (1)$$

where **M** e **K** are the mass and stiffness matrix, respectively, $\mathbf{C}_s$ is the structural viscous damping matrix, $\mathbf{f}_d$ is the control force vector (which in principle can be a nonlinear function of the displacement and velocity vectors), and **r** is the allocation vector of the scismic forces. Among different constitutive laws, describing the constitutive behavior of the control device, it is possible to approximate it through an equivalent Kelvin-Voight linear model.

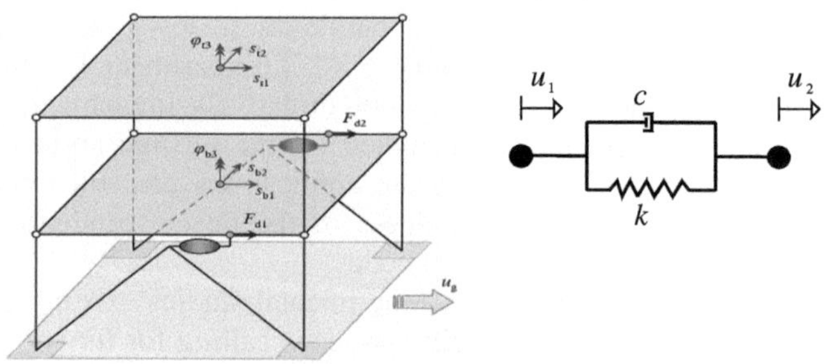

Fig. 1 Structural schemes: (a) seismic protected frame; (b) Kelvin-Voight dissipative devices

Adopting the Kelvin-Voight model for all the control devices in the structural system, and assuming known their placement according to a certain design strategy, the control force vector obeys to the force-velocity-displacement relationship

$$\mathbf{f}_{\mathrm{d}}(\dot{\mathbf{u}},\mathbf{u}) = \mathbf{C}_{\mathrm{d}}\dot{\mathbf{u}} + \mathbf{K}_{\mathrm{d}}\mathbf{u} \tag{2}$$

So that the equation of motion (1) can be rearranged as

$$\mathbf{M}\ddot{\mathbf{u}} + (\mathbf{C}_{\mathrm{s}} + \mathbf{C}_{\mathrm{d}})\dot{\mathbf{u}} + (\mathbf{K} + \mathbf{K}_{\mathrm{d}})\mathbf{u} = -\mathbf{M}\mathbf{r}\,\ddot{u}_{\mathrm{g}}(t) \tag{3}$$

where the additional damping matrix $\mathbf{C}_{\mathrm{d}}$ is non-proportional in the general case. Then, defining a state vector as $\mathbf{x}=\{\mathbf{u}^{\mathrm{T}}, \dot{\mathbf{u}}^{\mathrm{T}}\}^{\mathrm{T}}$, equation (3) can be rearranged as

$$\dot{\mathbf{x}} = \mathbf{A}\mathbf{x} + \mathbf{H}\ddot{u}_g \tag{4}$$

where the state matrices **A** and **H** are

$$\mathbf{A} = \begin{bmatrix} \mathbf{0} & \mathbf{I} \\ -\mathbf{M}^{-1}(\mathbf{C}+\mathbf{C}_{\mathrm{d}}) & -\mathbf{M}^{-1}(\mathbf{K}+\mathbf{K}_{\mathrm{d}}) \end{bmatrix}, \quad \mathbf{H} = \begin{bmatrix} \mathbf{0} \\ -\mathbf{r} \end{bmatrix} \tag{5}$$

The dynamic structural response, also in the case of seismic excitation, is strictly dependent on the input-output transfer functions, which are expression of the system spectral properties. Therefore, analyzing the frequency and mode dependence on the stiffness and viscosity properties of the devices may be a matter of theoretical and technical interest. The frequency and modal damping of the system ensue from the complex roots of the characteristic equation

$$\det[\mathbf{A}(\mathbf{C}_{\mathrm{d}}, \mathbf{K}_{\mathrm{d}}) - \lambda\mathbf{I}] = 0 \tag{6}$$

in which it is convenient to assume $\mathbf{C}_{\mathrm{d}} = c\boldsymbol{\Gamma}$ and $\mathbf{K}_{\mathrm{d}} = k\boldsymbol{\Gamma}$ for sake of simplicity. Therefore a parametric analysis can be carried out, tracking the equation root loci versus the independent variation of the only significant control parameters c and k

$$\lambda_{\mathrm{i}} = a_{\mathrm{i}}(c,k) \pm i b_{\mathrm{i}}(c,k) \tag{7}$$

where the real $a_{\mathrm{i}}(c,k)$ and the imaginary part $b_{\mathrm{i}}(c,k)$ of the i-th eigenvalue are found to be highly nonlinear function of the control parameters. Subsequently, it is possible to find the loci of optimal c- or k-values, imposing the condition

$$\frac{\partial a_{\mathrm{i}}(c,k)}{\partial c} = 0, \quad \text{or} \quad \frac{\partial a_{\mathrm{i}}(c,k)}{\partial k} = 0 \tag{8}$$

whose solution determines the maximum achievable real part of the i-th eigenvalue in the c (fixed k) or k (fixed c) parameter range. As the real part of the eigenvalue relates to the modal damping ξ_i, it is expected that the optimal values of the parameters, referred for instance to the principal structural mode, could ensure the best performance of the passively controlled structure.

2.1 Reference Linear Active Control Feedback

Active control strategies have been deeply investigated to enhance the performance of seismic protection systems [8,9]. In the case of classical linear quadratic regulator (LQR) in which the active control device is driven by optimal linear feedback, the dissipative force is yet a general linear function of displacement and velocity vectors, in the form of a non-collocated relationship, such as

$$\bar{\mathbf{f}}_d(\dot{\mathbf{u}},\mathbf{u}) = \mathbf{G}_1\dot{\mathbf{u}} + \mathbf{G}_2\mathbf{u} \tag{9}$$

where $\mathbf{G}_1$ and $\mathbf{G}_2$ are full gain matrices which determines the relation between the active force acting between two floors in an assigned direction and the whole displacement and velocity variables describing the frame dynamic motion (the dynamic state). The feedback is available from direct measures or reconstructed by a dynamic observer. The overall control can be designed according to the H2/LQG method. The LQG design provides both control feedback (LQR) and Kalman observer (Linear Gaussian). The linear control force in Equation (9) minimizes the cost functional

$$J(\mathbf{x},\mathbf{f}_d) = \int_0^F (\mathbf{x}^T\mathbf{Q}\,\mathbf{x} + \bar{\mathbf{f}}_d^T\mathbf{R}\,\bar{\mathbf{f}}_d)\,dt \tag{10}$$

where $\mathbf{Q}$ and $\mathbf{R}$ are weight matrices. Equation (4) representing the controlled system assumes now the following form

$$\begin{aligned}\dot{\mathbf{x}} &= \mathbf{A}\mathbf{x} + \mathbf{B}\bar{\mathbf{f}}_d + \mathbf{H}\ddot{u}_g \\ \bar{\mathbf{f}}_d &= \mathbf{G}\mathbf{x}\end{aligned} \tag{11}$$

Coherently with the passive case, it is possible to study the eigenvalue loci of the controlled system matrix $\mathbf{A}_c(r)=\mathbf{A}+\mathbf{B}\mathbf{G}(r)$ varying the cost parameter r, used to define the second weight matrix as $\mathbf{R} = r\mathbf{\Gamma}$. The frequency and modal damping of the system again ensue from the complex solutions of the characteristic equation

$$\det[\mathbf{A}_c(r) - \lambda\mathbf{I}] = 0 \tag{12}$$

Consequently, the root loci can be determined varying the parameter r, as

$$\lambda_i = a_i(r) \pm i b_i(r) \tag{13}$$

where $a_i = \mathrm{Re}(\lambda_i)$ and $b_i = \mathrm{Im}(\lambda_i)$ are nonlinear functions of the control parameter. Similarly to the passive case, it is possible to find the loci of optimal r-values, imposing the condition

$$\frac{\partial a_i(r)}{\partial r} = 0 \tag{14}$$

3 Semi-active Control Design

Recently, the semi-active control design has been fully exploited both for model prototypes [3] and real structures. The designers have to solve two principal issues: the device best placement and the optimal sizing of the mechanical device characteristics. In particular, in [10] it was evidenced that in the design process of semi-active protection system for full scale irregular building the lower and the higher force values are achieved when the minimum (OFF) or maximum voltage (ON) is supplied, respectively. A complete design process for semi-active seismic protection of frame structure includes the definition of the maximum and minimum device force (or maximum and minimum equivalent viscous damping). Consequently, to select the force capacity range of the physical semi-active device, a methodology based on equivalent optimal viscous damping may be pursued. In this respect, let assume that the semi-active device delivers a control force

$$\mathbf{f}_{\mathrm{d}}(\mathbf{u},\dot{\mathbf{u}},\boldsymbol{\upsilon})=\mathbf{f}_{\mathrm{d}}(\mathbf{u},\dot{\mathbf{u}},\mathbf{0})+\Delta\mathbf{f}(\mathbf{u},\dot{\mathbf{u}},\boldsymbol{\upsilon}(t)) \tag{15}$$

where $\Delta\mathbf{f}$ is the force increment due to the voltage change $\boldsymbol{\upsilon}(t)$ with respective to the passive part. Consider Equation (2) as description of the passive (OFF) component, and Equations (7-8) as design criteria. In order to exemplify the criterion, Fig. 2 represents the root loci of a 2-dof and 4-dof frame structures [7], varying the c and k sizing coefficient representing the first term of Equation (15) for a semi-active damper. Fig. 2a shows the effects of increasing the viscous coefficient c of a dashpot placed at the first floor of a 2-dof frame structure on the system eigenvalues in the Argand plane. Increasing the parameter produces an increment of the modal damping up to a certain value (marked with a dot) through the nonlinear dependence of the real part $a_1(c,k)$ of the fundamental eigenvalue, while the natural frequencies, related to the imaginary part $b_i(c,k)$, flip to each other due to the increasing of the lower and decreasing of the higher one.

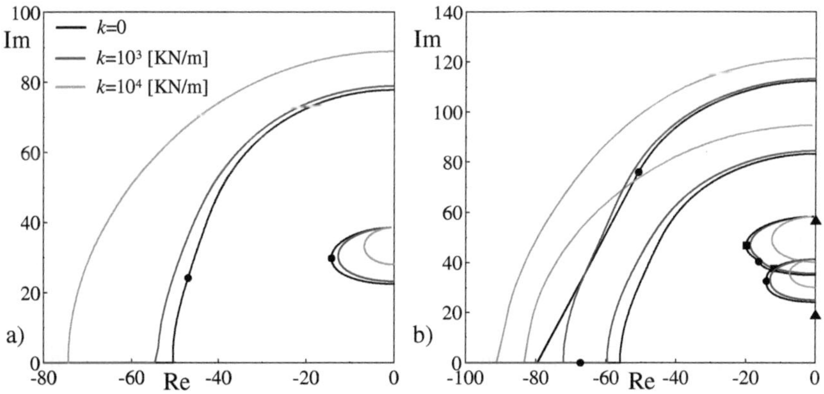

Fig. 2 Eigenvalue loci of dynamic systems varying the parameters c and k: (a) 2-dof symmetric, and (b) 4-dof non-symmetric frame structures

The device stiffness k produces a modal damping increment in the second mode and the opposite on the first mode. The effect of Kelvin-Voight devices on lateral-torsional coupling has been studied in the three-dimensional model of an asymmetric two-floor frame structure. In this case, depicted in Fig.2b, the coupled lateral-torsional modes are modified similarly to the previous planar case, with the only difference that the optimal c-values are, now, related to the two lateral-rotational modes and they are not equal (dot and square marks in Fig.2b).

A consistent procedure is here proposed as design criterion for the force device increment $\Delta\mathbf{f}$, regulated by the voltage $\mathbf{v}(t)$ in semi-active devices. In particular, the force increment is determined through the design of a reference active feed-back control which should be reproduced, as much as possible, by the semi-active strategy. Therefore, the reference active device follows the "constitutive" relation defined by Equation (9) and its effect on the dynamic system may be again represented by the root loci determined from the solution of Equation (12), obtained varying the design parameter r. It must be remarked that the reference active device is designed in a non-collocated configuration which permits the root loci to span a larger range of values. Fig. 3 presents the root loci for the 2-dof (Fig. 3a) and 4-dof (Fig. 3b) system varying the design parameter r. Looking at Fig.3a, it is evident that the design of the active device allows the increment of the second frequency, and consequently the avoidance of the flipping phenomenon noticed in the passive case, in which the second frequency had necessarily to decrease. Larger modal damping for the higher mode can be also achieved. A similar behavior is confirmed also in the 4-dofs case characterized by the latero-torsional modal coupling (Fig.3b).

Therefore, the semi-active control design is strongly conditioned by the actual possibility to simultaneous optimize the passive device characteristic (c and k) and the optimal reference active control intensity (r value).

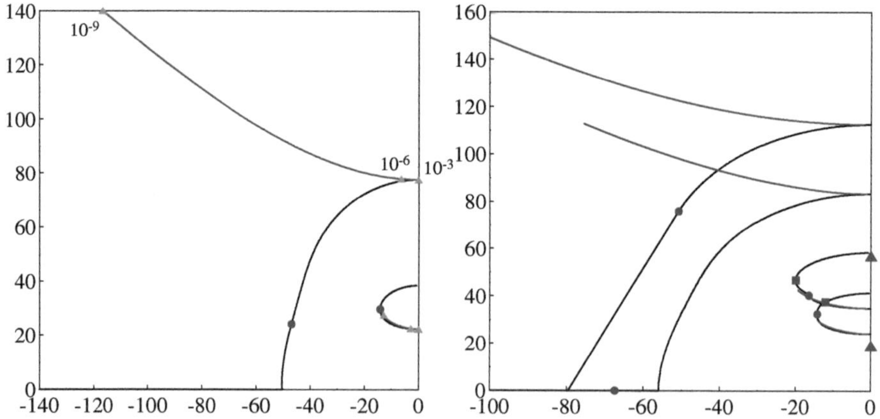

Fig. 3 Comparison between the root locus varying c and k and varying the r-parameter of the LQR: (a) 2-dof symmetric and (b) 4-dof non-symmetric frame structures

3.1 Semi-active Control Strategies

In recent years, different types of control algorithms for semi-active devices have been studied including *Lyapunov Stability Theory*, *Decentralized Bang-Bang Control*, *Maximum Energy Dissipation* and *Clipped-Optimal Control* [1].

In particular, the clipped-optimal control has been used in simulating the possible efficacy for real buildings [10] and implemented in prototype experiment [5]. Here, the clipped strategy logic is discussed on the basis of the overall design procedure. In particular, the controller is designed to perform as closely as possible to the linear optimal controller which defines the desired control force vector $\bar{\mathbf{f}}_{\mathrm{d}} = \{\bar{F}_{\mathrm{d1}}, \bar{F}_{\mathrm{d2}}\}^{\mathrm{T}}$. To force the i-th damper to generate approximately the corresponding desired optimal control force F_{ci}, the command signal υ_{i} is selected as follows. When the damper is providing the desired optimal force, the voltage applied to the damper should remain unchanged. If the force produced by the damper is lower than the desired optimal force and the two forces have the same sign, the voltage applied to the current driver is instantaneously increased to the maximum level admitted $\upsilon_{\mathrm{i}}^{\max}$, in order to increase the force produced by the damper aiming to match the desired control force. Otherwise, the commanded voltage is set to zero. Therefore the command signal follows the law

$$\upsilon_{\mathrm{i}} = \upsilon_{\mathrm{i}}^{\max} H\left((\bar{F}_{\mathrm{di}} - F_{\mathrm{di}})\, F_{\mathrm{di}}\right) \tag{16}$$

where H is the Heaviside function.

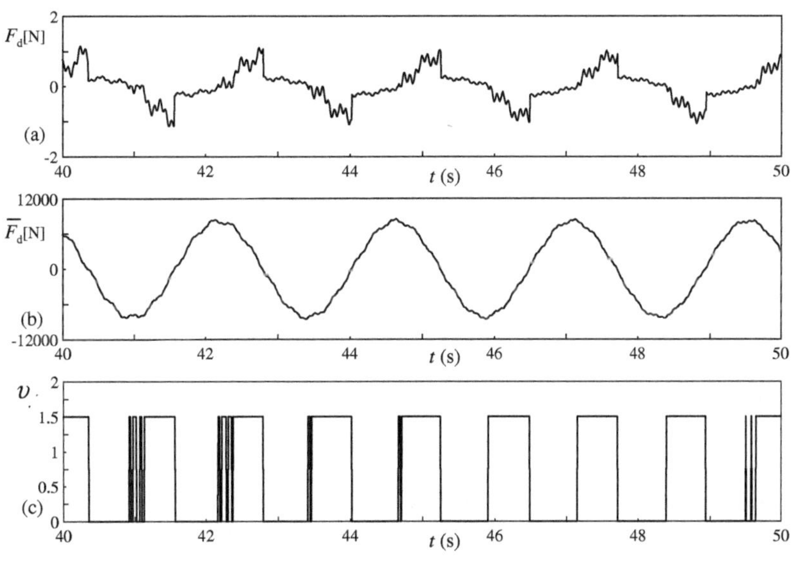

Fig. 4 Comparison between the semi-active force determined by the clipped-optimal algorithm and the reference active force in harmonic motion of the 2-dof system; (a) Kelvin-Voight model semi-active force, (b) reference active force; (c) applied voltage.

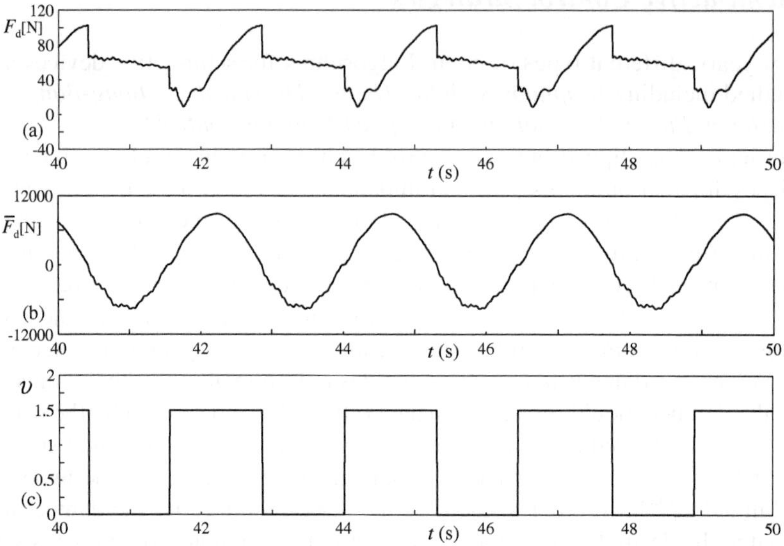

Fig. 5 Comparison between semi-active force determined by clipped-optimal algorithm and reference active force in harmonic motion of the 2-dof system: (a) Bouc-Wen model semiactive force, (b) reference active force; (c) applied voltage.

The clipped-optimal control has been tested applying a harmonic excitation to the 2-dof frame structure, equipped with a semi-active device modeled by a Kelvin-Voight device with viscous and stiffness coefficient varying in time, depending on the supplied voltage. The passive behavior of the device (voltage OFF) is selected on the basis of the optimal condition, Equation (8), while the voltage-depending part is requested to follow the reference active force according to the clipped optimal law (16). In Fig. 4 the comparison between the two forces is presented. It should be noted that most of the experimental studies available in the literature of semi-active control of prototypal structure employ magnetorheological (MR) dampers [3-5,7]. On this respect, the nonlinear behavior of the MR dampers may be described by the 9-parameter phenomenological model proposed by Spencer [5], in which the Bouc-Wen block is combined with a series dashpot and a parallel spring. The equation governing the relationship between the damper force F_d and the application point displacement u_d and velocity $\dot{u}_d$ is

$$F_d(\dot{u}_d, u_d) = c_1 \dot{\upsilon}_d + k_1 (u_d - u_{d0}) \tag{12}$$

where the evolution of the displacement variable υ_d and the internal auxiliary variable ζ is governed by a couple of differential equations

$$\dot{\upsilon}_d = (c_0 + c_1)^{-1} \left[k_0 (u_d - \upsilon_d) + c_0 \dot{u}_d + \alpha \zeta \right] \tag{13}$$

$$\dot{\zeta} = A(\dot{u}_d - \dot{\upsilon}_d) - \beta(\dot{u}_d - \dot{\upsilon}_d)|\zeta|^n - \gamma \zeta |\dot{u}_d - \dot{\upsilon}_d| |\zeta|^{n-1} \tag{14}$$

The coefficients k_0 and c_0 in the Bouc-Wen block assess the stiffness and damping at higher velocities, the stiffness k_1 of the spring accounts for the damper accumulator, while the series dashpot with viscosity c_1 reproduces the roll-off phenomenon. The parameters defining the Bouc-Wen model of the MR dampers are purposely tuned to simulate the experimental behavior of the commercial device Lord RD1005-3, as experimentally identified by dynamic tests [7]. In particular, the voltage-dependence of the significantly-varying coefficients $c_1(v)$, $c_0(v)$ and $A(v)$ has been described through a polynomial function interpolating to the identified results at different voltage amplitudes. In Fig. 5 the behavior of the available damper with respect the designed one is represented. Even if the clipped-optimal algorithms perform in the desired manner, a general deficiency of the device in terms of available deliverable force can be noted.

3.2 Implementation on a Prototype Structure

The results of an experimental campaign are here summarized. The project aimed to exploit the available technology in the wide area of earthquake engineering in developing design methods and implementation guidelines to improve civil construction code. To this end, a prototype frame structure was used as a benchmark study for different types of earthquake protection systems (Fig. 6). For this prototype, equipped by two magneto-rheological damper, acting in the direction of the column's minimum flexibility to the first floor, as first step, has been defined the analytical model, describing the three-dimensional motion formulated according to the direct displacement method [7]. To obtain a representative and reliable model, dynamical tests have been done for the updating of the parameters characterizing the mass and stiffness matrices.

Fig. 6 Prototypal building: (a) frame, (b) MR damper, (c) actuator, (d) accelerometers

Using the updated model, the performance of the semi-active control according to the clipped optimal strategy have been tested both numerically and experimentally. The modified Bouc-Wen block [5,7] has been implemented in the model to describe the nonlinear constitutive relationship for the MR dampers (Lord RD 1005-3) used in the experimental tests. The optimal control forces have been designed according to the H2/LQG method. Further insights and a detailed discussion of the numerical and experimental results are presented in [7].

4 Conclusion

The paper deals with the sizing of semi-active device for seismic protection of frame structures. The issue plays a fundamental role in the design process of enhanced dissipative bracings. The root loci of the controlled systems are used to determine both minimum (OFF) and maximum (ON) semi-active device characteristics. Last ones are searched looking at reference linear active control demand for the device force. The method is completed by a clipped-optimal non-collocated feedback used to change the applied voltage of magneto-rheological dampers. An experimental investigation has evidenced the needs of a clear design procedure.

References

[1] Soong, T.T., Spencer Jr., B.F.: Supplemental energy dissipation: state-of-the-art and state-of-the practice. Engineering Structures 24, 243–259 (2002)

[2] Yang, G., Spencer Jr., B.F., Carlson, J., Sain, M.: Large-scale MR fluid dampers: modelling and dynamic performance considerations. Engineering Structures 24, 309–323 (2002)

[3] Li, H.N., Li, X.L.: Experiment and analysis of torsional seismic responses for asymmetric structures with semi-active control by MR dampers. Smart Materials & Structures 18(7) (2009)

[4] Shook, D.A., Roschke, P.N., Lin, P.N.: Semi-active control of a torsionally-responsive structure. Engineering Structures 31(1), 57–68 (2009)

[5] Dyke, S.J., Spencer, B.F., Sain, M.K., Carlson, J.D.: Modeling and control of magnetorheological dampers for seismic response reduction. Smart Materials & Structures 5, 565–575 (1996)

[6] Ying, Z.G., Zhu, W.Q., Soong, T.T.: A stochastic optimal semiactive control strategy for er/mr dampers. Sound & Vibration 259, 45–62 (2002)

[7] Gattulli, V., Lepidi, M., Potenza, F.: Seismic protection of frame structures via semi-active control: modeling and implementation issues. Earthquake Engineering & Engineering Vibration 8, 627–645 (2009)

[8] Soong, T.T.: Active Structural Control: theory and practice. Wiley, New York (1990)

[9] Gattulli, V., Lin, R.C., Soong, T.T.: Nonlinear control laws for enhancement of structural control effectiveness. In: Proceedings of 5th U.S. National Conference for Earthquake Engineering, Chicago, Luglio 10-14, pp. 971–980 (1994)

[10] Yoshida, O., Dyke, S.J.: Response control of full-scale irregular buildings using magnetorheological dampers. Journal of Structural Engineering- ASCE 131, 734–742 (2005)

Iterative Procedures in Application of the LQGP Approach to the Duffing Oscillator

Piotr Kaczyński and Lesław Socha

Department of Mathematics and Natural Sciences,
Cardinal Stefan Wyszyński University in Warsaw, Poland

Abstract. The problem of the determination of the quasi-optimal control for the Duffing oscillator using the LQGP technique and a linearization method is considered. A few cases of these oscillators are considered including Gaussian and Poisson excitations both additive and multiplicative. Some sufficient conditions of convergence for the considered iterative procedure used in the evaluation of quasi-optimal control are derived. Obtained results are illustrated by a numerical example.

Keywords: Duffing oscillator, Gaussian and Poisson excitations, LQ Problem, Iterative methods.

1 Introduction

Nonlinear control problem for stochastic systems does not have analytical, exact solution. One of approximate methods is a combination of statistical linearization technique and LQG optimal control theory for linearized systems introduced by Wonham and Cashman [10]. This approach was developed due to its simplicity and easy applicability for systems with Gaussian excitations; see for instance [1] and [4, 11]. Such combination leads to an iterative procedure including consecutive solving of both Riccati and Lyapuov equations.

To obtain a quasi-optimal control strategy one can use iterative methods described above. Application of this method for the Duffing oscillator gives very good results for both Gaussian and non-Gaussian excitations [9].

Despite the fact that such iterative procedures are commonly used in applications, there are very few publications dealing with the convergence of such procedures. One can find sufficient conditions of the convergence in [5] for vector and scalar systems with Gaussian excitations. However, condition for vector case is very difficult to verify.

In this paper we consider the problem of the determination of the quasi-optimal control for the Duffing oscillator with Gaussian additive excitation and Gaussian and Poisson multiplicative excitations, using a combination of Gaussian statistical linearization and the LQGP technique (an extension of LQG technique to dynamic systems with Gaussian and Poisson excitations). The main goal is to derive sufficient conditions for convergence of the iterative procedures applied to the determination of control.

W.Q. Zhu et al. (Eds.): Nonlinear Stochastic Dynamics & Control, IUTAM BOOK SERIES 29, pp. 251–260.
springerlink.com

2 Problem Formulation

We will now present a general problem formulation with which we deal in this paper. Consider a standard stochastic optimal control problem for a polynomial dynamic system described by the Itô vector differential equation

$$dx(t) = [Ax(t) + \Phi(x(t)) + Bu(t)]\,dt + G_0 d\xi_0 + \sum_{k=1}^{M} G_k x(t) d\xi_k(t) + \int_{\mathbb{R}^n} D(v)x(t)\bar{\nu}(dt,dv) \quad (1)$$

where $x(t)$ is n-dimensional state vector, $u(t)$ is m-dimensional control vector and $A \in \mathbb{R}^{n\times n}$, $B \in \mathbb{R}^{n\times m}$ and $G_0 \in \mathbb{R}^n$, $G_k = [G_k^{pi}] \in \mathbb{R}^{n\times n}, k = 1,\ldots,M \in \mathscr{N}$ are time invariant system coefficients matrices, $\Phi : \mathbb{R}^n \to \mathbb{R}^n$ is a polynomial vector function, $C(v)\colon \mathbb{R} \to \mathbb{R}^n$ and $D(v)\colon \mathbb{R} \to \mathbb{R}^{n\times n}$ are matrix functions, $\xi_k(t), k = 0,\ldots,M$ denote mutually independent, standard scalar Wiener processes, each of which does not depend on the centered Poisson measure $\bar{\nu}(t,v)$, where $\bar{\nu}(t,v) = \nu(t,v) - E[\nu(t,v)] = \nu(t,v) - t\pi(v)$,

$$\pi(v) = \begin{cases} \lambda & \text{for } \{(1,\ldots,1)\} \subseteq v \\ 0 & \text{for } \{(1,\ldots,1)\} \nsubseteq v \end{cases}. \quad (2)$$

The steady state control strategy minimizes the criterion

$$I = E\left[\bar{x}^T Q \bar{x} + \bar{u}^T R \bar{u}\right], \quad (3)$$

where $\bar{x}$ and $\bar{u}$ denote stationary values of the state and control vectors, respectively; $Q \in \mathbb{R}^{n\times n}$ and $R \in \mathbb{R}^{m\times m}$ are time invariant positive definite matrices.

We assume that the stationary solution of equation (1) exists and that the polynomial vector function can be approximated by the linearized form $\Phi(x) \approx A_e x$, where A_e is a $n \times n$ matrix of linearization coefficients. One can find different criteria and methods for determining A_e in the literature [8]. In this paper the mean-sqare criterion is used. Substituting $\Phi(x)$ by $A_e x$ we obtain the linearized system corresponding to (1) which has the following form

$$dx_L(t) = [(A + A_e)x_L(t) + Bu(t)]\,dt + G_0 d\xi_0 + \sum_{k=1}^{M} G_k x_L(t) d\xi_k(t) + \int_{\mathbb{R}^n} D(v)x_L(t)\bar{\nu}(dt,dv) \quad (4)$$

where x_L is the n-dimensional state vector of the linearized system. The problem of finding optimal steady-state control for system (4) with criterion (3) is well established in the literature [7]. If $(A + A_e, B)$ pair is controllable, then the optimal control strategy is given by

$$\hat{u} = -R^{-1}B^T P \bar{x}_L, \quad (5)$$

where $\bar{x_L}$ is the stationary value of the linearized system and $P = [P_{pq}]$ is the positive definite solution of the Algebraic Riccati Equation (ARE)

$$(A+A_e)^T P + P(A+A_e) - PSP + L + Q + \lambda D^T PD = 0 \tag{6}$$

with $S = BR^{-1}B^T$ and

$$L = [L_{ij}] = \sum_{k=1}^{M} \sum_{p=1}^{n} \sum_{q=1}^{n} R_i^{pk} R_j^{qk} P_{pq},\ R_i^{pk} = G_k^{pi},\ i,j = 1,\dots,n. \tag{7}$$

The corresponding covariance matrix of the system can be evaluated using the Itô formula applied to system (4) and the averaging operation. It can be found by solving the Lyapunov algebraic matrix equation

$$(A+A_e-SP)V_L + V_L(A+A_e-SP)^T + G_0G_0^T + \sum_{k=1}^{M} G_k V_L G_k^T + \lambda D V_L D^T = 0, \tag{8}$$

where V_L denotes the $n \times n$ covariance matrix of the linearized system stationary response $V_L = E\left[\bar{x}_L \bar{x}_L^T\right]$ with control strategy (5) applied.

3 Quasi-Optimal Nonlinear Control Problem

A general solution of the problem of finding the optimal steady-state control strategy for system (1) with criterion (3) is unknown. However, one can use the Gaussian statistical linearization method to obtain linearized form of (1) and then find the optimal control strategy for this linearized system solving (6). There are numerous literature positions which contain methods and algorithms for finding the solution of ARE, for example [2, 3]. The only problem is, that the linearization coefficients matrix A_e is dependent on the variance of the controlled system. Since finding variance matrix for the nonlinear system (1) is also difficult, one can use solution of the Lyapunov equation (8) for the linearized system as an approximation of the variance of the nonlinear system. Assuming that the approximate response of nonlinear system (1) is a Gaussian process, we denote the general dependency of A_e on the variance matrix V_L as

$$A_e = F(V_L), \tag{9}$$

where $F\colon \mathbb{R}^{n\times n} \to \mathbb{R}^{n\times n}$ is a (possibly nonlinear) vector function.

Evaluations described above and their results strictly depend on each other: optimal solution for the linearized system (4) depends on linearization coefficients matrix A_e, but this matrix is a function of linearized system's variance, which in fact, is dependent on the control problem solution. This leads to the following iterative procedure for finding quasi-optimal solution of control problem for the nonlinear system (1) with criterion (3) [10, 1].

QOC Procedure. (Quasi-optimal control procedure in a nonlinear quadratic problem)

Step 1 Set $V_0 = 0$ and evaluate $A_{e_0} = F(V_k)$, $k = 0$.

Step 2 Find the positive definite solution of the ARE (6) using linearization matrix A_{e_k} and denote the solution as P_{k+1}

Step 3 Substitute the solution from Step 2 into the algebraic Lyapunov equation (8) for covariance matrix V of the linearized system (4) and find its solution denoting it as V_{k+1}.

Step 4 Using the covariance matrix V_{k+1} find the linearization coefficients matrix $A_{e_{k+1}}$ given by the function $A_{e_{k+1}} = F(V_{k+1})$, where F is implied by the criterion used for determination of linearization coefficients.

Step 5 If the deviation of V_{k+1} with respect to V_k or P_{k+1} with respect to P_k (i.e. using matrix norms $||V_{k+1} - V_k||$ and $||P_{k+1} - P_k||$, where $||X|| = \max_{i,j} |x_{ij}|$) is greater than a given precision parameter ε, set $k = k+1$ and go back to Step 2, otherwise the procedure is finished and V_{k+1} and P_{k+1} are the quasi-optimal solutions of the polynomial control problem.

Procedures like QOC Procedure are commonly used in applications and in theoretical papers [10, 1, 9, 4]. However, there are very few publications which deal with convergence of those procedures. One can find some sufficient conditions for convergence in [5] derived for simplified system (1) (additive Gaussian excitation only) in scalar and vector case.

4 Procedure Convergence for the Duffing Oscillator

In this section we find sufficient conditions for the convergence of QOC Procedure when applied to the Duffing oscillator with Gaussian and Poisson additive and multiplicative noises. We consider system (1) for n=2, M=2 and the following parameters

$$A = \begin{bmatrix} 0 & 1 \\ -\omega_0^2 & -2\zeta\omega_0 \end{bmatrix}, B = \begin{bmatrix} 0 \\ b \end{bmatrix}, \Phi(x) = \begin{bmatrix} 0 \\ -\varepsilon x_1^3 \end{bmatrix}, G_0 = \begin{bmatrix} 0 \\ g_0 \end{bmatrix}$$

$$G_1 = \begin{bmatrix} 0 & 0 \\ g_1 & 0 \end{bmatrix}, G_2 = \begin{bmatrix} 0 & 0 \\ 0 & g_2 \end{bmatrix}, D(v) = \begin{bmatrix} 0 & 0 \\ d_1 & d_2 \end{bmatrix}, \tag{10}$$

where $x(t) = \begin{bmatrix} x_1(t) \; x_2(t) \end{bmatrix}^T$ is 2-dimensional state vector, $u(t) \in \mathbb{R}$ is the scalar control variable, $\omega_0, \zeta, b, g_0, g_1, g_2, d_1, d_2$ and ε are all positive real values.

The control strategy minimizes stationary criterion (3) with

$$Q = \begin{bmatrix} q_1 & 0 \\ 0 & q_2 \end{bmatrix}, \; R = r, \tag{11}$$

where $q_1, q_2, r \in \mathbb{R}$. In order to find quasi-optimal solution of the control problem, we use the mean-square criterion to linearize function Φ as follows [8]

$$\Phi(x) \approx A_e x = \begin{bmatrix} 0 & 0 \\ -3\varepsilon v_{11} & 0 \end{bmatrix} \begin{bmatrix} x_1 \\ x_2 \end{bmatrix} \tag{12}$$

The corresponding Riccati equation for the linearized system is simplified to the following set of three equations

$$-2\omega_0^2 p_{12} - 6\varepsilon p_{12} v_{11} - s p_{12}^2 + (g_1^2 + \lambda d_2^2) + q_1 = 0 \tag{13}$$

$$p_{11} - 2\zeta\omega_0 p_{12} - \omega_0^2 p_{22} - 3\varepsilon p_{22} v_{11} - s p_{12} p_{22} = 0 \tag{14}$$

$$2p_{12} - 4\zeta\omega_0 p_{22} - s p_{22}^2 + g_2^2 p_{22} + q_2 + \lambda d_2^2 p_{22} = 0, \tag{15}$$

where $s = \frac{b^2}{r}$. The Lyapunov equation is simplified to

$$2v_{12} = 0 \tag{16}$$

$$v_{22} - \left(\omega_0^2 + s p_{12} + 3\varepsilon v_{11}\right) v_{11} - (2\zeta\omega_0 + s p_{22}) v_{12} = 0 \tag{17}$$

$$-2\left(\omega_0^2 + 3\varepsilon v_{11} + s p_{12}\right) v_{12} - 2(2\zeta\omega_0 + s p_{22}) v_{22} + g_0^2 + g_1^2 v_{11} + g_2^2 v_{22} + \lambda d_1^2 v_{11} + \lambda d_2^2 v_{22} = 0 \tag{18}$$

We will now deal with a specific case of system (10) with $g_1 = 0$ and $d_1 = 0$. The following theorem gives sufficient conditions of convergence of the QOC Procedure in this specific case.

Theorem 1. *Consider the Duffing oscillator system* (1) *with parameters* (10) *with* $g_1 = 0, d_1 = 0$ *and control criterion* (3) *with parameters* (11). *If there exists* α_0, $0 < \alpha_0 < 1$ *such, that the following inequalities are satisfied:*

$$3\varepsilon < s \tag{19}$$

$$s\sqrt{\left(g_2^2 + \lambda d_2^2 - 4\zeta\omega_0\right)^2 + 4sq_2} > 1 \tag{20}$$

$$2sg_0^2 < (4\zeta\omega_0 - g_2^2 - \lambda d_2^2)^2 \tag{21}$$

$$\frac{s}{6\varepsilon} + \frac{1}{\omega_0^2} < \alpha_0 \tag{22}$$

where $s = \frac{b^2}{r}$, *then QOC Procedure applied to system* (10) *is convergent.*

Proof. *See Appendix.*

Note that inequalities (20) and (21) can be replaced by the following condition

$$\left(g_2^2 - 4\zeta\omega_0 - \lambda d_2^2\right)^2 > \max\left\{\left(\frac{1}{s^2} - 4sq_2\right),\ 2sg_0^2\right\} \tag{23}$$

We will now deal with a general case. Consider system (1) with $M = 2, n = 2$ and parameters g_1 and d_1 not equal to zero. Their presence allows us to take into account noises not only in dumping, but also in stiffness.

We will now present the theorem which shows, that for a special set of parameters, the QOC Procedure is convergent.

Theorem 2. *Consider Duffing oscillator system* (1) *with parameters* (10) *and control criterion* (3) *with parameters* (11). *For sufficiently small value of* g_0^2 *the QOC Procedure is convergent.*

Proof. *Similarly to the proof of Theorem 1 we evaluate the derivative of* $v_{11}^{(k+1)}$ *as a function of* v_{11}^k, *which can be presented in the following form*

$$\frac{\partial v_{11}^{(k+1)}}{\partial v_{11}^k} = -\frac{g_0^2 \frac{\partial \mathscr{F}(v_{11}^k)}{\partial v_{11}^k}}{\left(\mathscr{F}(v_{11}^k)\right)^2} \tag{24}$$

where $\mathscr{F} : \mathbb{R} \to \mathbb{R}$ *is a scalar, Lipschitz function. We can also show that both* p_{12} *and* p_{22} *are positive and bounded. Finally we can state that for sufficiently small* g_0^2 *there exists* $0 < \alpha_0 < 1$ *for which*

$$\frac{\partial v_{11}^{(k+1)}}{\partial v_{11}^k} < \alpha_0 \tag{25}$$

5 Example

In this section we present some applications of theorems we derived in this paper. We also make a practical comparison of the proposed QOC Procedure and classical algorithms for finding function zeros.

In this example we will show the application of the Theorem 1. Consider the Duffing oscillator, that is system (1),(2) with parameters (10) and control criterion (3). Assume $d_1 = 0$ and $g_1 = 0$.

To find the sufficient condition for convergence of the QOC Procedure we use conditions (19)-(22). Note, that these conditions are only sufficient. That means, that some systems may not satisfy those conditions, however the QOC Procedure may be convergent. Especially condition (19) is very restrictive.

We will now deal with specific set of parameters. Assume $\alpha_0 = 0.99$ and $\varepsilon = 0.3, \zeta = 1, \omega_0 = 2, r = b = 1, \lambda = 1, g_0 = 1, g_2 = 2, q_1 = q_2 = 1, d_2 = 1, g_1 = d_1 = 0$. It is easy to show, that conditions (19)-(22) are satisfied. Therefore, applying Theorem 1 we find, that the QOC Procedure is convergent. Numerical tests showed, that subsequent values of v_k evaluated using QOC Procedure converged in 6 iterations.

Further numerical tests showed, that it is very difficult to find a set of parameters which would make the QOC Procedure non convergent. In fact, we did not find such set of parameters. It is worth noting, that this is not a general case for all systems – this depicts only the Duffing oscillator. One can find examples of other systems (even scalar systems) [5], for which the QOC Procedure is not convergent. Further work should be done to find less restrictive sufficient conditions than those in Theorem 1.

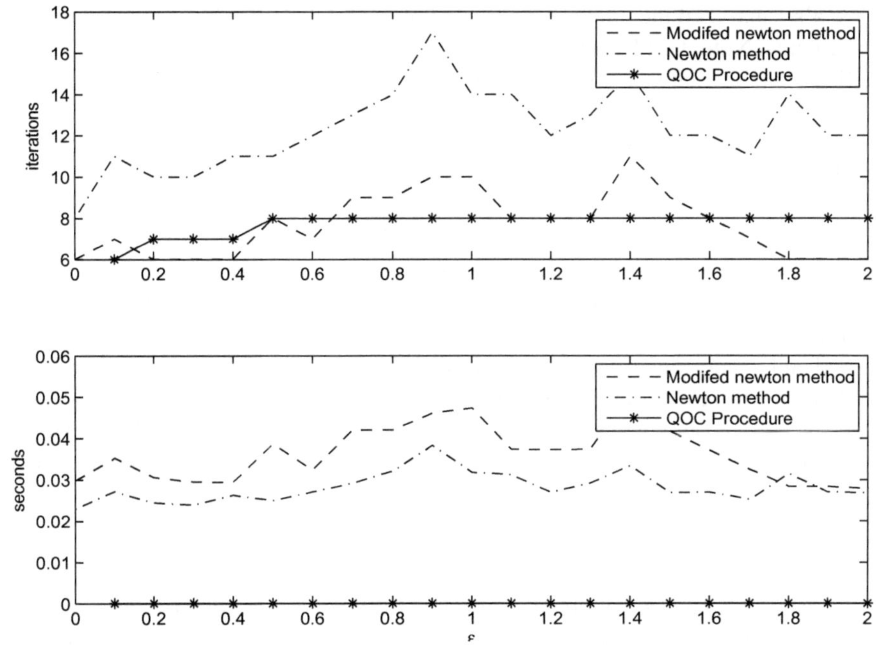

(a) Convergence comparison versus parameter ε

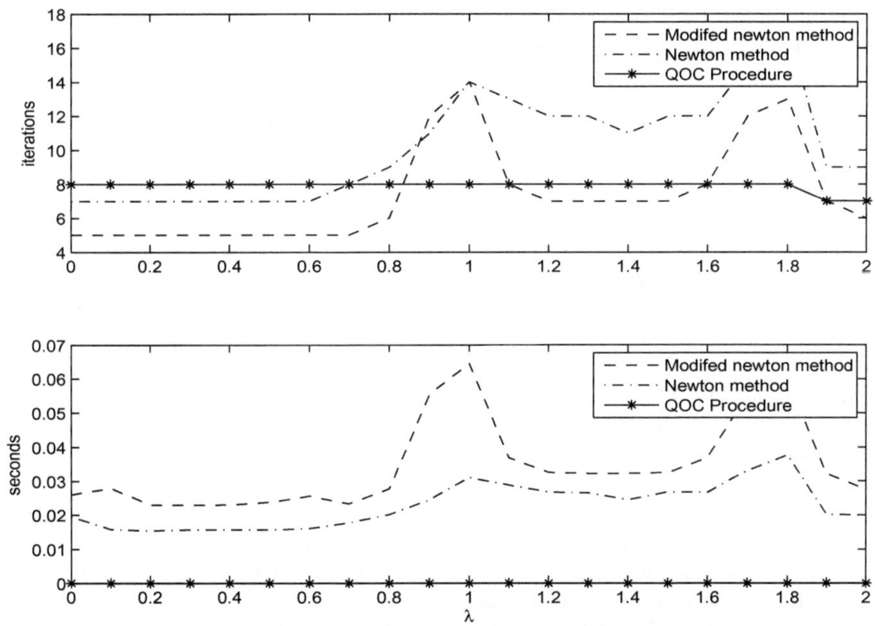

(b) Convergence comparison versus parameter λ

Fig. 1 Convergence speed comparison with precision $\varepsilon = 10^{-8}$

We also analyzed the convergence speed of the QOC Procedure using numerical experiments. Note, that Lyapunov equation (8) and Riccati equation (6) for Duffing oscillator can be treated as a set of 6 equations. That means, that actually, they can be solved using classic methods. We have compared the QOC Procedure (when using equations (13)-(18)) with classic Newton method [3] and its modification, which was described in recent work [6].

Comparison results are roughly presented in Figure 5. The number of iterations required to achieve selected precision is usually the highest for classic Newton's method. Modified Newton's method [6] and evaluation using QOC Procedure are more or less equal, depending on the set of parameters. However, if we compare execution times, not the iterations, proposed method is far better, while Newton's method and its modification are comparable. The execution time was measured on a standard desktop PC (with Intel Core 2 Duo 2GHz processor). The QOC Procedure achieves assumed precision about ten times faster than two other methods. This is due to the fact, that Newton methods need to evaluate the gradient of the function.

6 Conclusions

In this paper we have considered the problem of the determination of quasi-optimal control for the Duffing oscillator with parametric and external Gaussian and Poisson excitations. To solve this problem we have used the iterative procedure, called QOC Procedure, that combines Gaussian statistical linearization and LQGP technique. Furthermore, we have presented two theorems depicting sufficient conditions for convergence of the QOC Procedure. An example, which proved applicability of those theorems and compared the convergence speed of proposed procedure with classic Newton's method and its recent modification from [6], was presented.

References

1. Beaman, J.: Nonlinear quadratic Gaussian control. Int. J. Cont. 39, 343–361 (1984)
2. Bittani, S., Laub, A., Willems, J.C.: The Riccati Equation. Springer, Berlin (1991)
3. Golub, G.H., Van Loan, C.F.: Matrix Computations. North Oxford Academic, London (1986)
4. Han, S.I., Kim, J.S.: Nonlinear quadratic Gaussian control with loop transfer recovery. Mechatronics 13, 273–293 (2003)
5. Kaczyński, P., Socha, L.: Iterative procedures in application of lqg approach to control problems for nonlinear stochastic systems. In: Proceedings of the American Control Conference, ACC 2007, pp. 1753–1758 (2007)
6. Noor, M.A., Waseem, M.: Some iterative methods for solving a system of nonlinear equations. Computers and Mathematics with Applications 57, 101–106 (2009)
7. Paraev, J.: Introduction to Statistical Dynamics of Control and Filtering Processes. Sovetskoe Radio, Moscow (1976) (in Russian)
8. Socha, L.: Linearization Methods for Stochastic Dynamic Systems. Springer, Heidelberg (2008)
9. Socha, L., Proppe, C.: Control of the Duffing oscillator under non-Gaussian external excitation. Europ. J. Mech. A/Solid 6, 1069–1082 (2002)

10. Wonham, W.M., Cashman, W.F.: A computational approach to optimal control of stochastic saturating systems. Int. J. Cont. 10, 77–98 (1969)
11. Ying, Z.G., Ni, Y.Q., Ko, J.M.: Semi–active optimal control of linearized systems with multi–degree of freedom and applications. J. Sound Vib. 279, 373–388 (2005)
12. Zeidler, E.: Applied Functional Analysis, Applications to Mathematical Physics. Springer, New York (1995)

Appendix: Proof of Theorem 1

In order to show the convergence we consider only the variance v_{11}. We shall denote v_{11} evaluated in the k-th iteration of the procedure as v_{11}^k. We derive the element p_{12} of the Riccati matrix P using (13). Since (13) is the second order algebraic equation with respect to p_{12} it is easy to find the positive solution. Similarly we can derive p_{22} from equation (15) assuming that p_{12} is known. Note, that we don't have to evaluate p_{11} in order to derive v_{11} from equations (17)-(18). The next step in the proof is to find the solution of the Lyapunov equation (8). We derive v_{22} from (18) and finally we derive the solution of (17), which is also the second order algebraic equation with respect to v_{11} and we denote this solution as the concurrent iteration's solution $v_{11}^{(k+1)}$. We then find the derivative of p_{12}

$$\frac{\partial p_{12}}{\partial v_{11}^k} = -\frac{6\varepsilon}{2s} f_1(v_{11}^k),\ f_1(v_{11}^k) = \left(1 - \frac{(2\omega_0^2 + 6\varepsilon v_{11}^k)}{\sqrt{(2\omega_0^2 + 6\varepsilon v_{11}^k)^2 + 4sq_1}}\right). \tag{26}$$

Since $v_{11} \geq 0$ and $s, q_1, \omega_0, \varepsilon > 0$ and using both (19) and the fact that $f_1(v_{11}^k) < 1$ for all $v_{11}^k \geq 0$, we have $-1 < \frac{\partial p_{12}}{\partial v_{11}^k} < 0$.

We perform a similar calculation of the derivative of p_{22}

$$\frac{\partial p_{22}}{\partial v_{11}^k} = \frac{4}{4s\sqrt{(g_2^2 + \lambda d_2^2 - 4\zeta\omega_0)^2 + 4s(q_2 + 2p_{12})}} \frac{\partial p_{12}}{\partial v_{11}^k}. \tag{27}$$

Using (20) and the fact, that $p_{12} \geq 0$ we can easily show, that $-1 < \frac{\partial p_{22}}{\partial v_{11}^k} < 0$. Again we use similar calculation of the derivative of v_{22}. We get

$$\frac{\partial v_{22}}{\partial v_{11}^k} = -\frac{2sg_0^2}{(4\zeta\omega_0 + 2sp_{22} - g_2^2 - \lambda d_2^2)^2} \frac{\partial p_{22}}{\partial v_{11}^k}. \tag{28}$$

Using (21), derivative bounds of p_{22} and the fact, that $p_{22} \geq 0$, it is straightforward to show, that $0 < \frac{\partial v_{22}}{\partial v_{11}^k} < 1$. Finally we evaluate the derivative of $v_{11}^{(k+1)}$ as a function of v_{11}^k. We get

$$\frac{\partial v_{11}^{(k+1)}}{\partial v_{11}^k} = -\frac{s}{6\varepsilon}\left(1 - \frac{\omega_0^2 + sp_{12}}{f_2(p_{12}, v_{22})}\right) \frac{\partial p_{12}}{\partial v_{11}^k} + \frac{1}{f_2(p_{12}, v_{22})} \frac{\partial v_{22}}{\partial v_{11}^k} \tag{29}$$

where $f_2(p_{12}, v_{22}) = \sqrt{(\omega_0^2 + s p_{12})^2 + 12\varepsilon v_{22}}$. Assuming that there exists a constant $0 < \alpha_0 < 1$ which satisfies (22) and taking into account bounds of derivatives of p_{12} and v_{22}, nonnegativity of both p_{12} and v_{22} and finally the condition (22) we can state, that

$$0 < \frac{\partial v_{11}^{(k+1)}}{\partial v_{11}^k} < \alpha_0. \tag{30}$$

Application of Banach's fixed point theorem [12] gives the convergence of the procedure because the function which evaluates $v_{11}^{(k+1)}$ in terms of v_{11}^k is contractive.

Stochastic Optimal Time-Delay Control and Stabilization of Quasi-Integrable Hamiltonian Systems

Z.H. Liu[1] and W.Q. Zhu[2]

[1] Department of Civil Engineering, Xiamen University, Xiamen, Fujian 361005, China
[2] Department of Mechanics, Zhejiang University, Hangzhou, Zhejiang 310027, China

Abstract. Innovative procedures for the stochastic optimal time-delay control and stabilization are proposed for a quasi-integrable Hamiltonian system subject to Gaussian white noises. First, the problem of stochastic optimal control with time delay is formulated. Then, the problem is converted into a stochastic optimal control without time delay, and the converted control problems are then solved by applying the stochastic averaging method and the stochastic dynamical programming principle. The optimal time-delay stabilization of quasi-integrable Hamiltonian systems is formulated as an ergodic control with a cost function determined by minimizing the largest Lyapunov exponent of the controlled system. As an example, a two-degree-of-freedom quasi-integrable Hamiltonian system with time delay in feedback control forces is investigated in detail to illustrate the procedures and their effectiveness.

Keywords: Time-delay feedback control, Stochastic averaging method, Stochastic optimal control, Feedback stabilization.

1 Introduction

The time delay in feedback control can be caused by physical properties of control equipments, measurements of system states, filtering and data processing, calculating and executing control forces, etc.. The time delay in feedback control may not only deteriorate the performance of controlled systems but also destabilize the controlled systems. The dynamics of time-delay feedback controlled systems under stochastic excitation has been analyzed by using several methods. Recently, a stochastic averaging procedure was proposed for quasi-integrable Hamiltonian systems with time-delayed feedback control, and has been applied to predict the response, stochastic stability, stochastic Hopf bifurcation [1-3]. In principle, the dynamical programming principle can be extended to the stochastic optimal control problems with time delay [4]. However, only a few problems are solvable and most problems are practically intractable since the resulting Hamilton-Jacobi-Bellman (HJB) equation is of infinite-dimensions [5]. In the present paper, innovative procedures for the stochastic optimal time-delay control and stabilization of quasi-integrable Hamiltonian systems are proposed and illustrated with an example.

W.Q. Zhu et al. (Eds.): Nonlinear Stochastic Dynamics & Control, IUTAM BOOK SERIES 29, pp. 261–270.
springerlink.com

2 Formulation of Stochastic Optimal Time-Delay Control

Consider an n-degree-of-freedom quasi-Hamiltonian system with time-delay feedback control, governed by the following equations:

$$\dot{Q}_i = \frac{\partial H'}{\partial P_i}$$
$$\dot{P}_i = -\frac{\partial H'}{\partial Q_i} - \varepsilon c_{ij}\frac{\partial H'}{\partial P_j} + \varepsilon u_i(\mathbf{Q}_\tau, \mathbf{P}_\tau) + \varepsilon^{1/2} f_{ik} W_k(t) \tag{1}$$
$$i, j = 1, 2, \ldots, n; \quad k = 1, 2, \ldots, m.$$

where Q_i and P_i are generalized displacements and momenta, respectively; $H' = H'(\mathbf{Q}, \mathbf{P})$ is a twice differentiable Hamiltonian; ε is a small parameter; $\varepsilon c_{ij} = \varepsilon c_{ij}(\mathbf{Q}, \mathbf{P})$ represent the coefficients of quasi-linear dampings; $\varepsilon^{1/2} f_{ik} = \varepsilon^{1/2} f_{ik}(\mathbf{Q}, \mathbf{P})$ represent the magnitudes of stochastic excitations; $\varepsilon u_i(\mathbf{Q}_\tau, \mathbf{P}_\tau)$ with $\mathbf{Q}_\tau = \mathbf{Q}(t-\tau)$ and $\mathbf{P}_\tau = \mathbf{P}(t-\tau)$ denote feedback control forces with time delay τ; $W_k(t)$ are Gaussian white noises in the sense of Stratonovich with zero mean and correlation functions $E[W_k(t)W_l(t+T)] = 2D_{kl}\delta(T)$.

The objective of stochastic optimal control is to minimize a performance index

$$J(\mathbf{u}) = \lim_{T\to\infty}\frac{1}{T}\int_0^T L[\mathbf{Q}, \mathbf{P}, \mathbf{u}(\mathbf{Q}_\tau, \mathbf{P}_\tau)]dt \tag{2}$$

for semi-infinite time-interval ergodic control. In Eq.(2), $L[\mathbf{Q}, \mathbf{P}, \mathbf{u}(\mathbf{Q}_\tau, \mathbf{P}_\tau)]$ is the cost function, which is a continuous, differentiable and convex function. Eqs.(1) and (2) constitute a stochastic optimal time-delay control problem of quasi-integrable Hamiltonian system.

3 Conversion to Stochastic Optimal Control Problems without Time-Delay

Assume that Hamiltonian H' in Eq.(1) is separable and of the form

$$H' = \sum_{i=1}^{n} H_i'(q_i, p_i), \quad H_i' = \frac{1}{2}p_i^2 + G(q_i) \tag{3}$$

where $G(q_i) \geq 0$ is symmetric with respect to $q_i = 0$ and with minimum at $q_i = 0$. The conservative system corresponding to (1) has a family of periodic solutions. For weak dampings, weak excitations, and small time delay τ, the following approximate expressions can be used

$$\begin{aligned} P_i(t-\tau) &\doteq P_i(t)\cos\omega_i\tau + Q_i(t)\omega_i\sin\omega_i\tau \\ Q_i(t-\tau) &\doteq Q_i(t)\cos\omega_i\tau - [P_i(t)\sin\omega_i\tau]/\omega_i \end{aligned} \tag{4}$$

Thus the time-delay feedback control forces can be expressed approximately in terms of the state variables without time delay, i.e., $u_i(\mathbf{Q}_\tau, \mathbf{P}_\tau) = u_i(\mathbf{Q}, \mathbf{P}; \tau)$. By using this approximation and considering possible Wong-Zakai correction terms, Eq.(1) can be rewritten as

$$\begin{aligned} dQ_i &= \frac{\partial H}{\partial P_i}dt \\ dP_i &= \left[-\frac{\partial H}{\partial Q_i} - \varepsilon m_{ij}\frac{\partial H}{\partial P_j} + \varepsilon u_i(\mathbf{Q}, \mathbf{P}; \tau)\right]dt + \varepsilon^{1/2}\sigma_{ik}(\mathbf{Q}, \mathbf{P})dB_k(t) \end{aligned} \tag{5}$$

where $H = H(\mathbf{Q}, \mathbf{P}; \tau)$ is a modified Hamiltonian, which is still assumed to be separable, i.e., $H = H_1 + H_2 + \cdots + H_n$, $B_k(t)$ are unit Wiener processes, and $\sigma_{ik}\sigma_{jk} = 2D_{kl}f_{ik}f_{jl}$. Accordingly, the performance index in Eq.(2) is modified to

$$J(\mathbf{u}) = \lim_{T\to\infty}\frac{1}{T}\int_0^T L[\mathbf{Q}, \mathbf{P}, \mathbf{u}(\mathbf{Q}, \mathbf{P}; \tau)]dt \tag{6}$$

Eqs.(5) and (6) constitute a stochastic optimal control problem without time delay, converted from the original problem (1) and (2).

Applying the stochastic averaging method for quasi-integrable Hamiltonian systems [6] to system(5), in the non-resonance case, the following partially averaged Itô stochastic differential equations are obtained:

$$dH_r = [\bar{m}_r(\mathbf{H}) + \left\langle u_i\frac{\partial H_r}{\partial P_i}\right\rangle]dt + \bar{\sigma}_{rk}(\mathbf{H})dB_k(t) \tag{7}$$

where $\mathbf{H} = [H_1, H_2, \cdots, H_n]^{\mathrm{T}}$, $\langle\cdot\rangle$ denotes averaging operation; $\bar{m}_r(\mathbf{H})$ and $\bar{\sigma}_{rk}(\mathbf{H})$ are, respectively, the averaged drift coefficients and diffusion coefficients.

Eq.(7) implies that $\mathbf{H}$ is a controlled diffusion vector process. Correspondingly, partial average performance index (6) becomes

$$J(\mathbf{u}) = \lim_{T\to\infty}\frac{1}{T}\int_0^T L(\mathbf{H},\langle\mathbf{u}\rangle)dt \tag{8}$$

By applying the stochastic dynamical programming principle [7] to Eq.(7) and (8), the following dynamical programming equation can be established:

$$\inf_{\mathbf{u}}\{\frac{1}{2}\bar{\sigma}_{rk}\bar{\sigma}_{sk}\frac{\partial^2 V}{\partial H_r \partial H_s} + [\bar{m}_r(\mathbf{H}) + \left\langle u_i \frac{\partial H_r}{\partial P_i}\right\rangle]\frac{\partial V}{\partial H_r} + L(\mathbf{H},\langle\mathbf{u}\rangle)\} = \gamma \tag{9}$$

where $V(\mathbf{H})$ is called the value function, and

$$\gamma = \lim_{T\to\infty}\frac{1}{T}\int_0^T L(\mathbf{H}(t),\langle\mathbf{u}^*(t)\rangle)dt \tag{10}$$

is the optimal average cost function and $\mathbf{u}^*(t)$ is the optimal control.

Let L be quadratic with respect to $\mathbf{u}$, i.e.,

$$L(\mathbf{H},\langle\mathbf{u}\rangle) = f_1(\mathbf{H}) + \langle\mathbf{u}^{\mathrm{T}}\mathbf{R}\mathbf{u}\rangle \tag{11}$$

where $f_1(\mathbf{H}) \ge 0$ and $\mathbf{R}$ is a diagonal matrix with positive elements R_i. Minimizing the left-hand side of Eq.(9) with respect to $\mathbf{u}$ yields

$$u_i^* = -\frac{1}{2R_i}\frac{\partial V}{\partial H_r}\frac{\partial H_r}{\partial P_i} \tag{12}$$

Substituting Eq.(12) into Eq.(9) and averaging the terms involving u_i^* lead to the final dynamical programming equation. Solving the equation and substituting the resultant $V(\mathbf{H})$ into Eq.(12) yield the optimal control. Notc that $\mathbf{u}^*$ are generally nonlinear in Q_i and P_i. The reverse of Eq.(4) is

$$\begin{aligned} Q_i(t) &\doteq Q_i(t-\tau)\cos(\omega_i\tau) + P_i(t-\tau)\sin(\omega_i\tau)/\omega_i \\ P_i(t) &\doteq P_i(t-\tau)\cos(\omega_i\tau) - Q_i(t-\tau)\omega_i\sin(\omega_i\tau) \end{aligned} \tag{13}$$

Substituting Eq.(13) into Eq.(12), the following time-delay optimal control can be obtained:

$$u_i^*(\mathbf{Q}_\tau,\mathbf{P}_\tau) = -\frac{1}{2R_i}\frac{\partial V}{\partial H_r}\frac{\partial H_r}{\partial P_i}\bigg|_{\mathbf{Q}=\mathbf{Q}(\mathbf{Q}_\tau,\mathbf{P}_\tau),\ \mathbf{P}=\mathbf{P}(\mathbf{Q}_\tau,\mathbf{P}_\tau)} \tag{14}$$

Substituting $u_i^*(\mathbf{Q}_\tau,\mathbf{P}_\tau)$ in Eq.(14) into Eq.(7) and averaging the terms involving u_i^* yield

$$dH_r = [\bar{\bar{m}}_r(\mathbf{H})]dt + \bar{\sigma}_{rk}(\mathbf{H})dB_k(t),\ \ \bar{\bar{m}}_r(\mathbf{H}) = \bar{m}_r(\mathbf{H}) + \left\langle u_i^*\frac{\partial H_r}{\partial P_i}\right\rangle \tag{15}$$

The response of the controlled quasi-integrable Hamiltonian system can be predicted by solving the Fokker-Planck-Kolmogorov (FPK) equation corresponding to the fully averaged Itô Eq.(15). The root mean square displacements σ_h^u, σ_h^c of uncontrolled and controlled systems and the root mean square control forces σ_{u^*} can also be calculated. To characterize the performance of a controller, two quantities are introduced. One is the control effectiveness, defined as

$$K = \frac{\sigma_h^u - \sigma_h^c}{\sigma_h^u} \tag{16}$$

K indicates the relative reduction of the root mean square displacement due to the control. The other is control efficiency, which is defined as the ratio of reduction to root mean-square control force, i.e.,

$$\mu = K / \sigma_{u^*} \tag{17}$$

Obviously, the higher K and μ are, the better the controller is.

4 Time-Delay Feedback Stabilization

Now consider the case in which the stochastic excitations in Eq.(1) are pure parametric, the trivial solution is an equilibrium, and it may be unstable without control. Following the same derivation, we obtain Eq.(15) for the controlled system. Since the stochastic excitations are pure parametric, the drift and diffusion coefficients of Eq.(15) satisfy the following conditions:

$$\bar{\bar{m}}_r(0) = 0, \bar{\sigma}_{rk}(0) = 0 \tag{18}$$

indicating that $\boldsymbol{H}$=0 is the trivial solution for the controlled system. Since the stability of the trivial solution is considered, the system can be linearized about the trivial solution, and $\bar{\bar{m}}_r$ and $\sigma_{rk}\sigma_{sk}$ are homogeneous in H_s of degree one. We assume that diffusion process $\mathbf{H}(t)$ is nonsingular. Introduce the following new variables

$$\rho = (\ln H)/2, \quad \alpha_r = H_r / H\ . \tag{19}$$

The Itô equations for ρ and α_r are obtained as:

$$d\rho = Q^c(\boldsymbol{\alpha})dt + \Sigma_k(\boldsymbol{\alpha})dB_k(t) \tag{20a}$$

$$d\alpha_r = m_r^c(\boldsymbol{\alpha})dt + \sigma_{rk}(\boldsymbol{\alpha})dB_k(t) \tag{20b}$$

where $\boldsymbol{\alpha} = [\alpha_1, ..., \alpha_{n-1}, \alpha_n]^{\mathrm{T}}$, $r = 1,2,...,n; k = 1,2,...,m.$ Note that only n-1 equations for α_r in Eq.(20b) are independent; thus we can use $\boldsymbol{\alpha}' = [\alpha_1, ..., \alpha_{n-1}]^{\mathrm{T}}$ to replace $\boldsymbol{\alpha}$ in (20a) and (20b). It is noted that Eq. 20(b) is independent on Eq. (20a), and the stationary probability density of $\boldsymbol{\alpha}'$ can be obtained.

Define the Lyapunov exponent of averaged system (15) as the asymptotic rate of the exponential growth of the square root of H , i.e.,

$$\lambda = \lim_{t\to\infty}\frac{1}{t}\ln H^{1/2} \tag{21}$$

The largest Lyapunov exponent of controlled system (15) can be obtained as

$$\lambda_1^c = \int Q^c(\boldsymbol{\alpha}')p^c(\boldsymbol{\alpha}')d\boldsymbol{\alpha}' \tag{22}$$

where $Q^c(\boldsymbol{\alpha}')$ is obtained from $Q^c(\boldsymbol{\alpha})$ in Eq.(20a). and $p^c(\boldsymbol{\alpha}')$ is the stationary probability density of $\boldsymbol{\alpha}'$ obtained from solving the reduced FPK equations associated with Itô Eq.(20b). The necessary and sufficient condition for asymptotic Lyapunov stability with probability one of the trivial solution of Eq.(15) is $\lambda_1^c < 0$. It can also be considered as the approximate condition for asymptotic Lyapunov stability with probability one of the trivial solution of original system (1). Let $u_i^* = 0$, the largest Lyapunov exponent λ_1^u of the uncontrolled system can also be obtained.

The difference of the Lyapunov exponent between controlled and uncontrolled system is $\lambda^c - \lambda^u$. If it is negative, the system is more stable (or less unstable). Furthermore, if λ_1^c is negative, the trivial solution is asymptotic stable. Thus, the feedback stabilization can bc achieved by proper setting the cost function in Eq.(2) so that λ_1^c is negative and minimized.

Apply the dynamical programming Eq.(9) and also assume the cost function of the form of (11). Design of the feedback stabilization is actually to select $f_1(\mathbf{H})$ and $\mathbf{R}$ to make λ_1^c negative and minimized. In the following, an example is given to illustrate the designing procedure in detail.

5 Example

As an example, consider system

$$\ddot{X}_i + \alpha_{ij}\dot{X}_j + \dot{X}_i(\beta_j X_j^2) + \omega_i^2 X_i = u_{i\tau} + k_{ij}(X)W_j(t)\ , \ \ i,j=1,2 \tag{23}$$

where $W_j(t)$ are uncorrelated Gaussian white noises with small intensities $2D_i$; $u_{i\tau}$ are time-delay feedback control forces. The Hamiltonian system associated with Eq.(23) is linear and integrable.

First consider the case of purely external stochastic excitations, i.e., $k_{ij}(X)=k_{ij}$. Assume that $\omega_1/\omega_2=r/s$, where r,s are prime integers. The partially averaged performance index is of the form of Eq.(8) and the cost function is of the form of Eq.(11) with $\mathbf{H}=[H_1,H_2]^T$, $\mathbf{u}=[u_1,u_2]^T$, $R=diag[R_1,R_2]^T$, and

$$f_1(\mathbf{H})=s_{00}+s_{10}H_1+s_{01}H_2+s_{20}H_1^2+s_{11}H_1H_2+s_{02}H_2^2+s_{30}H_1^3 \\ +s_{21}H_1^2H_2+s_{12}H_1^2H_2+s_{03}H_2^3 \tag{25}$$

The solution of the dynamic programming equation for ergodic control is assumed to be of the form

$$V(\mathbf{H})=c_1H_1+c_2H_2+c_3H_1^2+c_4H_1H_2+c_5H_2^2 \tag{26}$$

Following the procedure described in section 3, the time-delay optimal control forces can be obtained as

$$u_{i\tau}^*=-\frac{1}{2}R_i^{-1}\frac{\partial V}{\partial H_i}\left(\cos(\omega_i\tau)P_{i\tau}-\omega_i\sin(\omega_i\tau)Q_{i\tau}\right) \tag{27}$$

Table 1 Results for the first degree of freedom of the system

τ	0	0.2	0.4	0.6	0.8	1.0
$\sigma(x_1)$	0.385	0.390	0.396	0.403	0.408	0.411
$\sigma(u_1)$	0.125	0.133	0.141	0.151	0.163	0.176
K_1	0.635	0.631	0.625	0.619	0.614	0.611
μ_1	5.068	4.758	4.425	4.094	3.765	3.470
$\bar{\sigma}(x_1)$	0.385	0.391	0.401	0.412	Unstable	
$\bar{\sigma}(u_1)$	0.125	0.133	0.145	0.164		
$\bar{K}_1$	0.635	0.629	0.622	0.610		
$\bar{\mu}_1$	5.068	4.728	4.302	3.719		

Numerical calculations are carried out for system (23) with the following parameters: $\alpha_{11}=\alpha_{22}=0.01$, $\alpha_{12}=\alpha_{21}=0$, $\beta_1=\beta_2=0.01$, $\omega_1^2=1.0$, $\omega_2^2=2.0$, $k_{11}=k_{22}=1.0$, $k_{12}=k_{21}=0$, $D_1=D_2=0.05$, $R_1=R_2=10$, $s_{30}=s_{03}=1$, and $s_{10}=s_{01}=0$. The results are shown in Table 1, where $\sigma(x_1)$, $\sigma(u_1)$, K_1, μ_1 denote root mean square of displacement x_1, root mean square of control force $u_{1\tau}^*$, control effectiveness and control efficiency, respectively, by using the proposed optimal time-delay control, while $\bar{\sigma}(x_1)$, $\bar{\sigma}(u_1)$, $\bar{K}_1$, $\bar{\mu}_1$ are those by using a control without considering time-delay effect, i.e., letting cos($\omega_1\tau$)=1 and sin($\omega_1\tau$)=0 in

Eq. (27). It is seen from Table 1 that the proposed controller is better in terms of control effectiveness and efficiency. Furthermore, the system with the proposed controller is stable even for larger time delay in contrast with the instability by using the control without considering time delay effect.

Now consider another case in which the stochastic excitations are pure parametric, i.e., $k_{ij}(X)=k_{ij}X_j$. The partially averaged performance index is of the form of Eq.(8) with cost function of the form of Eq.(11), where f_1(H) and R are to be determined. For stabilization, only the asymptotic behaviors near H=0 are of interest, and the linearized averaged Itô equations are

$$\begin{aligned} dH_1 &= [F_1(\mathbf{H})+\left\langle u_1\frac{\partial H_1}{\partial P_1}\right\rangle]dt+G_{11}(\mathbf{H})dB_1(t)+G_{12}(\mathbf{H})dB_2(t) \\ dH_2 &= [F_2(\mathbf{H})+\left\langle u_2\frac{\partial H_2}{\partial P_2}\right\rangle]dt+G_{21}(\mathbf{H})dB_1(t)+G_{22}(\mathbf{H})dB_2(t) \end{aligned} \tag{28}$$

The optimal control strategy is of the form in Eq.(12). The value function $V(\mathbf{H})$ should be linear function of H_1 and H_2. Then it is seen from dynamical programming equation that $f_1(H)-\gamma$ should also be linear function of H_1 and H_2. Let

$$V(H)=C_1H_1+C_2H_2,\quad f_1(H)-\gamma=k_1H_1+k_2H_2 \tag{29}$$

Substituting Eq.(12) and (29) into final dynamical programming principle equations lead to the following equations:

$$\begin{aligned} k_1+F_{11}C_1+F_{21}C_2-C_1^2/4R_1 &= 0 \\ k_2+F_{12}C_1+F_{22}C_2-C_2^2/4R_2 &= 0 \end{aligned} \tag{30}$$

C_1 and C_2 can be solved for given k_i and R_i. The optimal control u_i^* are then obtained from Eq.(12) as follows:

$$u_i^* = -\frac{C_i}{2R_i}P_i \tag{31}$$

Finally, the expression for the largest Lyapunov exponent of the controlled system(23) is

$$\lambda_1^c = \int_0^1 Q^c(\alpha_1)p^c(\alpha_1)d\alpha_1 \tag{32}$$

where $Q^c(\alpha_1)$ and $p^c(\alpha_1)$ are of the controlled system. When $u_i=0$, the largest Lyapunov exponent λ_1^u of uncontrolled system is also obtained.

The objective of the feedback stabilization of system(23) is to determine k_i, R_i such that the largest Lyapunov exponent determined by Eq.(28) is negative and minimized. After k_i and R_i are determined, the optimal time-delay control is

$$u_{i\tau}^* = -\frac{C_i}{2R_i}\left(\cos(\omega_i\tau)P_{i\tau} - \omega_i \sin(\omega_i\tau)Q_{i\tau}\right) \tag{33}$$

Numerical results have been obtained for system (23) with the following parameters $\alpha_{11} = -0.02, \alpha_{22} = -0.01$, $\alpha_{12} = \alpha_{21} = 0.01$, $\beta_1 = 0.01$, $\beta_2 = 0.01$, $\omega_1^2 = 1, \omega_2^2 = 2$, $k_{11} = k_{12} = k_{21} = k_{22} = 1.0$, $D_1 = D_2 = D$, $k_1 = k_2 = 0.0005$, $R_1 = R_2 = 1$. The initial condition for simulations are: $X_1 = X_2 = 0.01$ and $\dot{X}_1 = \dot{X}_2 = 0$. The largest Lyapunov exponents are shown in Fig.1, where the symbol US denotes the uncontrolled system ($u_i = 0$), WTD the controlled system without time-delay ($u_i^* = -(C_i / 2R_i)P_i$), NCT the controlled system without considering time-delay effects ($u_{i\tau} = -(C_i / 2R_i)P_i(t-\tau)$), and PCS the controlled system using proposed control strategy (Eq.(33)). It is seen from Fig.1 that the largest Lyapunov exponent of uncontrolled system (λ_1^u) is positive, indicating that the trivial solution is unstable. When time delay occurs in the controlled system, the largest Lyapunov exponents by using PCS are generally less than those by using NCT control strategy. It is also shown that the stabilization effect of the proposed control strategy is quite obvious especially for weak noise excitations.

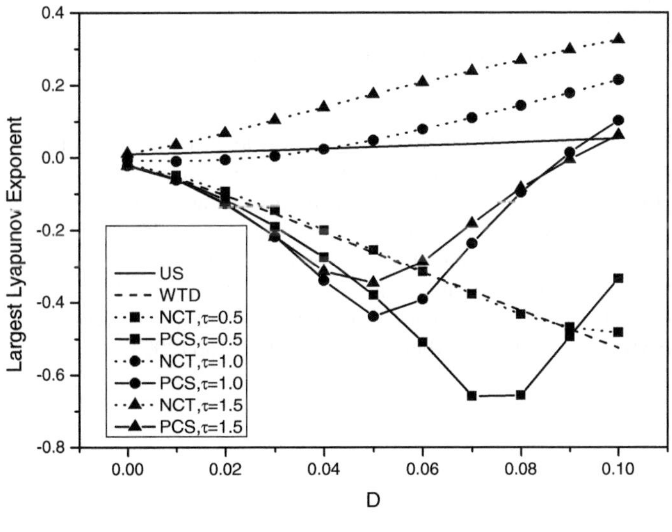

Fig. 1 The largest Lyapunov exponents λ_1 versus stochastic excitations intensity D

6 Conclusions

In the present paper, innovative procedures for stochastic optimal time-delay control and stabilization of quasi-integrale Hamiltonian systems have been proposed. The time-delayed feedback control forces were approximated by control forces without time-delay and the original control problem was converted into a stochastic optimal control problem without time-delay, which was solved by applying the stochastic averaging method and the stochastic dynamical programming principle. One example has been worked out in detail to illustrate the procedure. The result showed that the proposed control strategy procedures are effective.

Acknowledgements. This study was supported by the National Natural Science Foundation of China under Grant No. 10772195, 10932009, the Zhejiang natural Science Foundation of China under Grant No. 7080070 and the Special Foundation for Young Scientists of Fujian Province of China under Grant No. 2008F3100.

References

[1] Liu, Z.H., Zhu, W.Q.: Stochastic averaging of quasi-integrable Hamiltonian systems with delayed feedback control. Journal of Sound and Vibration 299, 178–195 (2007)
[2] Liu, Z.H., Zhu, W.Q.: Asymptotic Lyapunov stability with probability one of quasi-integrable Hamiltonian systems with delayed feedback control. Automatica 44, 1923–1928 (2008)
[3] Liu, Z.H., Zhu, W.Q.: Stochastic Hopf bifurcation of quasi-integrable Hamiltonian systems with delayed feedback control. Journal of Theoretical and Applied mechanics 46, 531–550 (2008)
[4] Kolmanovskii, V.B., Shaikhet, L.E.: Control of Systems with Afterffects. In: Translation of Mathematical Monographs, vol. 157. American Mathematical Society, Providence (1996)
[5] Elsanousi, I., Oksendal, B., Sulem, A.: Some solvable stochastic control problems with delay. Stochastics and Stochastics Reports 71, 69–89 (2000)
[6] Zhu, W.Q., Huang, Z.L., Yang, Y.Q.: Stochastic averaging of quasi integrable Hamiltonian systems. ASME Journal of Applied Mechanics 64, 975–984 (1997)
[7] Zhu, W.Q., Ying, Z.G., Soong, T.T.: An optimal nonlinear feedback control strategy for randomly excited structural systems. Nonlinear Dynamics 24, 31–51 (2001)
[8] Zhu, W.Q., Huang, Z.L.: Feedback stabilization of quasi-integrable Hamiltonian systems. ASME Journal of Applied mechanics 70, 129–136 (2003)

Part 5
Modeling of Stochastic Dynamical Systems and Stochastic Excitations

Modeling of Stochastic Dynamic Excitations and the Probability Density Evolution Theory for Nonlinear Stochastic Dynamics

J. Li, Q. Yan, and J.B. Chen

State Key Laboratory of Disaster Reduction in Civil Engineering & School of Civil Engineering, Tongji University,1239 Siping Road, Shanghai 200092, China

Abstract. Basic thoughts of physical stochastic systems are delineated. Different from traditional modeling where the measured data are statistically analyzed to obtain second-order characteristics, e.g covariance function or power spectral density, in the present framework, physical mechanism of stochastic dynamic excitation is firstly studied and used as a basis to construct a random function model with random parameters, of which the probability distributions are then identified via measured data. Modeling of fluctuating wind via physical stochastic model is exemplified. Stochastic response analyses of nonlinear structures by incorporating physical stochastic models of dynamic excitations and the probability density evolution methods are implemented. Investigation results demonstrate that this is a promising way.

Keywords: Physical stochastic modeling, Fluctuating wind, Probability density evolution method, Nonlinear structure.

1 Introduction

The past decades have seen great development in stochastic dynamics, not only in modeling of stochastic excitations but also in analysis theory of stochastic dynamical systems [11,17]. Interestingly, when we examine from these two fundamental aspects, it is easy to find that they are essentially coordinative. Researchers have devoted many endeavors to modeling of stochastic dynamic excitations, resulting in e.g. the Kanai-Tajimi spectrum for ground motion [5], the Davenport spectrum for fluctuating wind [3] and the Neumann spectrum for sea waves [14], etc. These models set the foundation for classical random vibration based on moments and power spectral density. Nonetheless, it has been recognized that via this path it is very hard to solve problems of response analysis and dynamic reliability evaluation, particularly of MDOF nonlinear systems [17].

Critical revisiting may find that this path of modeling stochastic process is essentially a phenomenological methodology, which might induce some shortages. To overcome these shortages, physical mechanism or background of dynamic excitations should be introduced as a basis, leading to physical stochastic models. In this framework, a stochastic dynamic excitation is described by a random

W.Q. Zhu et al. (Eds.): Nonlinear Stochastic Dynamics & Control, IUTAM BOOK SERIES 29, pp. 273–282.
springerlink.com

function, of which the shape is determined from the physical mechanism or empirical physical background. Probability distributions of the random variables involved in the random function are identified from measured data.

2 Physical Stochastic Modeling of Dynamic Excitations

2.1 *Basic Thoughts of Physical Stochastic Systems*

To describe a stochastic process $X(t),\ t \in [0,T]$, theoretically the finite-dimensional distributions, $f(x,t), f(x_1,t_1;x_2,t_2),\cdots, f(x_1,t_1;x_2,t_2;\cdots;x_n,t_n),\cdots$, are complete. However, in this description exist some disadvantages, e.g.: (1) to obtain and calibrate the finite-dimensional distributions for a stochastic process, prohibitively huge data and computational efforts are needed which makes it practically impossible [1]; and (2) even this is possible the relationship between the ensemble information (finite-dimensional distributions) and the information of the samples, are indirect, i.e. to generate a sample from the finite-dimensional distributions is not straightforward. This induces inconsistency in modeling and generating a stochastic process. To overcome the first disadvantage, the second-order stochastic processes are studied in depth, say, via covariance function or power spectral density function instead of finite-dimensional distributions. However, this reduction from the complete description does not change the essence of phenomenological description. Actually, in traditional modeling the measured data are statistically analyzed directly, without considering the embedded physical background that will shape the measured data, to construct covariance function or power density spectrum [1]. Here, a stochastic process $X(t)$ was regarded as an abstract function of time *t*, dependence of $X(t)$ on the random events (or more rigorously, on the point in probability space) is not explicitly involved.

An alternative complete description is to expose the stochastic process $X(\varpi,t)$ as a function of basic random event ϖ and time *t* simultaneously. For a stochastic process $X(t)$, $t \in [0,T]$ with an embedded physical background, a function can be found such that

$$X(t) = g(\eta, t),\ \ t \in [0,T] \tag{1}$$

where $g(\cdot)$ is a function determined by physical laws or empirical physical relations, η is the random variable(s) embedded in the system that induce randomness of the process. Generally, if a set of samples are available, then distributions and/or characteristic values of the random variable(s) η can be identified via appropriate criteria. Clearly, once information of the random variable(s) η are known, then the stochastic process specified by Equation (1) is completely described. This description of stochastic process, via a random function of basic random variable(s) based on physical background/mechanism, might be referred to a physical stochastic process description. Compared to traditional phenomenological description, relationship between the ensemble and the sample is straightforward as in Equation (1). In addition, the data needed to identify information of the random variable(s) are greatly reduced.

A convenient tool to the physical stochastic process description is the random Fourier function, which can be defined as [7]

$$F_X(\eta,\omega) = \frac{1}{\sqrt{T}} \int_0^T g(\eta,t) e^{-\mathrm{i}\omega t} \mathrm{d}t \tag{2}$$

The random Fourier function is usually a complex function that can be rewritten as

$$F_X(\eta,\omega) = \left|F_X(\eta,\omega)\right| e^{\mathrm{i}\varphi(\eta,\omega)} \tag{3}$$

where $\left|F_X(\eta,\omega)\right|$ is the Fourier amplitude spectrum and $\varphi(\eta,\omega)$ is the Fourier phase spectrum. Generally the amplitude spectrum is a kind of spectral decomposition of kinetic energy (Li and Chen, 2009 [11]) and the phase spectrum is the leading factor controlling the shape of the time history (Seong and Peterka, 2001[15]).

It is easy to prove that

$$S_X(\omega) = E\left[\left|F_X(\eta,\omega)\right|^2\right] = \int_0^\infty \left|F_X(\mathbf{z},\omega)\right|^2 p_\eta(\mathbf{z}) d\mathbf{z} \tag{4}$$

where $S_X(\omega)$ is the power spectral density function if $X(t)$ is stationary. However, we should note that the model in Equation (2) is not confined to stationary processes.

For many engineering dynamic excitations, it is relatively easy to find the expression of $F_X(\eta,\omega)$. With this background, probabilistic information of the random variable(s) η can be identified from the measured data [7]. The proposed modeling process will be exemplified in the following section by modeling of fluctuating wind.

2.2 Physical Stochastic Models: Exemplified by Fluctuating Wind

Existence of eddies is one of the main characters of turbulence. The fluctuating part of wind speed could be regarded as superposition of a series of eddies with different scale and frequency. The eddy scale in atmosphere is in a wide range from over one kilometer to several millimeters. It is recognized that turbulent eddy motion in the atmospheric boundary layer exhibits three spectral ranges: energy-containing sub-range, inertial sub-range and dissipation range. In wind engineering, the first two sub-ranges are mostly concerned. For the energy spectrum in inertial sub-range, "-5/3" power law has been reported both theoretically and in atmospheric boundary-layer measurements. The analysis of large-scale turbulence is more complex because of their strong anisotropy. Despite the complexity, several laboratory and field experiments reported "-1" power law at production scales for the longitudinal velocity spectrum in boundary-layer flows (Katul and Chu, 1998[6]). Employing wind speed records in our research group, we also find the "-1" power law in energy-containing sub-range of both three-dimensional spectrum and longitudinal spectrum.

The wave-length l, reciprocal of the wave-number k, could give a straight view of the eddy scale. According to Taylor's "frozen" hypothesis, the wave-number $k = 2\pi n / \overline{U}$, where $n = \omega/2\pi$ is the frequency. Thus the stochastic Fourier wave-number spectrum could be defined, following Equation (2), as

$$F(\eta, k) = \sqrt{\overline{U}/2\pi} F(\eta, n) \tag{5}$$

A specified wind speed record is the joint effect of many eddies in different scales. As mentioned in the energy spectrum (power spectral spectrum) there are energy-containing sub-range and equilibrium sub-range in which the "-1" law and "-5/3" law are obeyed, respectively. Therefore, the boundary position k_c between the two sub-ranges might be called as "boundary wave-number" and the reciprocal of k_c marked as l_c might be called as "boundary wave-length". Thus, considering Equation (4), the wave-number Fourier spectrum is defined as a logarithmically bilinear function

$$|F(\eta, k)| = \begin{cases} Ak^{-1/2}, & k_{\text{low}} < k \le k_c \\ Bk^{-5/6}, & k_c < k < k_{\text{up}} \end{cases} \tag{6}$$

where k_{low} and k_{up} are the lower and upper limits of the wave number of interest. Generally 10-min duration and 10Hz sampled wind speed records are adopted. In this case $k_{\text{up}} = 2\pi \times 5 / \overline{U} = 10\pi / \overline{U}$, $k_{\text{low}} = 2\pi \times (1/600) / \overline{U} = \pi / (300\overline{U})$. A and B are dependent parameters related to total energy,

$$A = \sigma^2 / \left(\ln k_c - \ln k_{\text{low}} - 1.5 k_c^{2/3} k_{\text{up}}^{-2/3} + 1.5\right), \; B = A k_c^{1/3} \tag{7}$$

where σ is the standard deviation of the fluctuating wind speed. The only independent parameter in Equation (6) is k_c.

According to Tchen (1953[16]), the leading factor affecting the value of k_c is the ratio γ between the main-flow vorticity V_m and turbulence vorticity V_t

$$\gamma = V_m / V_t \tag{8}$$

where V_m is the shear-ratio and V_t could be expressed as (Hinze, 1975[4])

$$V_t = \left[\int_0^\infty k^2 S(k)\, dk \right]^{1/2} = \left[\int_0^\infty k^2 |F(\eta, k)|^2\, dk \right]^{1/2} \tag{9}$$

Substituting Equations (6) and (7) in (9) yields

$$V_t = \sigma\left[-0.25k_c^2 - 0.5k_{\text{low}}^2 + 0.75k_c^{2/3}k_{\text{up}}^{4/3}\right]^{1/2} / \left[\ln k_c - \ln k_{\text{low}} - 1.5k_c^{2/3}k_{\text{up}}^{-2/3} + 1.5\right]^{1/2} \quad (10)$$

To calculate the value of γ, 2000 sets of 10-min fluctuating wind speed records are applied. The correlation coefficient between γ and $d\overline{U}(z)/dz$ is 0.8288, thus it can be assumed that γ is approximately linearly variant against $d\overline{U}(z)/dz$. Through curve-fitting, the relationship can be written as:

$$\gamma = 1.19 d\overline{U}(z)/dz + 0.0637 \quad (11)$$

Basic random variables in the model are the ground roughness z_0 and 10-min mean wind speed $\overline{U}(z)$ at a certain height. It is found that z_0 can be characterized by lognormal distribution and $\overline{U}(z)$ can be characterized by Gumbel Distribution (Li and Yan, 2009[13]). Fig. 1 displays comparison of the mean and standard-deviation of the amplitude spectrum between the measured data and the presented model. It is seen that the model spectra are almost identical with the measured spectra at all four heights.

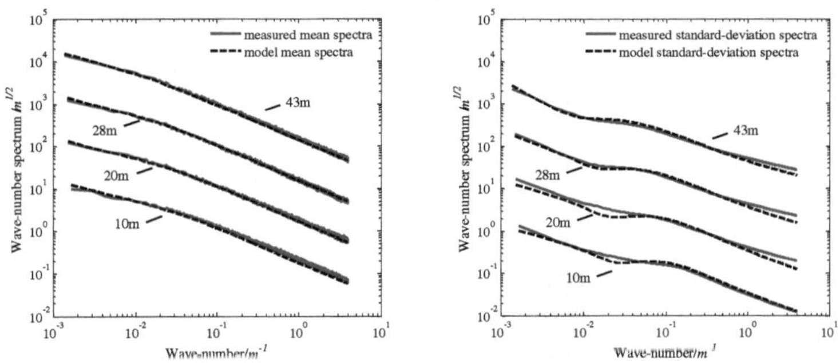

Fig. 1 Comparison of the mean and standard-deviation of the spectra between model and measurements

The characteristic speed of different eddies could be expressed as (Hinze, 1975[4]):

$$v(n) = \sqrt{|F(n)|^2 \Delta n} \quad (12)$$

The ratio between the distance that the eddy goes forward in a time interval and the wave length characterizes the changed circle numbers of the eddy. Each circle corresponds to a phase change of 2π. The phase evolution speed of eddies with different frequency could be expressed as:

$$\Delta\dot{\varphi}(n) = 2\pi v(n) k(n) \tag{13}$$

Different characteristic speed leads to different phase evolution speed. Generally speaking, large scale eddies possess slower phase evolution speed than small scale eddies do. It is reasonable to suppose that the measured fluctuating wind speed is the superposition of a series of harmonic waves which evolved from identical initial phase at time T_e, as shown in Fig. 2. For simplicity, we can assume the initial phases are all zero. T_e is named phase evolution time with the unit second.

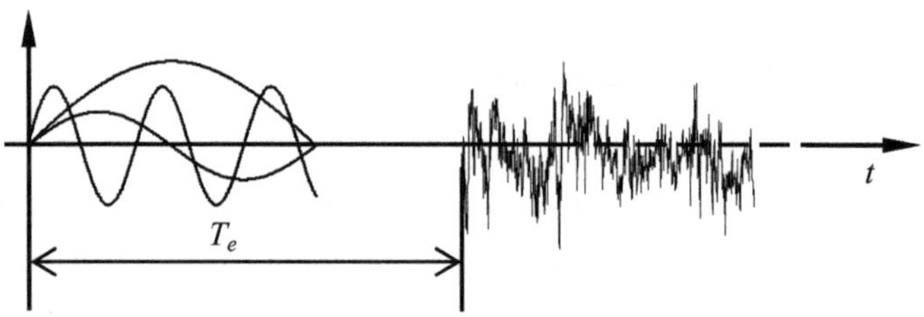

Fig. 2 Schematic diagram of phase evolution time *Te*

The value of T_e can be identified from the fluctuating wind speed samples with relaxed criterion. 800 sets of wind record are adopted. The identified values of T_e can be fitted by Gamma distribution, as shown in Fig. 3.

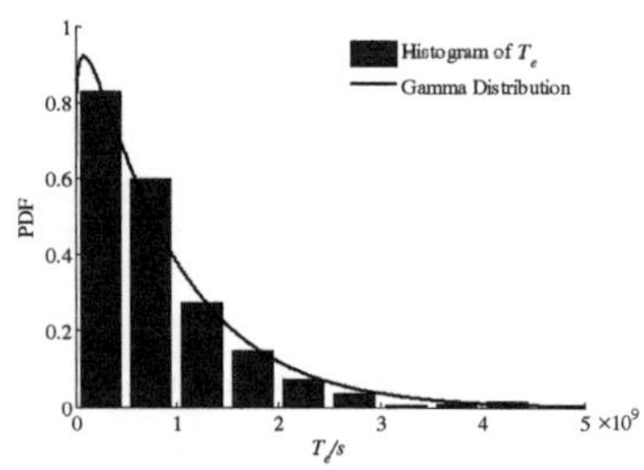

Fig. 3 Probability distribution fitting of phase evolution time *Te*

Thus, both the amplitude and the phase spectrum are specified, where totally 3 random variables, ground roughness z_0, 10-min mean wind speed at a certain height $\bar{U}(z)$, and phase evolution time T_e, are involved.

Likewise, physical stochastic models can also be studied for earthquake ground motion and sea wave. For details, refer to [8,12].

3 Stochastic Responses of Nonlinear Structures via Probability Density Evolution Equation

3.1 Fundamentals of the Probability Density Evolution Method

Consider a generic nonlinear MDOF structure subjected to physically modeled stochastic dynamic excitations. The equation of motion reads

$$\mathbf{M}\ddot{\mathbf{X}} + \mathbf{C}\dot{\mathbf{X}} + \mathbf{f}(\mathbf{X}) = \mathbf{DF}(\boldsymbol{\Theta}, t) \tag{14}$$

where $\mathbf{M}$ and $\mathbf{C}$ are the n by n mass and damping matrices, respectively, $\mathbf{f}$ is the nonlinear restoring forces, $\mathbf{D}$ is the n by r load location matrix, $\mathbf{F}$ is the r by 1 stochastic external forces, $\boldsymbol{\Theta}$ is the random variable(s) with known joint PDF $p_{\boldsymbol{\Theta}}(\boldsymbol{\theta})$ involved in the physical stochastic process. For fluctuating wind, the forces are specified by the models outlined in the foregoing sections.

Denote by $\mathbf{Z}(t) = (Z_1, Z_2, \cdots, Z_m)^{\mathrm{T}}$ the physical quantities of interest in the system (14). The randomness of $\mathbf{Z}(t)$ comes entirely from $\boldsymbol{\Theta}$ and thus the augmented system $(\mathbf{Z}(t), \boldsymbol{\Theta})$ is a probability preserved system [10,11]. According to the random event description of the principle of preservation of probability, we have

$$\frac{D}{Dt}\int_{\Omega_t} p_{\mathbf{Z}\boldsymbol{\Theta}}(\mathbf{z}, \boldsymbol{\theta}, t) d\mathbf{z}\, d\boldsymbol{\theta} = 0 \tag{15}$$

where $p_{\mathbf{Z}\boldsymbol{\Theta}}(\mathbf{z}, \boldsymbol{\theta}, t)$ is the joint probability density function of $(\mathbf{Z}(t), \boldsymbol{\Theta})$. Combining with the uncoupled physical equations, we have the following m-dimensional partial differential equation governing evolution of the joint probability density

$$\frac{\partial p_{\mathbf{Z}\boldsymbol{\Theta}}(\mathbf{z}, \boldsymbol{\theta}, t)}{\partial t} + \sum_{j=1}^{m} \dot{Z}_j(\boldsymbol{\theta}, t) \frac{\partial p_{\mathbf{Z}\boldsymbol{\Theta}}(\mathbf{z}, \boldsymbol{\theta}, t)}{\partial z_j} = 0 \tag{16}$$

Here $\dot{Z}_j(\boldsymbol{\theta}, t)$ is time rate of the physical quantity $Z_j(t)$ with specified $\{\boldsymbol{\Theta} = \boldsymbol{\theta}\}$. Once Equation (16) is solved, the instantaneous probability density can be obtained by

$$p_{\mathbf{Z}}(\mathbf{z}, t) = \int_{\Omega_{\boldsymbol{\Theta}}} p_{\mathbf{Z}\boldsymbol{\Theta}}(\mathbf{z}, \boldsymbol{\theta}, t) d\boldsymbol{\theta} \tag{17}$$

Equation (16) is the generalized density evolution equation, which reveals the intrinsic connections between a stochastic system and the deterministic counterpart. It tells us that evolution of the probability density in time is proportional to varying of probability density in space.

In most practical cases, $m = 1$ is adequate. Solving Equation (16) under appropriate initial and boundary conditions, the instantaneous probability density function of the responses can be captured [9]. Further, reliability evaluation of the nonlinear structures can be carried out [2,10].

3.2 Numerical Implementation Procedure

To perform probability density evolution analysis for responses of nonlinear structures, Equations (14), (16) and (17) should be solved in an incorporative manner, with the physical stochastic models embedded. For instance, if the response analysis of a nonlinear structure under strong wind is to be carried out, the first step is to employ the physical stochastic models outlined in the fore-going sections. To generate representative time histories, selection of representative points, denoted by $\boldsymbol{\theta}_q, q = 1, 2, \cdots, n_{\text{sel}}$, in the space of $\boldsymbol{\Theta}$ is needed. Then for each specified $\boldsymbol{\theta}_q$, a time history of dynamic excitation can be generated by the physical stochastic model. A deterministic dynamic response analysis can then be carried out for the system (14), resulting in responses of the system, including time rate $\dot{Z}(\boldsymbol{\theta}_q, t)$ of the physical quantities of interest. These quantities can then be substituted in Equation (16), which can therefore be solved by the finite difference method with appropriate difference scheme. Details of the numerical algorithm can be found in Li and Chen (2006) [9] and Li and Chen (2009) [11].

4 Numerical Examples

A steel TV tower, totally 388 m high with structure of 268m and shaft of 120m. In the analysis the P-Delta effect and geometric nonlinearity are taken into account. Parameters of the wind field are: the ground roughness z_0 obeys log-normal distribution with the mean 0.05 and coefficient of variation 0.38; the average wind speed at 10m high obeys the extreme type-I distribution with the mean 28.5m/s and coefficient of variation 0.1. Pictured in Fig.4 are three typical representative time histories of wind speed generated by the proposed physical stochastic model.

Some of the results are discussed here. Shown in Fig. 5 is information of the stochastic inter-story drift between the 9th and 10th story, including the PDF at three different time instants (Fig.5(a)), the instantaneous PDF evolving against time (Fig.5(b)) and the contour of the PDF evolution surface (Fig.5(c)). It is seen that the PDF evolves in a complex manner, generally non-stationary. In addition, the PDF is not symmetric to its center (Fig.5(a)), which can also be seen clearly from Fig.5(c).

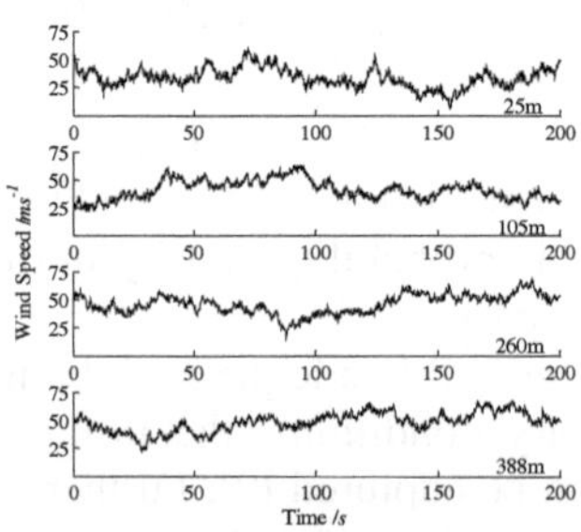

Fig. 4 Typical representative wind speed time history generated by the proposed model

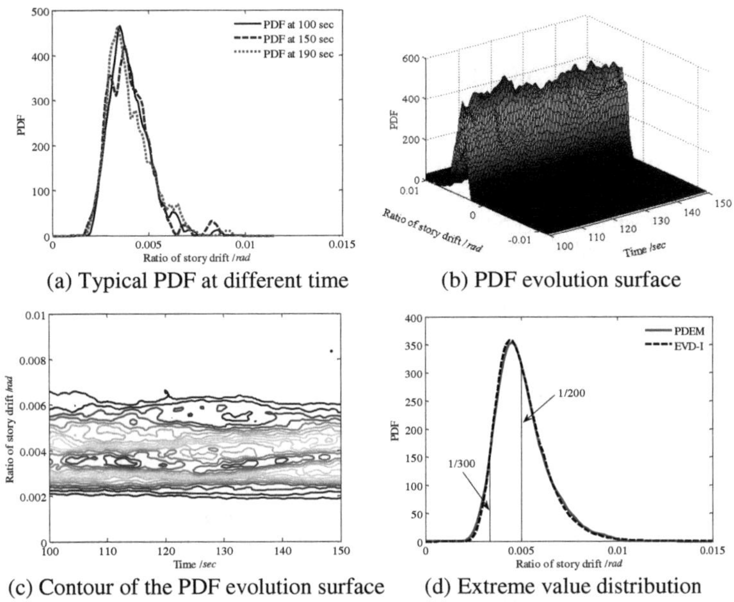

(a) Typical PDF at different time (b) PDF evolution surface

(c) Contour of the PDF evolution surface (d) Extreme value distribution

Fig. 5 Probabilistic information of inter-9-10-story drift angle

The extreme value distribution can also be obtained via the probability density evolution method. Fig.5(d) shows the extreme value distribution in this case. It is seen that now the extreme value distribution is close to the Extreme type-I distribution. But this accordance is rather accidental than often. Further, reliability of the tower can be obtained by integration of the extreme value distribution.

5 Conclusions

Methodology of modeling stochastic dynamic excitations and that of random vibration analysis are essentially cooperative. Phenomenological modeling of stochastic process is in consistency with classical random vibration, which leads to difficulties in tackling stochastic response analysis of nonlinear systems. The methodology of physical modeling of stochastic dynamic excitations is described in the present paper. Physical modeling of fluctuating wind is exemplified. Probability density evolution method is then incorporated with the physical stochastic excitations to implement response analysis of nonlinear structures. Investigations show that the proposed methodology is promising in stochastic dynamic response analysis and reliability evaluation of nonlinear structures under stochastic excitations.

References

[1] Bendat, J.S., Piersol, A.G.: Random Data: Analysis and Measurement Procedures, 3rd edn. John Wiley & Sons, Inc., New York (2000)
[2] Chen, J.B., Li, J.: The extreme value distribution and dynamic reliability analysis of nonlinear structures with uncertain parameters. Structural Safety 29, 77–93 (2007)
[3] Davenport, A.G.: The spectrum of horizontal gustiness near the ground in high winds. J. Royal Meteorol Soc. 87, 194–211 (1961)
[4] Hinze, J.Q.: Turbulence. McGraw-Hill, New York (1975)
[5] Kanai, K.: Semi-empirical formula for the seismic characteristics of the ground. Bull. Earthquake Research Institute 35, 309–325 (1957)
[6] Katul, G., Chu, C.R.: A theoretical and experimental investigation of energy-content scales in the dynamic sublayer of boundary-layer flows. Boundary-Layer Meteorology 86, 279–312 (1998)
[7] Li, J.: Physical stochastic models for dynamic excitations of engineering structures. In: Li, J., Chen, J.B. (eds.) New Advances in Random Vibration and Applications, pp. 119–132. Tongji University Press, Shanghai (2009)
[8] Li, J., Ai, X.Q.: Study on random model of earthquake ground motion based on physical process. Earthquake Engineering and Engineering Vibration 26(5), 21–26 (2006) (in Chinese)
[9] Li, J., Chen, J.B.: The probability density evolution method for dynamic response analysis of non-linear stochastic structures. International Journal for Numerical Methods in Engineering 65, 882–903 (2006)
[10] Li, J., Chen, J.B.: The principle of preservation of probability and the generalized density evolution equation. Structural Safety 30, 65–77 (2008)
[11] Li, J., Chen, J.B.: Stochastic Dynamics of Structures. John Wiley & Sons, Chichester (2009)
[12] Xu, Y.Z., Li, J.: Stokes model of wind-wave interaction. Advances in Water Science 20(2), 281–286 (2009) (in Chinese)
[13] Li, J., Yan, Q.: Physical models for the stochastic dynamic excitations of structures: in the case of fluctuating wind speed. Engineering Mechanics 26(sup II), 175–183 (2009) (in Chinese)
[14] Neumann, G.: On wind generated ocean waves with special reference to the problem of wave forecasting. NYU, College of Eng., Res. Div., Dept. of Meteor and Oceanogr, 136 (1952)
[15] Seong, S.H., Peterka, J.A.: Experiments on Fourier phases for synthesis of non-Gaussian spikes in turbulence time series. Journal of Wing Engineering and Industrial Aerodynamics 89, 421–443 (2001)
[16] Tchen, C.M.: On the spectrum of energy in turbulent shear flow. Journal of Research of the National Bureau of Standards 50, 51–62 (1953)
[17] Zhu, W.Q.: Nonlinear stochastic dynamics and control in Hamiltonian formulation. Applied Mechanics Reviews 59, 230–248 (2006)

On the Consideration of Model Uncertainties in Model Updating of Dynamic Systems

G.I. Schuëller and B. Goller

Institute of Engineering Mechanics, University of Innsbruck, Austria, EU

Abstract. Model updating procedures are applied in order to improve the match between experimental data and corresponding model output. The updated, i.e. improved, finite element (FE) model can be used for more reliable predictions of the structural performance in the target mechanical environment. The discrepancies between the output of the FE-model and the results of tests are due to the uncertainties that are involved in the modeling process. These uncertainties concern the structural parameters, measurement errors, the incompleteness of the test data and also the FE-model itself. The latter type of uncertainties is often referred to as model uncertainties and is caused by simplifications of the real structure that are made in order to reduce the complexity of reality. Several approaches have been proposed for taking model uncertainties into consideration, where the focus of this manuscript will be set on the updating procedure within the Bayesian statistical framework. A numerical example involving different degrees of non-linearity will be used for demonstrating how this type of uncertainty is considered within the Bayesian updating procedure.

Keywords: Bayesian statistics, model updating, stochastic analysis, model uncertainties, non-linear dynamics.

1 Introduction

The topic of model updating has been in the focus of intensive research now for over four decades and it continues to be a topic of high importance for the accurate prediction of structural performance of dynamic systems [1]. The need for taking uncertainties into account within the model updating process has been widely recognized and it has led to the development of several approaches for performing model updating under the consideration of uncertainties. The thereby involved spectrum of uncertainties is interpreted in different ways by the two main schools for probability interpretations, namely the frequentist and Bayesian interpretation.

The frequentist interpretation of probability leads to a differentiation of uncertainties into two categories: the first category comprises the uncertainty in the parameters and is denoted aleatoric uncertainty. Its source is seen to be the inherent randomness of physical parameters. Model uncertainties (or epistemic uncertainties) on the other hand arise from the complexity of physical processes that have not been understood sufficiently enough in order to be explicitly modelled. The uncertainties in the modelling process must therefore form another category since

W.Q. Zhu et al. (Eds.): Nonlinear Stochastic Dynamics & Control, IUTAM BOOK SERIES 29, pp. 283–292.
springerlink.com

the probability of a model does not make sense if probability is interpreted as the relative frequency of a random event in the long run.

In order to consider the whole spectrum of uncertainties in the analysis, different approaches have been proposed. One way to treat epistemic uncertainties consists in the shift of model uncertainties to parameter uncertainties and in considering them as variables describing events in the long run, i.e. in a frequentist interpretation of probability (see [2, 3]). Another way to treat model uncertainties is given by the non-parametric approach [4]. Within this approach, the relaxation of the topological connectivity of the structural matrices aims at a consideration of the uncertainties in processes that are not modelled explicitly by structural parameters. This approach broadens the set of structural models (i.e. all stiffness matrices which are symmetric and positive definite) and uses the Principle of Maximum Entropy to construct a PDF over this set. Applications of the non-parametric model in context with structural model updating are shown in e.g. [5]. Alternative approaches for taking epistemic uncertainties into account consist in the use of possibility theory and fuzzy sets. In e.g. [6, 7] it is discussed how this method is fitted into the robust updating process with the aim of damage detection.

The Bayesian interpretation of probability does not distinguish between these two categories, since all uncertainties are seen as epistemic uncertainties [8]. In this context, probability is not interpreted as the relative occurrence of a random event in the long run, but as the plausibility of a hypothesis. Probability quantifies the uncertainty about propositions and therefore its domain contains both physical variables and models by themselves. The wider scope of the interpretation of probability in the Bayesian sense leads to the fact that the reason of uncertainty of both parameters and models is seen in the incomplete available information.

In this manuscript, the approach for considering the entire spectrum of uncertainties within the Bayesian statistical framework is discussed. First, the basic principles of Bayesian updating procedures are summarized (Sec. 2), where the prediction error, which takes into account the discrepancies between model output and measurements, is subject of a thorough discussion in Sec. 3. Finally, in Sec. 4, a linear beam model is updated where the reference data derives from non-linear models involving different degrees of non-linearity. This provides a means for investigating quantitatively the effect of model uncertainties.

2 Bayesian Model Updating

The concept of the Bayesian statistical framework is to embed a deterministic model in a class of probability models as introduced in [9, 10]. Each probability model in the chosen model class is described by probability distributions of the unknown parameters and the prediction error. Based on the available data, the initial knowledge of the range of the unknown parameters is updated, making some parameter ranges more plausible if the data provide the necessary information.

This embedment of the deterministic model in a model class is performed by the use of the Bayes' Theorem which is given by

$$p(\theta \mid \mathcal{D},\mathcal{M}) = c^{-1} p(\mathcal{D} \mid \theta,\mathcal{M}) p(\theta \mid \mathcal{M}), \tag{1}$$

where θ is the vector of the unknown (adjustable) parameters, $\mathcal{D}$ denotes the set of available data points and $\mathcal{M}$ is the chosen model class. The constant c is a normalizing constant given by $c = \int p(\mathcal{D} \mid \theta,\mathcal{M}) p(\theta \mid \mathcal{M}) d\theta$. This constant c is actually $p(\mathcal{D} \mid \mathcal{M})$, which is called the *evidence* of the model class $\mathcal{M}$ and which is used in model class comparison and selection which is however not in the focus of the present manuscript.

The term $p(\mathcal{D} \mid \theta,\mathcal{M})$ is called *likelihood* and expresses the probability of the data conditional on the structural parameters, i.e. a probability model for the measured data. This term describes the discrepancies between model output and measurements through the prediction error, which is introduced in order to bridge the gap between model output, and measurements and which will be discussed in Sec. 3. The factor $p(\theta \mid \mathcal{M})$ is the *prior PDF*, which quantifies the initial plausibility of each model defined by the parameters θ within the model class $\mathcal{M}$. The product of these two terms determines the shape of the *posterior PDF* $p(\theta \mid \mathcal{D},\mathcal{M})$, which reflects the updated, relative plausibility of each model within the model class after incorporating the information contained in the data $\mathcal{D}$.

The analytical or also the numerical solution of Eq. (1) is only feasible for low dimensional problems. For the case of large number of uncertain parameters θ, efficient sampling algorithms have been developed which are based on Markov Chain Monte Carlo algorithms, like the multi-level Metropolis-Hastings algorithm (the so-called Transitional Markov Chain Monte Carlo algorithm) in [11], Gibbs sampler in [12] and Hybrid Monte Carlo in [13].

3 The Prediction Error in Bayesian Updating

Due to the complexity of real systems and the therefore arsing necessity for reducing the complexity of the model, the established numerical model can not predict exactly reality. Within the Bayesian updating procedure, the parameter values are updated in order to better represent the real structure, where the updating process is directed by the prior information and the information contained in the measurements of the investigated structure. However, since the model does not represent an exact picture of reality, there is no true parameter value and there remains a gap between model output and measurements which is taken into account by the so-called prediction error. The prediction error therefore makes it possible to go outside of the domain of the model class. As already pointed out, in the Bayesian sense, uncertainty in model parameters and in the model itself are interpreted as a lack of knowledge and therefore they both fall into the category of epistemic uncertainty. If model uncertainty is interpreted as the type of uncertainty that can not be considered within the structural parameters, the prediction error can be understood as an approach for considering this type of uncertainties. Hence, the prediction error provides a means for considering those uncertainties that cause the remaining lack of knowledge which prohibits a perfect matching between model and real system.

In general terms, the connection between the analytical output of the system $y(\theta)$ and the corresponding test value q is given by

$$q = y(\theta) + e \tag{2}$$

The choice of the PDF for the prediction is based on the maximum entropy principle. Based on the given knowledge that on average model and measurements agree (i.e. zero mean) and that the variance is finite, a Gaussian probability density function maximizes the uncertainty. It should be noted that the prediction-error variance is not taken as a known value, but it is included in the vector of uncertain parameters and it is updated based on the data.

In the present manuscript, model updating is performed using modal data. The formulation of the likelihood function using modal data is derived in [14] and is summarized in the following. The experimental data $\mathcal{D}$ from the structure is assumed to consist of N_s sets of modal data $\mathcal{D}$, comprised of N_m modal frequencies $\hat{\omega}_{1,j}$ and N_m incomplete mode-shape vectors $\hat{\psi}_{1,j} \in R^{N_0}$ where N_o is the number of observed degrees of freedom. The model output $q(\theta)$ is then the corresponding modal properties of the structural model defined by the parameter vector $\theta \in \Theta \in R^{Np}$, that is, eigenfrequencies $\omega_r(\theta)$ and partial eigenvectors $\psi_r(\theta)$.

First, the use of Eq. (2) for the modeshape vectors yields

$$\hat{\psi}_{r,j} = a_r \psi_r(\theta) + e_{\psi_r}, \tag{3}$$

where a_r is a scaling factor as defined in [14] to relate the scaling of the model mode shape vector $\psi_r(\theta)$ to that of the experimental mode shape vector $\hat{\psi}_{r,j}$.

Assuming a Gaussian distribution for the probabilistic characterization of the prediction error variance and with the choice of the scaling factor a_r as defined in [15], the likelihood function for the mode shape vector is given by

$$p(\hat{\psi}_{r,j} \mid \theta, \mathcal{M}) = c_1 \exp\left(-\frac{\psi_r^T \left(I - \hat{\psi}_{r,j}\hat{\psi}_{r,j}^T\right)\psi_r}{2\delta^2 \|\psi_r\|^2} \right), \tag{4}$$

where I is the identity matrix of size N_m x N_m and δ^2 denotes the prediction error variance (assumed to be equal for all N_m modes).

Secondly, Eq. (2) is formulated for the squared **eigenfrequencies**, which yields

$$\hat{\omega}_{r,j}^2 = \omega_r^2 + e_{\omega_r^2} \tag{5}$$

Using again a Gaussian probability model for the statistical description of the discrepancies between analytical and experimental eigenfrequencies, Eq. (2) yields

$$p(\hat{\omega}_{r,j}^2 \mid \theta, \mathcal{M}) = c_2 \exp\left[-\frac{1}{2}\left(\frac{\omega_r^2 / \hat{\omega}_{r,j}^2 - 1}{\varepsilon}\right)^2\right] \tag{6}$$

where ε^2 denotes the prediction error variance of the normalized squared eigenfrequencies (again assumed to be equal for all N_m modes).

Due to the assumed statistical independence between the mode shape vectors and the modal frequencies, between the different modes and between one data set to another, the resulting likelihood function can be written as

$$p(\mathcal{D} \mid \theta, \mathcal{M}) = c_3 \exp\left(-\frac{1}{2}\sum_{r=1}^{N_m}\sum_{j=1}^{N_s}\left[\left(\frac{(\omega_r^2 / \hat{\omega}_{r,j}^2 - 1)^2}{\varepsilon^2}\right) + \frac{\psi_r^T \left(I - \hat{\psi}_{r,j}\hat{\psi}_{r,j}^T\right)\psi_r}{\delta^2 \left\|\psi_r\right\|^2}\right]\right). \tag{7}$$

In the following investigation, a quantitative assessment of those uncertainties that can not be captured by model parameters is carried out and it is shown how these uncertainties are taking into account within the Bayesian model updating procedure. The focus will thereby be set on the parameters taking into account these uncertainties, namely the prediction error variances corresponding to the eigenvectors and squared, normalized eigenfrequencies.

4 Numerical Example

As a numerical example a beam model as shown in Fig. 1 has been chosen in order to quantitatively analyze the uncertainty by which the updating procedure is affected. The model used for structural model updating is a linear model with a nominal Young's modulus of $E = 9.45 \cdot 10^7$ N/m^2, the density is given by ρ=1800N/m^3 and the stiffness of the springs modelling the supports is assumed to be c=3.0·10^4 N/m. The structure is clamped at one end (visualized in Fig. 1 by the red circles at the supporting DOFs) and 6 springs (where only the respective 3 front springs are visible) connect the structure to the ground.

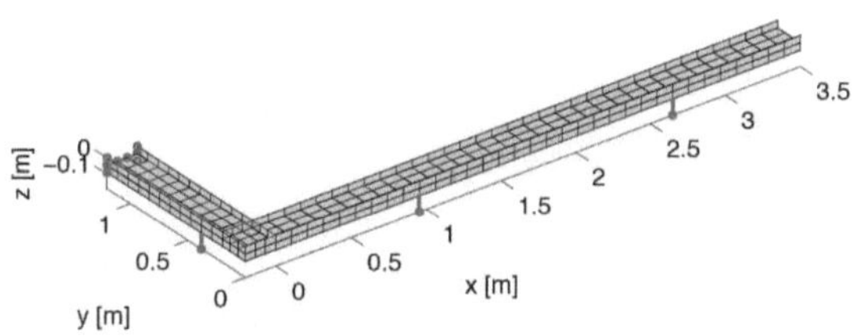

Fig. 1 FE-model of the beam used for model updating

Three cases of model updating are performed in the following which differ by the type of model which is used for simulating the reference data:

1. Case 1: In case 1, a linear model is used for generating the modal data set (N_s =1), which consists of N_m = 5 modes where the partial modeshape vectors have a length of N_o = 129. The parameter values are equal to E=9.0·10^7 N/m^2, ρ = 2000 N/m^3 and c = 2.0·10^4 N/m.
2. Case 2: In case 2, a slightly non-linear model is used, with the structural parameters chosen as in case 1, where however the supports involve non-linearities, i.e. the springs are only active when pressured. Modal properties are determined by using the structural response due to a sine-sweep excitation. Of course, the extracted modal properties are a function of the magnitude of the applied load.
3. Case 3: The reference model in case 3 is the same as in case 2 where additionally the width b= 4 cm of the supports is considered, which leads to the fact that a tilting movement is observable at the supports. The modal properties are determined as described above for case 2.

Table 1 Initial comparison of modal properties

		Case 1			*Case 2*			*Case 3*		
Mode	ω_r [Hz]	$\hat{\omega}_r$[Hz]	$\Delta\omega_r/\hat{\omega}_r$	MAC_{rr}	$\hat{\omega}_r$[Hz]	$\Delta\omega_r/\hat{\omega}_r$	MAC_{rr}	$\hat{\omega}_r$[Hz]	$\Delta\omega_r/\hat{\omega}_r$	MAC_{rr}
1	2.45	2.26	0.085	1.0000	2.24	0.092	0.9874	2.28	0.073	0.9977
2	3.62	3.37	0.074	0.9998	3.33	0.087	0.9625	3.35	0.081	0.9998
3	4.58	4.19	0.093	0.9995	3.95	0.161	0.8771	4.05	0.132	0.9245
4	5.57	5.10	0.092	0.9990	4.85	0.149	0.9422	4.76	0.170	0.9798
5	6.28	5.70	0.103	0.9980	5.70	0.102	0.9253	5.68	0.106	0.2212

In Tab. 1 the comparison of the initial modal properties of the 3 cases is shown. These modal data are compared by means of *i)* eigenfrequencies ω_r for r = 1... 5 obtained with the linear model and the frequencies at the resonance peaks of the reference models $\hat{\omega}_r$ and *ii)* the modal assurance criterion (MAC) of the first 5 modes defined by

$$MAC_{rr} = \frac{\left|\hat{\psi}_r^T \psi_r\right|^2}{\left|\hat{\psi}_r^T \hat{\psi}_r\right|\left|\psi_r^T \psi_r\right|}, \qquad 0.0 \leq MAC_{r,r} \leq 1.0 \tag{8}$$

where a MAC-value of 1.0 expresses full correlation and a MAC-value of 0.0 orthogonal vectors. The reference mode shapes are defined as the displacement of the structure at the resonance peaks due to a certain load magnitude.

In order to express the prior knowledge about the parameter values, Gaussian distributions with mean values equal to the nominal values and coefficients of variations of 10% are assigned to the Young's moduli and the densities, where for both properties 2 independent variables are used for the flange and the web of the beam. Uniform distributions within the bounds $c_i \sim U([1.5, 4.5]\cdot 10^4\,\text{N/m})$, $i = 1,2,3$ are used as prior PDFs for the stiffnesses of the supports, where the two springs constituting one support i are fully correlated. The transitional Markov Chain Monte Carlo algorithm is used for model updating.

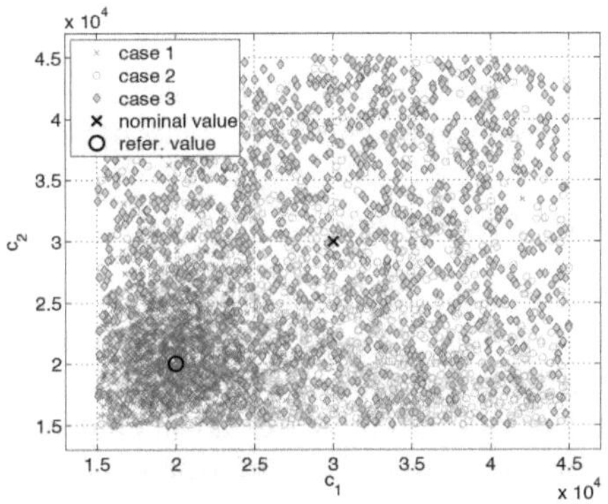

Fig. 2 Posterior samples of the spring stiffnesses c_1 and c_2 (cases 1-3)

4.1 Updated Structural Parameters

Fig. 2 shows the posterior samples of two stiffness values, where the red crosses derive from case 1, the green circles from case 2 and the blue diamonds from case 3. This figure points out that for case 1, which is the linear case, true parameter values exist and therefore the samples are concentrated around this reference point (depicted by a black circle), while for cases 2 and 3 no true values exist due to the involved non-linearity. In case 2, the degree of non-linearity affects only slightly the reference data used for model updating which results in posterior samples that express that values of these parameters in the upper, prior interval are less probable. However, the samples have a considerably higher dispersion if compared with case 1 which means that the posterior prediction error variance is higher than in case 1. The posterior samples of case 3 show no clear difference to the prior samples since they cover the entire support of the prior PDF. Hence, since the springs are only active when pressured and due to the additionally considered widths of the supports there is no information about the constant stiffness values $\{c_1, c_2\}$ in the reference data and no decrease of the prior uncertainty is obtained.

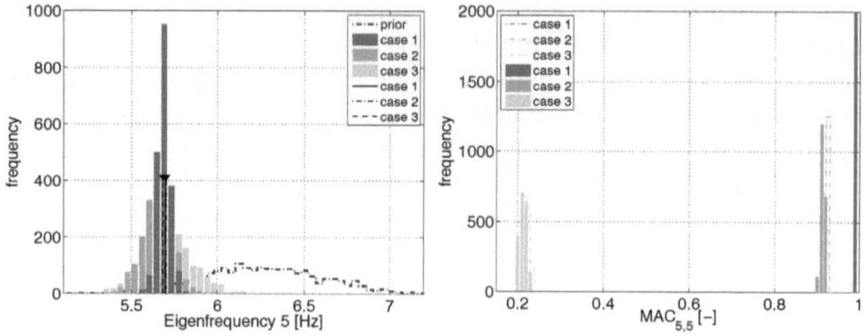

(a) (b)

Fig. 3 (a) Prior and posterior histograms of the 5th eigenfrequencies and corresponding reference values and (b) posterior histograms of the MAC-values of the 5th eigenvector (cases 1-3)

4.2 Posterior Modal Properties

As a next step, the effect of the updating procedure on the match of the modal properties is investigated. In Fig. 3(a), the prior (dashed-dotted line) and the posterior histograms (shaded bars) are shown exemplary for mode no. 5, where the measured eigenfrequencies of the three cases are included in the figure. It can be observed, that due the incorporation of the information contained in the modal data, the prior distributions are shifted towards the reference values leading to a considerably better match. As already observed for the posterior structural parameter values, the distributions of the three cases show a larger prediction error variance with increasing degree of non-linearity, which is visible through the scatter of the posterior histograms corresponding to the three investigated cases.

In Fig. 3(b) the correlation of the 5th mode with the respective reference mode shape is plotted by means of the MAC-value. The differences of the eigenvectors with the reference data becomes larger with increasing degree of non-linearity, whereby in case 3 this mode can almost not be identified ($MAC_{5,5} < 0.3$). In addition, it shall be pointed out that the posterior histograms do not show considerable changes with respect to the prior histograms. Hence, the prior space of the linear model does not span the full solution space and no better fit can be reached.

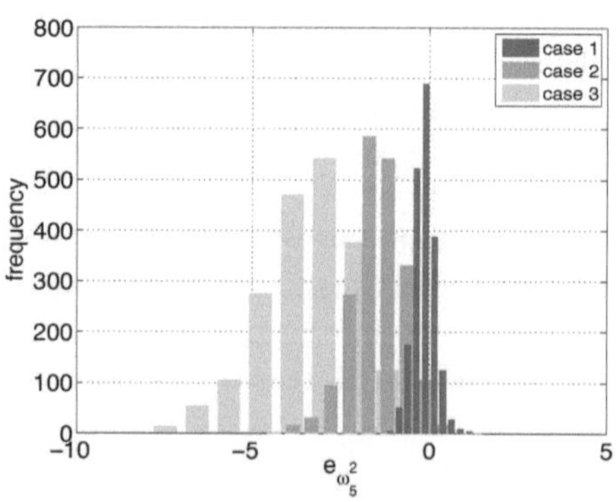

Fig. 4 Posterior histograms of the prediction errors of the 5th eigenfrequency (cases 1-3)

4.3 Posterior Prediction Error

As a last step, the distribution of the prediction error is discussed. In the Bayesian approach, the prediction error is modelled as a zero-mean Gaussian variable with an apriori unknown variance which is updated together with the structural parameters. The assumption, that model output and reference data agree on average, can be confirmed for the linear case, i.e. for case 1, as shown in Fig. 4. However, the figure also depicts that there is a shift in this distribution in case that the model used for generation of the data and the model employed for model updating show differences that are not defined by changes in the parameter values of the model (cases 2 and 3). Therefore, in such cases the model might be biased since the model class does not allow for changes in the parameter values to reach same mean values for the model output and the reference data. This situation requires a strategy for consideration of this shift. One approach could consists in an additive constant to be added to the model to allow for this bias, leading again to zero mean for the prediction error. However, this approach will not be further investigated here.

5 Conclusions

This manuscript has discussed the issues associated with the application of model updating for dynamic systems using modal properties where the underlying system shows slight non-linearities. It has been shown that these non-linearities which can not be captured by the structural parameters of the linear model used for model updating lead to larger prediction errors and hence larger prediction error variances. The results point out that an improvement of the prior model, which is conditional on the model class, could be achieved. However, the non-linearity of the reference structure leads to a bias in the updated model which evokes the need for an additional term considering this shift such that the assumption of a zero-mean Gaussian variable for the prediction error holds.

Acknowledgments. This research was partially supported by the European Space Agency (ESA) under Project No. 20829/07/NL/EM which is gratefully acknowledged by the authors. The second author is a recipient of a DOC-fForte-fellowship of the Austrian Academy of Science at the Institute of Engineering Mechanics (University of Innsbruck).

References

1. Friswell, M.I., Mottershead, J.E.: Finite Element Model Updating in Structural Dynamics. Kluwer Academic Publishers, Dordrecht (1995)
2. Menezes, R.C.R., Brenner, C.E.: On mechanical modeling uncertainties in view of failure data. In: Schuëller, G.I., Shinozuka, M., Yao, J.T.P. (eds.) Structural Safety and Reliability, ICOSSAR 1993, Balkema, Rotterdam, pp. 305–308 (1994)

3. Menezes, R.C.R., Schuëller, G.I.: On structural reliability assessment considering mechanical model uncertainties. In: Proceedings of the International Workshop Uncertainty: Models and Measures, Lambrecht, Germany, pp. 175–186 (1996)
4. Soize, C.: A nonparametric model of random uncertainties for reduced matrix models in structural dynamics. Probabilistic Engineering Mechanics 15, 277–294 (2000)
5. Capiez-Lernout, E., Soize, C.: Robust updating of uncertain damping models in structural dynamics for low- and medium-frequency ranges. Mechanical Systems and Signal Processing 22, 1774–1792 (2008)
6. Degrauwe, D., De Roeck, G., Lombaert, G.: Uncertainty quantification in the damage assessment of a cable-stayed bridge by means of fuzzy numbers. Computers and Structures 87, 1077–1084 (2009)
7. Altunok, E., Reda Taha, M.M., Ross, T.J.: Possibilistic Approach for Damage Detection in Structural Health Monitoring, vol. 133, pp. 1247–1256 (2007)
8. Cox, R.T.: Probability, frequency and reasonable expectation. American Journal of Physics 14, 1–13 (1946)
9. Beck, J.L., Katafygiotis, L.S.: Updating Models and their Uncertainties. I: Bayesian Statistical Framework. Journal of Engineering Mechanics 124, 455–461 (1998)
10. Katafygiotis, L.S., Beck, L.S.: Updating Models and their Uncertainties. II: Model Identifiability. Journal of Engineering Mechanics 124, 463–467 (1998)
11. Ching, J., Chen, Y.-C.: Transitional Markov Chain Monte Carlo Method for Bayesian Model Updating, Model Class Selection and Model Averaging. Journal of Engineering Mechanics 133, 816–832 (2007)
12. Ching, J., Muto, M., Beck, J.L.: Structural Model Updating and Health Monitoring with Incomplete Modal Data Using Gibbs Sampler. Computer-Aided Civil and Infrastructure Engineering 21, 242–257 (2006)
13. Cheung, S.H., Beck, J.L.: Bayesian model updating using Hybrid Monte Carlo simulation with application to structural dynamic models with many uncertain parameters. Journal of Engineering Mechanics 135, 243–255 (2009)
14. Vanik, M.W., Beck, J.L.: Bayesian probabilistic approach to structural health monitoring. Journal of Engineering Mechanics 126, 738–745 (2000)

Stochastic Reduced Order Models for Uncertain Infinite-Dimensional Geometrically Nonlinear Dynamical Systems- Stochastic Excitation Cases

X.Q. Wang[1], M.P. Mignolet[1], C. Soize[2], and V. Khanna[1]

[1] School of Mechanical, Aerospace, Chemical, and Materials Engineering, Arizona State University, USA
[2] Laboratoire Modélisation et Simulation Multi Echelle, Université Paris-Est, France

Abstract. The application of the nonparametric stochastic modeling technique to reduced order models of geometrically nonlinear structures recently proposed is here further demonstrated. The complete methodology: selection of the basis functions, determination and validation of the mean reduced order model, and introduction of uncertainty is first briefly reviewed. Then, it is applied to a cantilevered beam to study the effects of uncertainty on its response to a combined loading composed of a static inplane load and a stochastic transverse excitation representative of earthquake ground motions. The analysis carried out using a 7-mode reduced order model permits the efficient determination of the probability density function of the buckling load and of the uncertainty bands on the power spectral densities of the stochastic response, transverse and inplane, of the various points of the structure.

Keywords: Uncertainty, reduced order models, random matrices, geometrically nonlinear srructures, nonparametric stochastic modeling.

1 Introduction

The sharp increase in computational capabilities of the last 10-15 years has led to very satisfactory solutions for many complex structural dynamic problems *for given values of the structural parameters*. Further, these same analyses have also demonstrated that these solutions can be very sensitive to small variations of the structural parameters, thereby emphasizing the need to consider structural uncertainty. Several approaches have been devised to model this uncertainty and estimate its effects on the structural response; among those are the polynomial chaos methodology (e.g. Ghanem and Spanos [1]) and the nonparametric approach initially proposed by Soize [2,3]. The latter approach is particularly computationally attractive as it applies to reduced order models of the structure, seeking the distribution of the uncertain parameters that maximizes their statistical entropy under given physical constraints.

The nonparametric method has been applied to a broad class of problems including a recent extension to nonlinear geometric structural dynamic problems

W.Q. Zhu et al. (Eds.): Nonlinear Stochastic Dynamics & Control, IUTAM BOOK SERIES 29, pp. 293–302.
springerlink.com

[4] by relying on novel developments in the formulation of reduced order models for such structures (see Kim et al. [5] and references therein). According to these reduced order models, the structural response is expressed in a time-invariant basis with time-varying generalized coordinates satisfying coupled Duffing-type differential equations. Further, the parameters of this reduced order model, i.e. mass and linear, quadratic, and cubic stiffness coefficients, are identified directly from a full finite element model of the structure rendering the approach applicable to infinite-dimensional systems.

The earlier investigation of [4] is here further extended and validated to different infinite-dimensional structural models under stochastic excitations. A key element of the approach is the existence of a positive definite matrix $\mathbf{K_B}$ that regroups the linear, quadratic, and cubic stiffness coefficients. It is that matrix which is randomized in the nonparametric formulation while maintaining the positive definiteness so that the simulated stiffness properties, linear and nonlinear, are rendered uncertain in a physically admissible manner.

The complete process, reduced order modeling strategy and application of the nonparametric methodology, is presented on a cantilevered beam subjected to a static compressive load near the buckling limit and a transverse excitation corresponding to ground motions.

2 Reduced Order Modeling of Geometrically Nonlinear Structures

The formulation of reduced order model of geometrically nonlinear structures involves three specific issues: (i) the selection of the basis functions used to represent the motion of the structure, (ii) the determination of the form of the equations governing the generalized coordinates, and (iii) the determination of the coefficients of these equations. The resolution of these issues is briefly reviewed below.

2.1 Basis Functions Selection

In parallel with modal analysis of linear systems, the displacement field of the structure will be expressed in a modal expansion-type representation, i.e., as

$$u_i(\mathbf{X},t) = q_n(t)\, U_i^{(n)}(\mathbf{X}) \quad i = 1,2,3 \tag{1}$$

(summation over repeated indices, n here, is implied). In this representation, $U_i^{(n)}(\mathbf{X})$ denote time-invariant, spatially varying basis functions while $q_n(t)$ are the corresponding time-dependent generalized coordinates. Note that the spatial domain Ω to which $\mathbf{X}$ belongs is the undeformed configuration of the structure, see section 2.2.

In selecting the functions $U_i^{(n)}(\mathbf{X})$, it is first expected that the nonlinear reduced order model (1) should reduce naturally to a modal model in the limit of small motions. Thus, the nonlinear basis should completely include its linear counterpart, i.e. the set of linear modes significantly excited. This is however not enough and a complete representation of the structural response requires additional basis functions. As an example, consider the response to transverse loads of a flat, symmetric beam or plate subjected to a purely transverse loading. In the linear, infinitesimal case, the decoupling of the inplane and transverse modes implies that only the latter ones are necessary and no inplane motion takes place. However, finite deformations can only occur with a stretching of the beam or plate and, accordingly, with inplane deformations. Thus, the nonlinear basis required for a full representation must include both transverse linear modes and functions describing the inplane motions.

The basis functions selected here to complement the linear modes are the "dual" modes of [5], i.e. a set of static nonlinear displacement fields induced by external loads such that the response they would induce in the structure would be proportional to either one of the linear modes or a linear combination of two of them. Constructed in this manner, the dual modes capture the nonlinear effects corresponding to motions that would take place if the structure was behaving linearly.

2.2 Form of the Reduced Order Model Equations

The derivation of the form of the ordinary differential equations governing the evolution of the generalized coordinates $q_n(t)$ is next derived from the equations of finite deformation elasticity in a Galerkin procedure. To this end, note first that the time-invariance of the functions $U_i^{(n)}(\mathbf{X})$ is most easily achieved when the spatial domain Ω occupied by the structure is constant. This situation occurs when the displacement field $\mathbf{u}$ is expressed in the undeformed configuration in which the equations of elasticity are (see [6,7])

$$\frac{\partial}{\partial X_k}\left(F_{ij}\, S_{jk}\right) + \rho_0\, b_i^0 = \rho_0\, \ddot{u}_i \qquad \text{for } \mathbf{X} \in \Omega_0 \tag{2}$$

where $\mathbf{S}$ denotes the second Piola-Kirchhoff stress tensor, ρ_0 is the density in the reference configuration, and $\underline{b}^0$ is the vector of body forces. Further, in Eq. (2), the deformation gradient tensor $\mathbf{F}$ is defined by its components F_{ij} as

$$F_{ij} = \delta_{ij} + \frac{\partial u_i}{\partial X_j} \tag{3}$$

where δ_{ij} denotes the Kronecker symbol. Associated to Equation (2) are appropriate boundary conditions, e.g. specification of displacement and/or tractions on the boundary $\partial\Omega_0$ of the reference configuration domain.

To complete the formulation of the elastodynamic problem, it remains to specify the constitutive behavior of the material. In this regard, it will be assumed here that the second Piola-Kirchhoff stress tensors **S** is linearly related to the Green strain tensor **E**, i.e.

$$S_{ij} = C_{ijkl} E_{kl} \qquad \text{where} \qquad E_{ij} = \frac{1}{2}\left(F_{ki} F_{kj} - \delta_{ij}\right) \tag{4),(5}$$

where C_{ijkl} denotes the fourth order elasticity tensor.

Introducing the assumed displacement field of Equation (1) in Eqs (2)-(5) and proceeding with a Galerkin approach leads to the desired governing equations, i.e.

$$M_{ij}\ddot{q}_j + D_{ij}\dot{q}_j + K_{ij}^{(1)} q_j + K_{ijl}^{(2)} q_j q_l + K_{ijlp}^{(3)} q_j q_l q_p = F_i \tag{6}$$

in which a damping term $D_{ij}\,\dot{q}_j$ has been included to collectively represent various dissipation mechanisms. In Equation (6), M_{ij}, $K_{ij}^{(1)}$, and F_i denote the coefficients of the linear mass and stiffness matrices and the modal forces while $K_{ijl}^{(2)}$ and $K_{ijlp}^{(3)}$ are nonlinear stiffness coefficients.

2.3 *Identification of the Stiffness Coefficients*

The *form* of the reduced order model, derived in the previous section as Equation (6), involves a series of structure and loading dependent coefficients, i.e. M_{ij}, F_i, $K_{ij}^{(1)}$, $K_{ijl}^{(2)}$, and $K_{ijlp}^{(3)}$. While the modal masses (M_{ij}) and modal forces (F_i) can be evaluated as in linear modal models, the stiffness coefficients, linear, quadratic, and cubic, necessitate a dedicated identification strategy. The specific methodology used here was initially proposed in [8] and further modified in [5], it is based on the availability of a series of static nonlinear solutions (usually from a finite element of the structure) in which the (static) displacement field is imposed and the corresponding necessary forces are determined.

The identification procedure starts with the imposition of static displacement fields that are proportional to each of the basis functions $U_i^{(n)}(\mathbf{X})$, i.e.

$$u_i(\mathbf{X}) = q^{(j)}\ U_i^{(n)}(\mathbf{X}) \quad i = 1,2,3 \tag{7}$$

In fact, three such cases, with different values of the factor $q^{(j)} = q^{(1)}$, $q^{(2)}$ (typically - $q^{(1)}$), and $q^{(3)}$, are considered for each $U_i^{(n)}(\mathbf{X})$ and the corresponding necessary forces $\mathbf{F}^{(j)}(\mathbf{X})$ are determined from the finite element model and projected onto the basis functions $U_i^{(m)}(\mathbf{X})$ to yield the modal forces $F_m^{(j)}$. Introducing this data into the reduced order model equations (6) yields the conditions

$$K_{in}^{(1)}\, q^{(j)} + K_{inn}^{(2)}\left[q^{(j)}\right]^2 + K_{innn}^{(3)}\,\left[q^{(j)}\right]^2 = F_i^{(j)}. \tag{8}$$

Considering Equation (8) for j = 1, 2, and 3 leads, for each pair of indices i and n, to 3 linear algebraic equations in the 3 unknowns $K_{in}^{(1)}$, $K_{inn}^{(2)}$, and $K_{innn}^{(3)}$ from which these coefficients are determined.

The identification of the remaining stiffness coefficients proceed in a similar manner by imposing static displacement fields which are linear combinations of 2 and then 3 of the modal bases, see [4,5,8] for complete details.

The above identification procedure has successfully been applied to a variety of problems (e.g. see [5]) but was found to be too sensitive to small errors in the predicted modal forces in connection with cantilevered structures (see [10] for discussion). This difficulty led to a modification of the estimation procedure in which the linear and quadratic stiffness coefficients of the final model were indeed estimated as above but with cubic coefficients selected (the decondensation technique of [10]) to match the corresponding coefficients of a reduced order model in which only the transverse motions are used with the inplane ones condensed. This two-step approach was employed here for the cantilevered beam of section 3.

2.4 Nonparametric Uncertainty Modeling

Two different methodologies have been proposed for the consideration of uncertainty in linear structural dynamic systems. The first one, referred to here as parametric, introduces the uncertainty at the level of the full computational model (e.g. finite element model) through the randomization of some or all of its material properties (Young's modulus, Poisson's ratio, etc., e.g see [1]). This approach is particularly well suited for the consideration of *data uncertainty*, i.e. lack of knowledge or variability in the system properties but not for *model uncertainty* which is associated with deviations of the structure from its computational model. As example, for a beam that is nominally straight, such deviations include the presence of a curvature, a twist, or any other variation of geometry that would require a change of mesh in the finite element model.

A computationally efficient approach for the consideration of data and model uncertainty, referred to as the nonparametric method, has been proposed a few years ago, e.g. [2,3]. In this approach, the uncertainty is introduced directly at the level of the reduced order model by allowing the matrices it involves (e.g. mass, stiffness, and/or damping matrices) to be random. Further, the probability density functions of these matrices is derived (not chosen) to provide the maximum of its statistical entropy under mathematical and physical constraints, i.e. that it leads to a total unit probability, that all matrices be positive definite if physically required (for the mass, damping, and stiffness matrices), that their means be equal to the matrices of the mean reduced order model, and finally that the expected value of the square Frobenius norm of their inverse be finite, see [2,3] for complete discussion.

Key to the implementation of this approach is the simulation of random matrices according to the derived probability density function which is conveniently achieved as follows. Denote by $\overline{\mathbf{A}}$ the mean reduced order matrix considered and let $\overline{\mathbf{L}}$ be any decomposition (e.g. Cholesky) satisfying $\overline{\mathbf{A}} = \overline{\mathbf{L}}\,\overline{\mathbf{L}}^T$. Then, random matrices $\mathbf{A}$ may then be simulated as

$$\overline{\mathbf{A}} = \overline{\mathbf{L}}\,\mathbf{H}\,\mathbf{H}^T\,\overline{\mathbf{L}}^T \tag{9}$$

where the random matrix $\mathbf{H}$ is lower triangular. Further, its elements were shown [2,3] to be independent random variables with those located off diagonal being normally distributed with zero mean and common variance. Finally, the diagonal elements of $\mathbf{H}$ are proportional to the square root of Gamma distributed random variables [2,3]. A single free parameter exists in this strategy which can be selected to match a particular information on the level of variability, such as coefficient of variation of natural frequencies or the overall measure of uncertainty δ introduced in [2,3].

The above discussion was first carried out in the context of linear structural dynamic systems but it was recently extended [4] to reduced order models of nonlinear geometric problems of the form of Eq. (6). Pivotal in this extension is the property (e.g. see [4]) that the linear, quadratic, and cubic stiffness coefficients $K_{ij}^{(1)}$, $K_{ijl}^{(2)}$, and $K_{ijlp}^{(3)}$ can be combined to form a matrix $\mathbf{K_B}$ which is symmetric and positive definite. Then, random coefficients $K_{ij}^{(1)}$, $K_{ijl}^{(2)}$, and $K_{ijlp}^{(3)}$ can be obtained from random matrices $\mathbf{K_B}$ generated from their mean model counterpart $\overline{\mathbf{K}}_\mathbf{B}$ according to Eq. (6).

3 Effects of Uncertainty on a Cantilevered Structure

The methodology developed in the previous sections was applied to a cantilevered beam of length 0.2286m, width 0.0127m, and thickness 7.75×10^{-4}m which was discretized by the finite element method (with MSC NASTRAN) into 40 CBEAM elements of equal lengths. The beam material was high-carbon steel with a Young's modulus of 205,000 MPa, a shear modulus of 80,000 MPa, and a mass density of 7,875 kg/m^3 leading to natural frequencies of the first transverse modes of 12.4, 77.9, 218, and 427 Hz. A Rayleigh damping model was assumed that yielded damping ratios of 1.07%, 0.47%, 0.91%, and 1.69%, respectively. Finally, the beam was subjected to a combined loading: random time-varying transverse motions of its support and an inplane compressive static force.

The development of an accurate reduced order model represented the first step of the uncertainty analysis. The ground motions selected here exhibited a Kanai-Tajimi spectrum (see [10]) of characteristic frequency equal to 5Hz ($\omega_g = 10\pi$) and damping ratio of $\zeta_g = 0.3$. Given this low frequency excitation, only the first three linear, purely transverse modes were considered for the linear part of the basis. To these functions, 4 dual modes were added that exhibited only inplane motions thereby forming a 7 mode reduced order model.

The first validation of this reduced order model focused on the power spectral densities of the transverse and inplane relative displacements of the beam tip. A comparison of the spectra obtained from a full finite element computation (Nastran SOL 400) and from the reduced order model equations is presented in Figure 1 for two different standard deviations of the ground motions and a common compressive inplane force equal to 80% of its buckling limit (i.e. 4N). The standard deviations of the transverse tip deflections corresponding to this loading were found to be 10% and 18% of the beam length. Owing to the long Nastran computations, this comparison was achieved with records of 40 seconds from which the first 20 were removed as transient. The remaining 20 seconds of data may not be sufficient for an accurate capture of the low frequency response but are sufficient here for the validation of the reduced order model the response of which was similarly treated.

Clearly, the matching between full finite element and reduced order model predictions is excellent except at the very low frequencies for the inplane motions at the highest loading level. As the response levels increases, the Nastran and reduced order models will differ, see [9], because of the difference in the definitions of linear elasticity used in these methodologies, in a total Lagrangian in the latter while the former is believed to proceed in an updated Lagrangian framework. The results of Figure 1 demonstrate the appropriateness of the reduced order model for the prediction of the mean and uncertain beam responses.

The response of the uncertain beam to the specified combined loading was considered next using the nonparametric methodology of section 2.4. The free parameter was selected to achieve a coefficient of variation of 2% of the first transverse natural frequencies of the beam. With this level of uncertainty, the response of 300 random beams was computed using the stochastic reduced order model for 90 seconds with the first 20 seconds considered as transient. The uncertainty bands corresponding to the 2nd and 98th percentiles of the generated power spectra of the response were then determined and are shown in Figure 2 for the two excitation levels and for both transverse and inplane motions. Note that the power spectrum of the mean model is within the 2nd-98th percentiles band for the lowest excitation levels but it reaches the 98th percentile (for the transverse motions) or exceeds it (for the inplane motions) at the highest excitation level. This finding is justified by the inclusion in the nonparametric methodology of *model uncertainty*, i.e. the presence of coupling terms in the stochastic reduced order model which are not present for the mean model owing to its symmetry. Thus, the simulated reduced order models would be representative of typically curved beams for which the response is typically smaller than for the straight beam of the mean reduced order model. Uncertainty in the mass matrix was also considered with the non-parametric method but its effects appear very small and thus are not presented here for brevity.

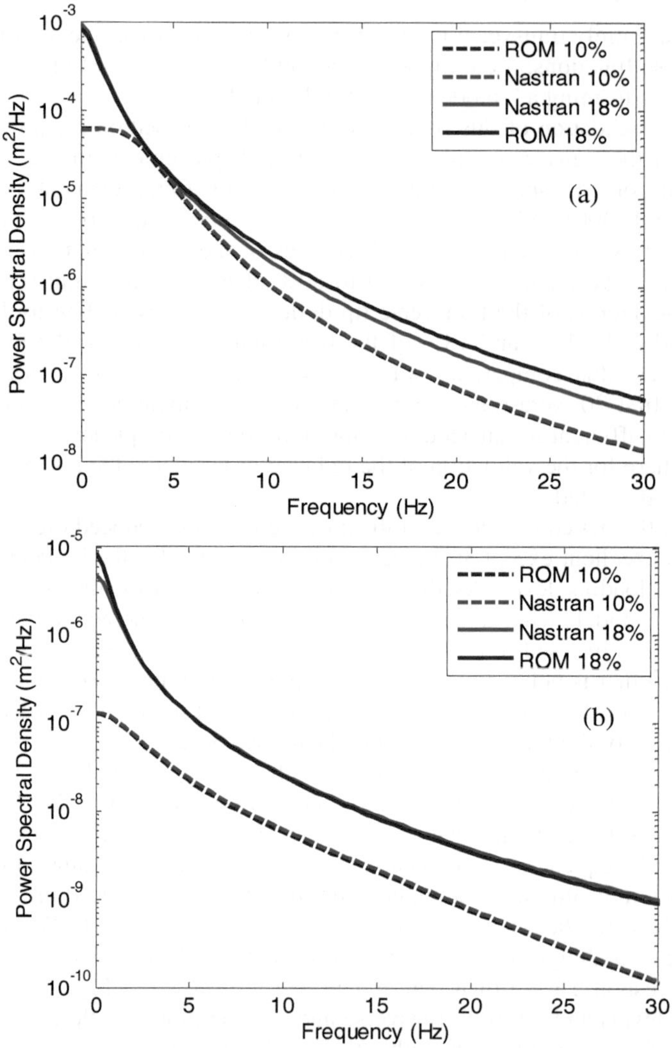

Fig. 1 Power spectra of the tip displacements, Nastran vs. reduced order model (ROM) for the two loading cases with standard deviation of transverse response = 10% and 18% of span (a) Transverse. (b) Inplane.

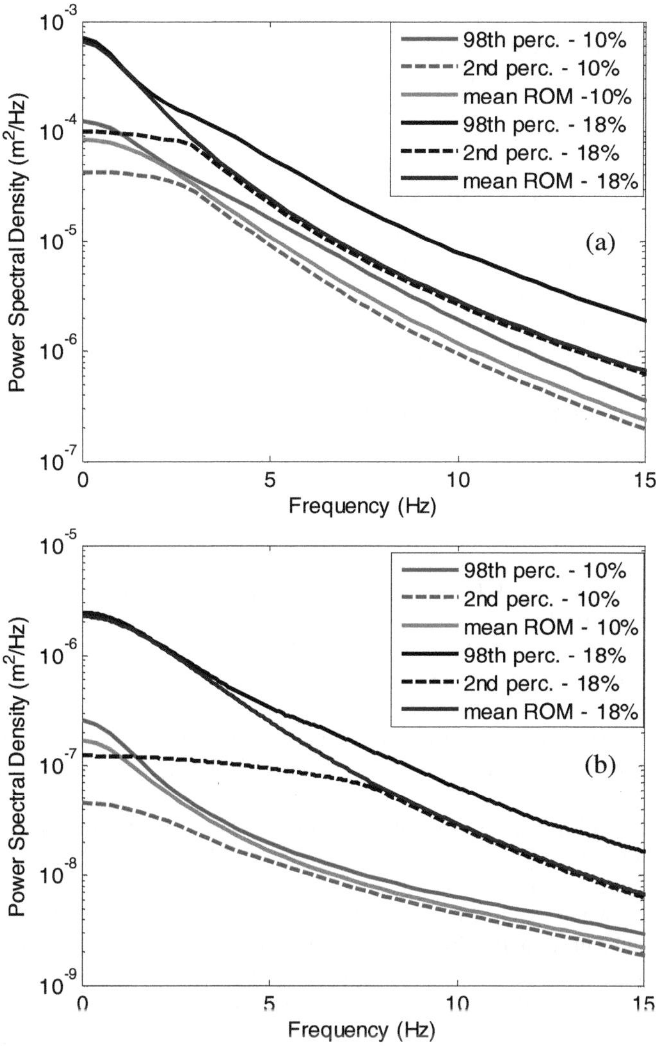

Fig. 2 Power spectra of the tip displacements, uncertainty in stiffnesses for the two loading cases with standard deviation of transverse response = 10% and 18% of span. Uncertainty bands corresponding to the 2nd and 98th percentiles and mean ROM. (a) Transverse. (b) Inplane.

4 Summary

The present paper reported on a continued investigation of the effects of uncertainty on the response of nonlinear geometric structures. Owing to the computationally expensive Monte Carlo simulations involved in such investigations, a nonlinear reduced order modeling strategy was adopted for the mean model, see sections 2.1-2.3 and references therein for details. Next, uncertainty was introduced in this mean model according to the nonparametric methodology (see section 2.4) which allows the consideration of uncertainty in both structural properties (data uncertainty) and geometry (as example of model uncertainty).

This framework was demonstrated on a cantilevered beam subjected to the combined action of a static compressive inplane load and a transverse random excitation typical of ground motions. The mean model was first constructed from a full finite element model and its predictive capabilities validated vs. this full model at significant displacement levels (tip deflections of up to 18% of beam length). Uncertainty was then introduced leading to a stochastic reduced order model the stationary response of which was determined. The uncertainty bands associated with the 2nd and 98th percentiles of the power spectrum of the tip displacements were determined and it was found that the mean model power spectrum fits well within these bands at lower response levels but shifts to the 98th percentile as the response level increases owing to model uncertainty effects.

References

1. Ghanem, R., Spanos, P.D.: Stochastic Finite Elements: A Spectral Approach. Springer, Heidelberg (1991)
2. Soize, C.: A Nonparametric Model of Random Uncertainties on Reduced Matrix Model in Structural Dynamics. Probabilistic Engineering Mechanics 15, 277–294 (2000)
3. Soize, C.: Maximum Entropy Approach for Modeling Random Uncertainties in Transient Elastodynamics. Journal of the Acoustical Society of America 109, 1979–1996 (2001)
4. Mignolet, M.P., Soize, C.: Stochastic Reduced Order Models for Uncertain Geometrically Nonlinear Dynamical Systems. Computer Methods in Applied Mechanics and Engineering 197, 3951–3963 (2008)
5. Kim, K., Wang, X.Q., Mignolet, M.P.: Nonlinear Reduced Order Modeling of Isotropic and Functionally Graded Plates. In: Proceedings of the 49th Structures, Structural Dynamics, and Materials Conference, AIAA Paper AIAA-2008-1873 (2008)
6. Fung, Y.C., Tong, P.: Classical and Computational Solid Mechanics. World Scientific, Singapore (2001)
7. Bonet, J., Wood, R.D.: Nonlinear Continuum Mechanics for Finite Element Analysis. Cambridge University Press, Cambridge (1997)
8. Muravyov, A.A., Rizzi, S.A.: Determination of nonlinear stiffness with application to random vibration of geometrically nonlinear structures. Computers and Structures 81, 1513–1523 (2003)
9. Kim, K., Khanna, V., Wang, X.Q., Mignolet, M.P.: Nonlinear reduced order modeling of flat cantilevered structures. In: Proceedings of the 50th Structures, Structural Dynamics, and Materials Conference, AIAA Paper AIAA-2009-2492 (2009)
10. Madsen, H.O., Krenk, S., Lind, N.C.: Methods of Structural Safety. Dover, New York (2006)

Part 6
Structural Health Monitoring

Decentralized Random Decrement Technique for Data Aggregation and System Identification in Wireless Smart Sensor Networks

Sung-Han Sim[1], B.F. Spencer, Jr.[1], Hongki Jo[1], and Juan Francisco Carbonell-Márquez[2]

[1] Department of Civil and Environmental Engineering, University of Illinois at Urbana-Champaign, USA
[2] Department of Structural Mechanics, University of Granada, Spain

Abstract. Smart sensors have been recognized as a promising technology with the potential to overcome many of the inherent difficulties and limitations associated with traditional wired structural health monitoring (SHM) systems. The unique features offered by smart sensors, including wireless communication, on-board computation, and cost effectiveness, enable deployment of the dense array of sensors that are needed for monitoring of large-scale civil infrastructure. Despite the many advances in smart sensor technologies, power consumption is still considered as one of the most important challenges that should be addressed for the smart sensors to be more widely adopted in SHM applications. Data communication, the most significant source of the power consumption, can be reduced by appropriately selecting data processing schemes and the related network topology. This paper presents a new decentralized data aggregation approach for system identification based on the Random Decrement Technique (RDT). Following a brief overview of RDT, which is an output-only system identification approach, a hierarchical approach is described and shown to be suitable for implementation in the intrinsically decentralized computing environment found in wireless smart sensor networks (WSSNs). RDT-based decentralized data aggregation is then implemented on the Imote2 smart sensor platform based on the Illinois Structural Health Monitoring Project (ISHMP) Services Toolsuite. Finally, the efficacy of the decentralized RDT method is demonstrated experimentally in terms of the required data communication and the accuracy of identified dynamic properties.

Keywords: wireless smart sensor, decentralized processing, Natural Excitation Technique, Random Decrement Technique, output-only system identification.

1 Introduction

Vibration-based structural health monitoring (SHM) can provide valuable information regarding the dynamic characteristics of structures. The identification process typically consists of measuring vibration responses from structures and analyzing the measured data to build a numerical model of the structure. Traditionally, the vibration responses are obtained using centralized data acquisition systems with wired sensors. However, the use of wired sensors has proven to be problematic, particularly for dense deployments of sensors on large-scale civil

W.Q. Zhu et al. (Eds.): Nonlinear Stochastic Dynamics & Control, IUTAM BOOK SERIES 29, pp. 305–314.
springerlink.com

infrastructure, primarily due to long setup times, difficulties in cabling, and high equipment costs.

Smart sensors provide a promising alternative to traditional wired sensor systems. Spencer et al. [20] defined smart sensors as having four features: (i) on-board computing capability, (ii) small size, (iii) wireless communication, and (iv) low cost. Recent studies [5, 8] have demonstrated the potential of such smart sensors to realize a dense array of sensors for monitoring large-scale civil infrastructure. However, challenges such as power consumption and long-term reliability still remain.

As smart sensors typically are battery-powered, power management is critical for long-term monitoring. Although several approaches for power harvesting have been reported (e.g., solar power, vibration power [8, 16]), power consumption must still be appropriately managed. Reducing wireless communication, the most significant source of power consumption for smart sensors, is an important goal.

Data communication in wireless smart sensor networks (WSSN) is intimately related to the data acquisition and processing schemes employed in a WSSN. In traditional centralized data collection approaches (see Fig. 1a), large amounts of data must be transferred to a central data repository; such approaches are not efficient for WSSNs due to problems with limited bandwidth, data congestion, and excessive power requirements. Thus, decentralized approaches that employ local data processing for data aggregation and condensation have been introduced. Independent processing, shown in Fig. 1b, utilizes the computational power of a local sensor node to process sensor data [4, 11, 12, 15, 19, 21]. While the amount of data wirelessly transferred in the network is significantly reduced, all spatial information (e.g., mode shapes) is lost in such approaches.

Gao et al. [7] proposed a coordinated computing strategy for damage detection that retains local spatial information, while concurrently reducing data communication in the network (see Fig. 1c). In this approach, the sensor network is divided into hierarchical sensor communities that consist of a limited number of sensor nodes in local proximity to each other. Nagayama and Spencer [13] implemented a coordinated computing strategy in a WSSN employing the Imote2 sensor platform for damage detection. An output-only identification approach, the Natural Excitation Technique (NExT) [10], was employed in conjunction with Eigensystem Realization Algorithm (ERA) [9]. The network topology employed in this approach consists of three types of sensor nodes: (a) gateways, (b) cluster-head, and (c) leaf node. The gateway node is directly linked to the base station, controlling operation of the sensor network and interfacing users with the WSSN. The use of NExT/ERA in the decentralized computing environment has been shown to be quite efficient from a data communication perspective.

The Random Decrement Technique (RDT) is an alternative output-only system identification method proposed by Cole [6] that has several attractive features. The decentralized implementation of NExT by Nagayama and Spencer [13] requires the complete time history data from the cluster-head in a group be transferred to the leaf nodes to calculate the correlation functions. In contrast, RDT only requires the trigger crossings be sent to the leaf nodes, which is typically much smaller in size than the raw sensor data. The output of the RDT is the random decrement (RD) function, which can be used for system identification.

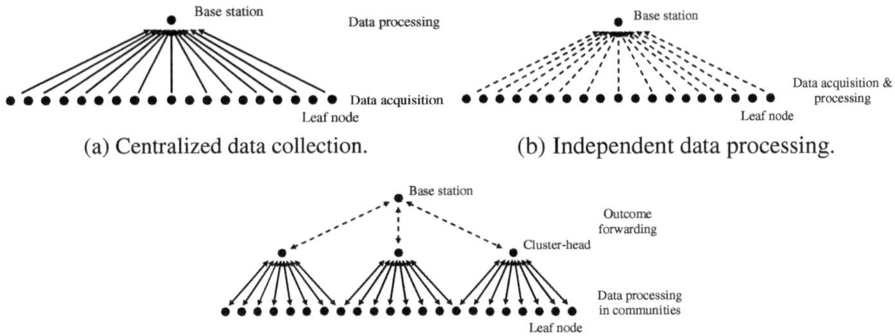

(a) Centralized data collection. (b) Independent data processing.

(c) Coordinated computing strategy.

Fig. 1 Data acquisition computing strategy [13]

This paper presents a new decentralized computing strategy for data aggregation and system identification based on RDT that is suitable for implementation in WSSNs. The implementation of decentralized NExT reported by Sim and Spencer [19] is used as a baseline for comparison. Finally, the decentralized RDT method is realized on the Imote2 smart sensor and experimentally verified. The efficacy of the RDT method is demonstrated in terms of required data communication and accuracy of identified dynamic models.

2 Random Decrement Technique

Cole [6] initially proposed RDT to estimate the dynamic properties of space structures excited by immeasurable ambient excitation. The basic assumption is that the dynamic response of a structure under ambient excitation is ergodic. From the structural response, n time history segments in the interval $[t_i, t_i+\tau]$ $(i = 1,\ldots,n)$are selected such that the displacement at t_i is equal to a specific trigger level. The response of a system at $t_i+\tau$ is then comprised of three components:

- o Deterministic response due to the initial displacement at time t_i
- o Deterministic response due to the initial velocity at time t_i
- o Random response due to the random excitation between t_i and $t_i+\tau$

If the average is taken over a sufficiently large number of the segments, the third part of the response due to the random excitation will tend toward zero. Furthermore, the velocity at time t_i is uncorrelated with the displacement and has zero mean. Thus, the part of the response due to the initial velocity also tends to zero. The resulting RD function is the free vibration caused by a nonzero initial displacement.

Vandiver et al. [22] provided a mathematical foundation for the random decrement function, showing that the RD function is proportional to the autocorrelation function for a linear, time-invariant system excited by a zero-mean, stationary, Gaussian random process. Later, Brincker et al. [2, 3] introduced a general triggering function and showed that RDT estimates a weighted sum of the auto- and cross-correlation functions and their time derivatives.

The RD functions can be estimated from data as follows:

$$D_{jk}(\tau) = \frac{1}{N}\sum_{i=1}^{N} x_j(t_i+\tau)\big|C_{x_k(t_i)} \tag{1}$$

where $D_{jk}(\tau)$ is the RD function obtained from $x_j(t)$ with respect to the reference $x_k(t)$, N is the total number of trigger events, $C_{x_k(t_i)}$ is the specified trigger condition, and t_i is the ith time obtained from the trigger event $C_{x_k(t_i)}$. When $j = k$, $D_{jj}(\tau)$ is an auto-RD function. In this study, the positive-point trigger condition [1] is considered:

$$C_{x_k(t_i)} = \left[\alpha_1\sigma_{x_k} \le x_1(t_i) < \alpha_2\sigma_{x_k}, -\infty \le \dot{x}_k(t_i) < \infty\right] \tag{2}$$

where $0 \le \alpha_1 < \alpha_2 \le \infty$. Once the RD functions are obtained, modal properties of the structure can be estimated using a wide range of system identification methods.

3 RDT-Based Decentralized Data Aggregation

RDT can significantly enhance the efficiency of data aggregation in the distributed computing environment in WSSNs. As previously described, central data collection and processing in WSSNs can cause severe data congestion due to limited communication bandwidth. The use of decentralized in-network processing condense the data can mitigate this problem.

To better understand the efficiency of decentralized processing, consider the centralized implementation of a community-wide data processing scheme, as shown in Fig. 2. For this community has n_s nodes, each sensor node measures data and transmits it to Node 1. For time history records of length N and n_d averages, the amount of transmitted data is $N \times n_d \times (n_s - 1)$.

Nagayama et al. [13] proposed a decentralized NExT implementation, taking advantage of each node's computing capability to reduce data communication (see Fig. 3). Node 1 sends a measured time history record as a reference signal to each node. Correlation functions are calculated in all nodes in the community and subsequently collected at Node 1. The amount of transmitted data is at most $N \times n_d + N/2 \times (n_s - 1)$. As the numbers of nodes or averages increase, the efficiency of the decentralized NExT implementation becomes clearer.

The decentralized RDT implementation shown Fig. 4 can further reduce data communication. In this approach, Node 1 sends the trigger information to all nodes in the community. Once the reference trigger data is received, each node calculates the RD functions that are subsequently collected at Node 1. Note that (1) the trigger information is in general much shorter than the time history record used as the reference for NExT, and (2) transmission of the reference takes a significant portion of the total communication in the decentralized NExT implementation, particularly when long records are used. Thus, the RDT-based decentralized data aggregation can considerably reduce data communication requirement.

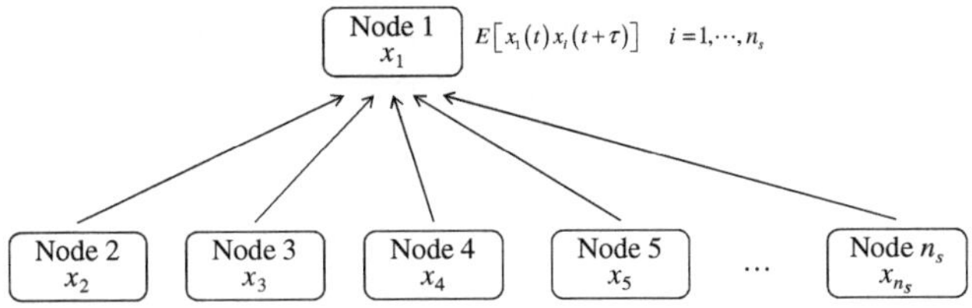

Fig. 2 Centralized NExT implementations [13]

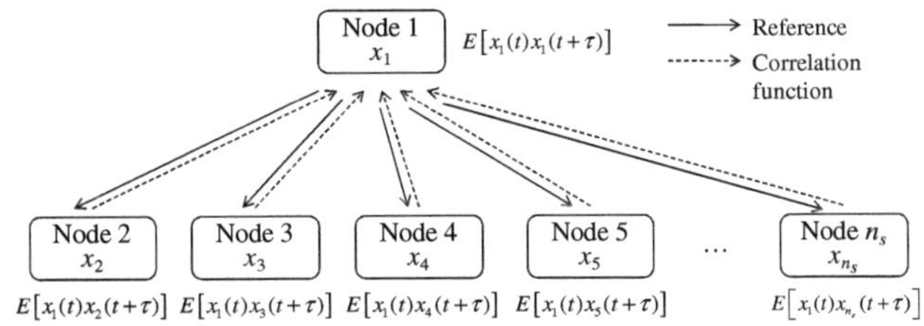

Fig. 3 Decentralized NExT implementation [13]

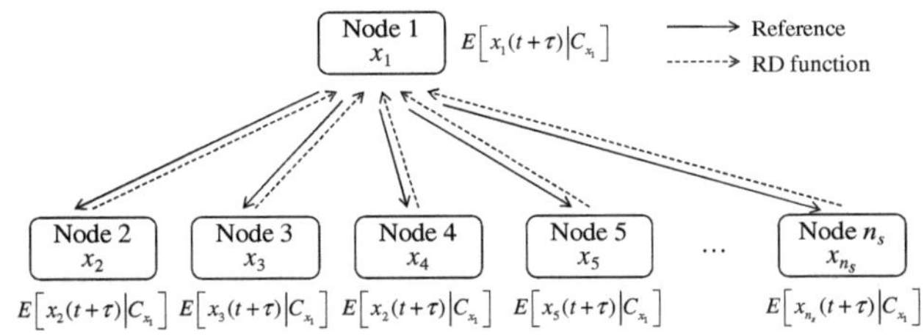

Fig. 4 Decentralized RDT implementation

Data communication required by the decentralized RDT implementation is closely related to the number of triggering points. For the positive-point trigger condition found in Eq. (2), the expected number of triggering points is [1]:

$$E\left[n(a_1,a_2)\right]=(N_X-N_\tau)\cdot\int_{a_1}^{a_2}p_X(x)dx \tag{3}$$

where $n(a_1, a_2)$ is the number of triggering points between a_1 and a_2, Δt is the sampling rate, $p_X(x)$ the probability density function of $X(t)$, and N_X and N_τ are the number points in $X(t)$ and the RD function, respectively. Thus, the total number of points to be wirelessly transferred in the decentralized RDT implementation is:

$$(N_X-N_\tau)\int_{a_1}^{a_2}p_X(x)dx+N_\tau(n_s-1) \tag{4}$$

Depending on the choice of a_1 and a_2, RDT can require substantially less data communication than NExT. The next section presents the implementation and experimental validation of RDT on the Imote2 wireless smart sensor platform.

4 Implementation and Experimental Validation of RDT-Based Decentralized Data Aggregation on WSSNs

RDT is implemented in the decentralized coordinated computing environment (see Fig. 1c) based on the Illinois Structural Health Monitoring Project (ISHMP) Services Toolsuite (http://shm.cs.uiuc.edu) for the Imote2 sensor platform (see Fig. 5). The ISHMP Services Toolsuite provides an open source library of services that are essential for developing SHM applications. The services, software components that perform specific tasks such as sensing, time synchronization, wireless communication, etc., can be assembled to develop customized SHM applications [17]. The ISHMP Services Toolsuite also contains SHM applications that can be readily used for monitoring and identification of structures using the Imote2 smart sensor platform.

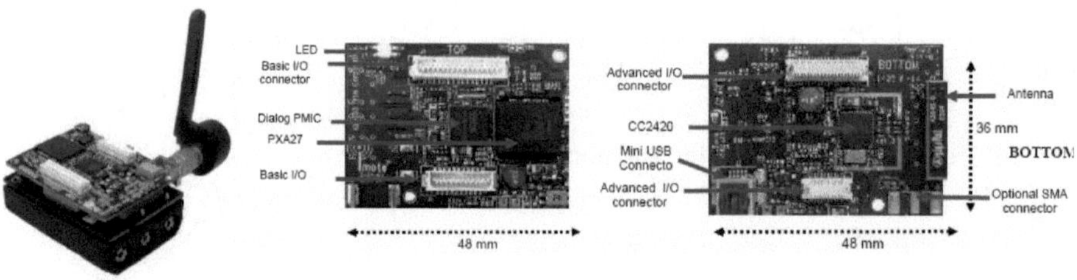

(a) Imote2 (b) Top view (c) Bottom view

Fig. 5 Imote2 sensor platform

DecentralizedDataAggregationRD is an implementation of RDT-based decentralized data aggregation in the hierarchical network shown in Fig. 1c. Note that overlapping nodes are allowed so that phase information from two overlapping local communities can be related to each other. The network is divided into local communities where RD functions are calculated at each node and gathered by the cluster-heads. To properly realize the decentralized implementation of RDT in the hierarchical network, the design of *DecentralizedDataAggregationRD* requires careful consideration of network topology, controlled network-wide flow, and fault tolerance. More details regarding the implementation can be found in [18, 19].

The performance of RDT-based decentralized data aggregation in WSSNs is experimentally investigated using the truss structure shown in Fig. 6. Herein, the estimation accuracy for global modal properties and data communication requirements are assessed. *DecentralizedDataAggregationRD* installed on the Imote2 sensors is employed to decentrally calculate the RD functions. *DecentralizedDataAggregation*, an implementation of NExT-based decentralized data aggregation found in the ISHMP Services Toolsuite [19], is examined for comparison.

The experimental structure considered is a simply supported truss that consists of steel hollow circular tubes with an inner diameter of 0.428 inches and an outer diameter of 0.612 inches (see Fig. 6). A shaker is used to excite vertically the truss with a band-limited white noise on the interval 0–100 Hz.

A total of 14 Imote2 sensors with SHM-A acceleration sensor boards [17] are installed on the bottom chord of the truss as shown in Fig. 6 and Fig. 7. The sensor network is divided into four local sensor groups that consist of four or six sensor nodes. Sensor nodes S2, S6, S10, and S14 serve as cluster-heads in each local sensor groups (see Fig. 7).

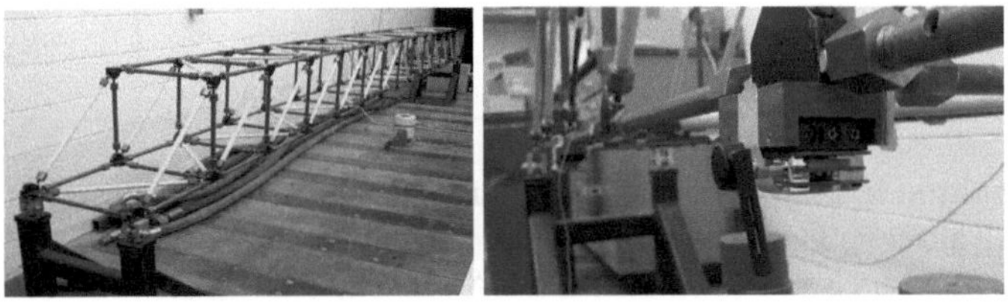

Fig. 6 Truss structure and an installed Imote2 sensor node

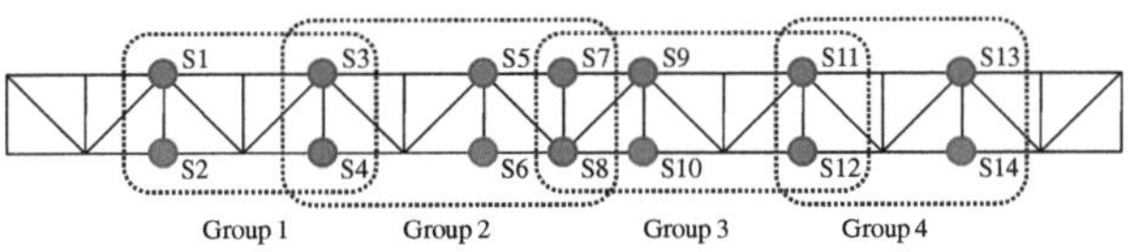

Fig. 7 Sensor topology (plan view) (S2, S6, S10, and S14 are cluster-heads)

Vertical accelerations are measured at each sensor node with a sampling rate of 280 Hz, with a 70 Hz cutoff frequency. The measured acceleration time histories are 10,752 points in length for both *DecentralizedDataAggregation* and *DecentralizedDataAggregationRD*. For correlation function estimation, a signal with 10,752 points allows 20 averages if 1,024 points of FFT and 50% overlap between windows are specified.

DecentralizedDataAggregation and *DecentralizedDataAggregationRD* are employed to estimate the correlation and RD functions. The positive-point trigger crossing with an interval of (σ, 2.5σ) was found to produce the best results. Local modal properties are estimated for both cases by ERA using the correlation and RD functions, and subsequently global modal properties are obtained. In addition, raw acceleration time history data from all sensor nodes are centrally collected to provide a reference for comparison of NExT and RDT in the distributed computing environment, i.e., the centrally collected accelerations, NExT/ERA and RDT/ERA are employed to estimate reference modal properties.

Table 3 summarizes the identified natural frequencies of the truss for system identification method. Compared to the cases of the centralized processing, both NExT/ERA and RDT/ERA based on the decentralized processing estimate natural frequencies with an excellent accuracy (see Table 3).

Table 3 Identified natural frequencies (Hz)

Mode	Centralized processing		Decentralized processing	
	NExT/ERA	RDT/ERA	NExT/ERA (case 1)	RDT/ERA (case 2)
1	20.59	20.68	20.77	20.73
2	32.88	32.87	32.88	32.92
3	41.42	41.38	41.30	41.43
4	63.81	63.70	63.80	63.85
5	69.08	69.04	68.99	69.08

The global mode shapes for each case are compared as shown in Fig. 8, which shows good agreements with the case of centralized processing. Note that the mode shapes for the centralized NExT/ERA and RDT/ERA are indiscernible from each other as shown in Fig. 8, because global mode shapes from Reference 1 and 2 are visibly indistinguishable. Both NExT/ERA and RDT/ERA in decentralized processing environment estimate global mode shapes accurately.

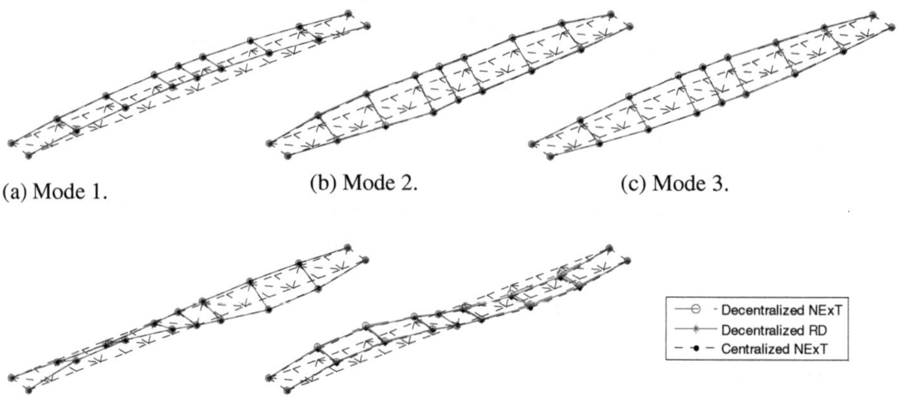

(a) Mode 1. (b) Mode 2. (c) Mode 3.

(d) Mode 4. (e) Mode 5.

Fig. 8 Global mode shapes

Wireless data communication required in correlation and RD function estimation is investigated to identify the efficiency of RDT-based decentralized data and is summarized in Table 5. RDT reduces wireless data communication to 29.04% of NExT.

The experiment using *DecentralizedDataAggregation* for NExT and *DecentralizedDataAggregationRD* for RDT shows both NExT and RDT are well-suited to the decentralized processing. RDT-based decentralized data aggregation is demonstrated to be more efficient with respect to data communication.

Table 5. Transferred data for NExT and RDT

Case 1 *DecentralizedDataAggregation*	Case 2 *DecentralizedDataAggregationRD*	RDT/NExT (%)
51,216 points	14,871 points	29.04

6 Conclusions

The RDT-based decentralized data aggregation approach was proposed for efficient data condensation and feature extraction, and verified experimentally. The performance of decentralized RDT was assessed in terms of (1) accuracy of the estimated modal properties and (2) efficiency in the wireless data communication. The NExT-based decentralized data aggregation approach was selected as a reference for comparison. *DecentralizedDataAggregationRD* has been developed as an implementation of RDT on the Imote2 wireless sensor platform and verified on a steel truss structure. From the experimental implementation, the efficacy of the RDT-based decentralized data aggregation strategy has been demonstrated.

Acknowledgments. This study is supported in part by the National Science Foundation Grants CMS 06-00433 (Dr. S.C. Liu, program manager). This support is gratefully acknowledged.

References

1. Asmussen, J.C.: Modal Analysis Based on the Random Decrement Technique – Application to Civil Engineering Structures. PhD Thesis. University of Aalborg, Denmark (1997)
2. Brincker, R., Krenk, S., Kirkegaard, P.H., Rytter, A.: Identification of the Dynamical Properties from Correlation Function Estimates, Bygningsstatikse Meddelelser. Danish Society for Structural Science and Engineering 63(1), 1–38 (1992)
3. Brincker, R.: Note about the Random Decrement Technique. Aalborg University (1995)
4. Caffrey, J., Govindan, R., Johnson, E., Krishnamachari, B., Masri, S., Sukhatme, G., Chitalapudi, K., Dantu, K., Rangwala, S., Sridharan, A., Xu, N., Zuniga, M.: Networked Sensing for Structural Health Monitoring. In: Proceedings of 4th International Workshop on Structural Control, New York, NY, June 10-11, pp. 57–66 (2004)
5. Cho, S., Jo., H., Jang, S.A., Park, J., Jung, H.J., Yun, C.B., Spencer Jr., B.F., Seo, J.: Structural Health Monitoring of a Cable-stayed Bridge Using Smart Sensor Technology: Data Analyses. Smart Structures and Systems: An International Journal (accepted) (2010)
6. Cole, H.A.: On-the-analysis of random vibrations. Paper No. 68-288. American Institute of Aeronautics and Astronautics (1968)
7. Gao, Y., Spencer Jr., B.F., Ruiz-Sandoval, M.: Distributed computing strategy for structural health monitoring. Journal of Structural Control and Health Monitoring 13(1), 488–507 (2006)

8. Jang, S.A., Jo, H., Cho, S., Mechitov, K.A., Rice, J.A., Sim, S.H., Jung, H.J., Yun, C.B., Spencer Jr., B.F., Agha, G.: Structural Health Monitoring of a Cable-stayed Bridge using Smart Sensor Technology: Deployment and Evaluation. Smart Structures and Systems: An International Journal (2010) (accepted)
9. Juang, J.N., Pappa, R.S.: An eigensystem realization algorithm for modal parameter identification and model reduction. Journal of Guidance, Control, and Dynamics 8(5), 620–627 (1985)
10. James, G.H., Carne, T.G., Lauffer, J.P.: Dynamic testing and system identification of a multispan highway bridge, SAND92-1666, UC-261. Sandia National Laboratories, Sandia (1993)
11. Lynch, J.P., Sundararajan, A., Law, K.H., Kiremidjian, A.S., Carryer, E.: Embedding Damage Detection Algorithms in a Wireless Sensing Unit for Operational Power Efficiency. Smart Materials and Structures 13(4), 800–810 (2004)
12. Lynch, J.P., Parra-Montesinos, G., Canbolat, B.A., Hou, T.C.: Real-time Damage Prognosis of High-performance Fiber Reinforced Cementitious Composite Structures. In: Proceedings of Advances in Structural Engineering and Mechanics (ASEM 2004), Seoul, Korea, September 2-4 (2004)
13. Nagayama, T., Sim, S.H., Miyamori, Y., Spencer Jr., B.F.: Issues in Structural Health Monitoring using Smart Sensors. Smart Structures and Systems 3(3), 299–320 (2007)
14. Nagayama, T., Spencer Jr., B.F.: Structural Health Monitoring Using Smart Sensors. Newmark Structural Engineering Laboratory (NSEL) Report Series, vol. 1. University of Illinois at Urbana-Champaign, Urbana (2007), http://hdl.handle.net/2142/3521
15. Nitta, Y., Nagayama, T., Spencer Jr., B.F., Nishitani, A.: Rapid damage assessment for the structures utilizing smart sensor MICA2 MOTE. In: Proceedings of 5th Int. Workshop on Structural Health Monitoring, Standford, CA, pp. 283–290 (2005)
16. Rahimi, M., Hardik, S., Sukhatme, G.S., Heideman, J., Deborah, E.: Studying the feasibility of energy harvesting in a mobile sensor networks. In: Proceedings of IEEE Int. Conference on Robotics and Automation, Taipai, Taiwan, vol. 1, pp. 19–24 (2003)
17. Rice, J.A., Spencer Jr., B.F.: Flexible Smart Sensor Framework for Autonomous Full-scale Structural Health Monitoring. Newmark Structural Engineering Laboratory (NSEL) Report Series, vol. 18. University of Illinois at Urbana-Champaign, Urbana (2009), http://hdl.handle.net/2142/13635
18. Sim, S.H., Spencer Jr., B.F., Zhang, M., Xie, H.: Automated Decentralized Modal Analysis using Smart Sensors. Journal of Structural Control and Health Monitoring (2010) (in press), doi:10.1002/stc.348
19. Sim, S.H., Spencer Jr., B.F.: Decentralized Strategies for Monitoring Structures using Wireless Smart Sensor Networks. Newmark Structural Engineering Laboratory (NSEL) Report Series, vol. 19. University of Illinois at Urbana-Champaign, Urbana (2009), http://hdl.handle.net/2142/14280
20. Spencer Jr., B.F., Ruiz-Sandoval, M.E., Kurata, N.: Smart Sensing Technology: Opportunities and Challenges. Journal of Structural Control and Health Monitoring 11, 349–368 (2004)
21. Tanner, N.A., Wait, J.R., Farrar, C.R., Sohn, H.: Structural Health Monitoring Using Modular Wireless Sensors. Journal of Intelligent Material Systems and Structures 14(1), 43–56 (2003)
22. Vandiver, J.K., Dunwoody, A.B., Campbell, R.B., Cook, M.F.: A Mathematical Basis for the Random Decrement Vibration Signature Analysis Technique. Journal of Mechanical Design 104, 307–313 (1982)

A Wave-Based Approach for Seismic Response Analyses of High-Rise Buildings

R. Zhang, S. Al Hilali, A. Abdulla, and M. Al Kurbi

Department of Mechanical Engineering, Petroleum Institute,
P.O. Box 2533, Abu Dhabi, United Arab Emirates

Abstract. This study examines one-dimensional wave propagation in a multi-story building with seismic excitation. In particular, the building is modeled as a series of shear beams for columns/walls and lumped masses for floors. Wave response at one location of the building is then derived to an impulse displacement at another location in time and frequency domains, termed here as wave-based or generalized impulse/frequency response function (GIRF/GFRF), which is dependent upon the building characteristics above the impulse location. Not only does this study illustrate features of GIRF/GFRF in terms of building properties, it also shows broad-based applications of the modeling. Two examples are presented with the use of the modeling. One is wave-based characterization of ten-story Millikan Library in Pasadena, California with the recordings of Yorba Linda earthquake of September 3, 2002. The other is analysis for influence of stochastic floor-to-column mass ratio, story-height and seismic input in seismic wave responses.

Keywords: Wave-based approach, Seismic responses of buildings, Wave propagation in buildings.

1 Introduction

Seismic response analyses of buildings are typically carried out within the framework of vibration theory applied to a discrete or a multi-degree-of-freedom model for building structure, in which structural dynamic properties are characterized with modal frequency and shape that are a function of physical parameters such as floor mass and column stiffness. This vibration-based approach builds on the belief that seismic responses are *synchronous* at different locations of the structure.

Alternatively, seismic response can also be viewed as the result of wave propagation, which is appropriate particularly for tall buildings. Accordingly, structural properties of a building can be characterized with wave-based indices such as wave speed, which is directly related to such physical parameters as mass density and shear modulus of the building materials. In fact, recent studies show the promising of the wave-based approach in better understanding of wave phenomena in seismic recordings and system identification (e.g., Safak, 1999; Todorovska et al., 2001; Snieder and Safak, 2006; Kohler et al., 2007). Building upon the aforementioned work, this study proposes seismic wave motion modeling in building structures and examines its effectiveness and broad-based applications.

W.Q. Zhu et al. (Eds.): Nonlinear Stochastic Dynamics & Control, IUTAM BOOK SERIES 29, pp. 315–325.
springerlink.com

2 Modeling of Wave Motion in Buildings

In this study, an *N*-story building is modelled as a series of shear beams for columns/walls and lumped masses for floors as shown in Fig. 1, in which one-dimension shear wave propagation in vertical direction is investigated. Each column/wall is characterized with shear wave speed $v = \sqrt{G/\rho}$ where G and ρ are respectively shear modulus and mass density, hysteretic damping ratio γ_c, story height h, and cross-sectional area A, while each floor with lumped mass m_f which excludes the mass overlapped with columns and walls, and damping γ_f (= c_f/m_f) where c_f is the hysteretic damping coefficient.

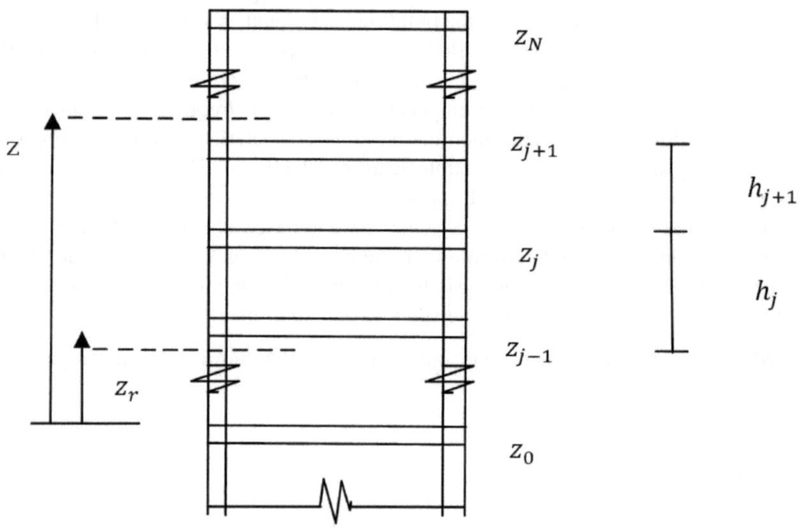

Fig. 1 A model for an *N*-story building subjected to seismic motion below z_0

For source-free, j^{th} column bounded with (z_{j-1}^+, z_j^-) and j^{th} floor bounded with (z_j^-, z_j^+) with $j=1,2...,N$, wave motion of shear displacement $u(z,t)$ is governed by

$$\frac{\partial^2 u(z,t)}{\partial z^2} = \frac{1}{v_j^2}\frac{\partial^2 u(z,t)}{\partial t^2} \tag{1}$$

$$G_{j+1}A_{j+1}\frac{\partial u(z_j^+,t)}{\partial z} - G_jA_j\frac{\partial u(z_j^-,t)}{\partial z} - c_{f_j}\frac{\partial u(z_j,t)}{\partial t} = m_{f_j}\frac{\partial^2 u(z_j,t)}{\partial t^2} \tag{2}$$

where superscripts + and – indicate respectively the positive and negative sides of height z. For convenience, height z indicates the positive side in this paper and thus superscript + can be dropped in later use.

Introducing Fourier transform representation of the wave motion

$$u(z,t) = \int_{-\infty}^{\infty} U(z,\omega)e^{i\omega t}d\omega, \qquad U(z,\omega) = \frac{1}{2\pi}\int_{-\infty}^{\infty} u(z,t)e^{-i\omega t}dt \tag{3a,b}$$

where i is imaginary unit and ω frequency. Inserting Eq. (3a) into Eqs. (1-2), one can solve for wave representation in frequency domain at z and wave relationship at z_l and z_m

$$U_z = U(z,\omega) = U_z^u + U_z^d = C_1 e^{-i\omega z/v_j} + C_2 e^{i\omega z/v_j} \tag{4}$$

$$\begin{Bmatrix} U_m^u \\ U_l^d \end{Bmatrix} = \begin{bmatrix} T_{ml} & R_{lm} \\ R_{ml} & T_{lm} \end{bmatrix} \begin{Bmatrix} U_l^u \\ U_m^d \end{Bmatrix} \tag{5}$$

where displacement U_z consists of up-going and down-going waves denoted with superscripts u and d, and transmission and reflection coefficients T_{ml} and R_{lm} (T_{lm} and R_{ml}) relate the outgoing waves $U_m{}^u$ and U_l^d to input waves U_l^u and $U_m{}^d$ for building segment bounded with (z_l, z_m).

For the jth column, the coefficients can be found

$$T_{j^-(j-1)} = T_{z_j^- z_{j-1}} = T_{(j-1)j^-} = e^{-i\omega h_j/v_j}, \qquad R_{j^-(j-1)} = R_{(j-1)j^-} = 0 \tag{6}$$

Wave attenuation in propagation due to damping can be taken into consideration by replacing real shear wave speed with complex one $v_j[1+i\gamma_{cj}\,|\,\omega|]$.

For the jth floor, one can find

$$T_{jj^-} = \frac{2}{1+r_{I_j} - r_{D_j} + ir_{M_j}} = B_{f_j} e^{-i\omega h_{ej}/v_j}$$

$$R_{jj^-} = T_{jj^-} - 1 \ , \quad T_{j^- j} = T_{jj^-} r_{I_j}, \quad R_{j^- j} = T_{j^- j} - 1 \tag{7}$$

where coefficients r_{Ij}, r_{Dj} and r_{Mj}, amplitude B_{fj} and equivalent floor-height h_{ej} can be found in terms of column impedance (ρv) ratio, cross-sectional area ratio, floor-to-column mass ratio (r_m), wave travel time for column length (h/v), i.e.,

$$r_{I_j} = \frac{\rho_{j+1}v_{j+1}}{\rho_j v_j}\frac{A_{j+1}}{A_j}, \quad r_{D_j} = \gamma_{f_j} r_{M_j}, \quad r_{M_j} = r_{m_j}\frac{h_j}{v_j}\omega, \quad r_{m_j} = \frac{m_{f_j}}{m_j}$$

$$B_{f_j} = \frac{2}{\sqrt{(1+r_{I_j} - r_{D_j})^2 + r_{M_j}^2}}, \quad h_{ej} = \frac{v_j}{\omega}\tan^{-1}\frac{r_{M_j}}{1+r_{I_j} - r_{D_j}} \tag{8}$$

Eqs. (7-8) indicates that lumped floor mass can be treated as a column-type continuum with equivalent height and damping but with non-zero reflection coefficients. For lumped mass at the building top, $\rho_{N+1}=v_{N+1}=A_{N+1}=0$ or $r_{I_N}=0$, which corresponds to the free-end boundary condition for the building. At the building lower end z_0 (or generally at referenced location z_r which could be selected as z_0), no segments below level z_0 are used in the model, yielding $r_{I_1}=\infty$. One can then find $T_{0^-0}=1$ and $R_{0^-0}=0$, suggesting down-going wave at z_0 is completely transmitted to the lower end z_0^- and no reflection to the up-going wave. This indicates the lower end z_0 has the fixed-end boundary condition.

For a composite building segment bounded with (z_l, z_n), or simply (l,n), with intermediate location z_m $(z_l<z_m<z_n)$ such as (z_{j-1}, z_j) with z_j^-, repeat use of Eq. (5) for (l,m) and (m,n) will lead to the representation of transmission and reflection coefficients in (l,n) in terms of those in two sub-segments in (l,m) and (m,n) as

$$T_{ln}=\frac{T_{lm}T_{mn}}{1-R_{nm}R_{lm}}, \qquad R_{ln}=R_{mn}+\frac{T_{nm}R_{lm}T_{mn}}{1-R_{nm}R_{lm}} \tag{9}$$

The above composition rule can be applied reversely for (n,l) and also repeatedly to find all the transmission and reflection coefficients between any two locations.

With the aforementioned coefficients R and T, wave response at z (or $z_R=z-z_r$) can then be related to those at referenced level z_r that could be at the bottom of the building or any other height as

$$D_{Rr}(\omega)=\frac{U_{z_R}}{U_{z_r}}=\frac{(1+R_{NR})T_{Rr}}{(1-R_{rR}R_{NR})(1+R_{Nr})}, \qquad d_{Rr}(t)=\int_{-\infty}^{\infty} D_{Rr}e^{i\omega t}d\omega \tag{10,11}$$

Eq. (10) indicates that D_{Rr} is dependent only upon R and T above z_r which are function of building properties and frequency. For $z=z_r$, Eqs. (10-11) lead to $D_{Rr}=1$ and $d_{Rr}=\delta(t)$, suggesting that D_{Rr} and d_{Rr} are respectively frequency and time displacement responses at z to displacement impulse at z_r. Subsequently, wave response representation in general, and displacement response at z to input displacement at z_r in particular, is then found as

$$u(z,t)=\int_{-\infty}^{\infty} D_{Rr}U_{z_r}e^{i\omega t}d\omega=\int_{-\infty}^{\infty} d_{Rr}(t-\tau)u(z_r,\tau)d\tau \tag{12}$$

which has the same mathematical form as traditional vibration response of Duhamel's or convolution integral with impulse response function.

While the aforementioned derivation is for displacement (u,U), it is straightforward to the extension to velocity $(v=du/dt, V=i\omega U)$ and acceleration $(a=d^2u/dt^2, A=-\omega^2 U)$ with D_{Rr} and d_{Rr} remaining the same. For acceleration input at z_r and displacement response at z, which is the typical case for displacement response to earthquake ground acceleration, Eq. (12) can be modified as

$$u(z,t) = \int_{-\infty}^{\infty} H_{Rr} A_{z_r} e^{i\omega t} d\omega = \int_{-\infty}^{\infty} h_{Rr}(t-\tau)a(z_r,\tau)d\tau \qquad (13)$$

where $H_{Rr}=-D_{Rr}/\omega^2$ and h_{Rr} have conventional meanings for frequency response function and impulse response function respectively. Because of the aforementioned generality in addition to the generalized impulse location at z_r which is not designated at z_0, D_{Rr} and d_{Rr} are referred to respectively as wave-based or generalized frequency response function (GFRF) and generalized impulse response function (GIRF).

3 Applications in Earthquake Engineering

For illustration, two examples are presented below to show the usefulness and effectiveness of the proposed modelling in system identification and seismic response analyses.

3.1 Uniform Shear-Beam Model

One can first examine a special case for the aforementioned model, i.e., uniform shear-beam model without lumped floor mass, which leads Eqs. (10-11) to

$$D_{Rr} = \frac{[1+e^{-i\omega(2H_r-2z_R)/v_c} e^{-\gamma_c|\omega|(2H_r-2z_R)/v_c}] e^{-i\omega z_R/v_c} e^{-\gamma_c|\omega| z_R/v_c}}{1+e^{-i\omega(2H_r)/v_c} e^{-\gamma_c|\omega|(2H_r)/v_c}} \qquad (14)$$

$$d_{Rr} = 8\omega_0 \sum_{j=1}^{\infty} (-1)^{j+1} e^{-\gamma_c \omega_j t} \cos\frac{\omega_j (H_r - z_R)}{v_c} \sin(\omega_j t) \qquad (15)$$

where $H_r=z_N-z_r$ and $z_R=z-z_r$ denote respectively the height and response location of the building portion bounded by (z_r,z_N), v_c is shear velocity of the building portion, and modal frequency ω_j is

$$\omega_j = (2j-1)\omega_0, \qquad \omega_0 = 0.5\pi v_c / H_r, \qquad j = 1,2\ldots,\infty \qquad (16)$$

Eq. (15) shows GIRF consists of infinite number of motion modes, each of which has exponentially decaying damping factor, modal shape and harmonic motion. The fundamental or first mode with $j=1$ has period $T_c=4H_r/v_c$ which is the travel time for waves to propagate up and down the building height (H_r) twice. Equations (14-16) are first derived by Snieder and Safak (2006).

To validate the uniform shear-beam model, this study follows Snieder and Safak (2006) to examine seismic recordings in ten-story Millikan Library after the Yorba Linda earthquake of September 3, 2002. Figure 2a shows the seismic acceleration recordings in the north-south component in the west side of the building at basement and the 1st to 10th floor. While features of wave propagation in the building can be seen from the arrival time of travelling shear waves from floor to floor in *10-11 s*, it can be observed more clearly through GIRF.

Figure 2b shows GIRF based on recordings at different floors ($j=1\text{-}10$) with respect to referenced motion at basement (b), denoted as $\tilde{d}_{jb}$, which is calculated from GFRF, i.e., $\tilde{D}_{jb} = (\tilde{U}_j \tilde{U}_b^*)/(|\tilde{U}_b|^2 + \varepsilon)$ where $\tilde{U}$ is recordings in frequency domain, superscript asterisk indicates the complex conjugate, and ε is a positive small number, implying the added white noise. The white noise is used primarily to avoid unstable calculation at some frequencies near the notches in the spectrum $\tilde{U}_b$, as suggested by Snieder and Safak (2006). As ε approaches zero, $\tilde{D}_{jb}$ is degenerated to the traditional expression of Eq. (10).

As shown in Fig. 2b, the GIRF at the basement is impulse acceleration with ε selected as 5% total power spectrum of basement motion. As a fictitious input or virtual source to the building, the impulse acceleration at basement is propagated upward at building shear velocity with gradually-increased amplitude. The increased wave amplitude as location approaching to the free top is due to the fact that transmission coefficient at the top ($T_{NN}^{-}=2$) results in the doubled up-going wave amplitude in comparison with the input at the basement if damping is not concerned. The travelling waves are then reflected after hitting the top with reflection coefficient ($R_{NN}^{-}=1$). They are then propagated downward with gradually-reduced amplitude, and zero amplitude at the basement. The decreased wavc amplitude as location moving downward is due to the fact that the basement with impulse displacement input is equivalent to the fixed basement end, which makes the wave motion disappear at time other than $t=0$. This phenomena can also be explained with transmission and reflection coefficients at the basement ($T_b^{-}{}_b=1$ and $R_b^{-}{}_b=0$), which indicates that all the down-going waves completely transmit through the basement and no up-going waves reflected from down-going waves at the basement, i.e., fixed basement boundary.

The aforementioned cycle of wave propagation continues as time goes on. For earlier time (*0-1 s*), the GIRF consists primarily of superposition of up-going and down-going travelling shear waves. For later time (*>1 s*) as the travelling waves can be regarded as standing waves, the GIRF develops the character of a resonance of the building, with the amplitude reduced as time goes (e.g., see GIRF at the 10th floor), which is typically the free-vibration or impulse response phenomenon. The increased amplitude of the GIRF at the fundamental modal frequency with the position changed from the 1st to the 10th floor is again attributed to the wave phenomena due to the fixed bottom and free top boundaries. While Fig. 2b also shows other higher-frequency vibration/wave motion modes, they are typically observed more clearly in frequency domain.

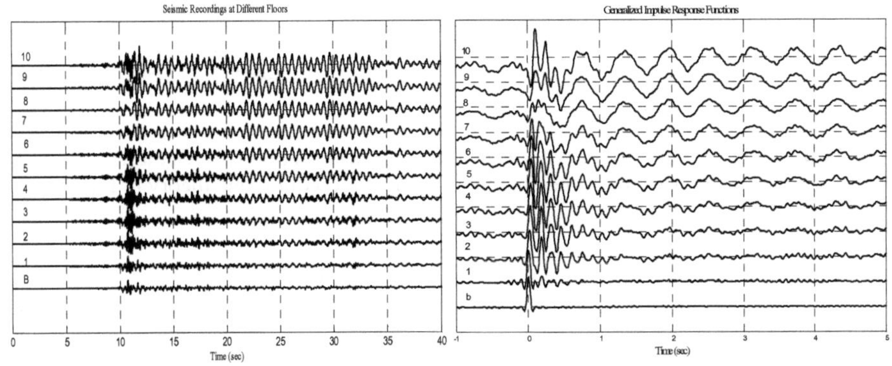

Fig. 2a,b Seismic recordings (Fig. 2a, left) and GIRF (Fig. 2b, right) at the different floors (indicated as 1-10) with respect to basement motion (indicated as *B* or *b*)

As an example for applications, this study shows system identification for shear wave speed and damping with the use of two sets of recordings. Figures 3a,b show the GFRFs of acceleration at the eighth floor with respect to referenced motion at basement and the 3rd floor respectively based on recordings (i.e., $\tilde{D}_{8b}$ and $\tilde{D}_{83}$) and the uniform shear-beam model (i.e., D_{8b} and D_{83}). The identified shear wave speeds for the whole building and the 3rd-floor-up building portion are 330 and 292 *m/s* respectively, indicating that shear modulus (G) of the 3rd-floor-up building portion is less than that of the whole building if mass density (ρ) remains the same. Alternatively, the lower portion of the building (i.e., from the basement to the 3rd floor) is more rigid in shear resistance than the upper portion (from the 3rd floor to the top). Typically, the stronger the shear rigidity of the building is, the less the corresponding damping ratio. While not universally correct, this phenomenon is also observed from the identified damping for the aforementioned case, i.e., 0.0187 and 0.0281 respectively for the whole and the 3rd-floor-up building portion. As shown in Table 1, the identified parameters are also compared with those using recordings at other floors (only one of them with D_{3b} and $\tilde{D}_{3b}$ is shown here due to the limited space) as well as those from Snieder et al., (2006) and Chopra (1995), indicating that the uniform shear-beam model is good enough to characterize the fundamentals of wave and vibration motion in buildings.

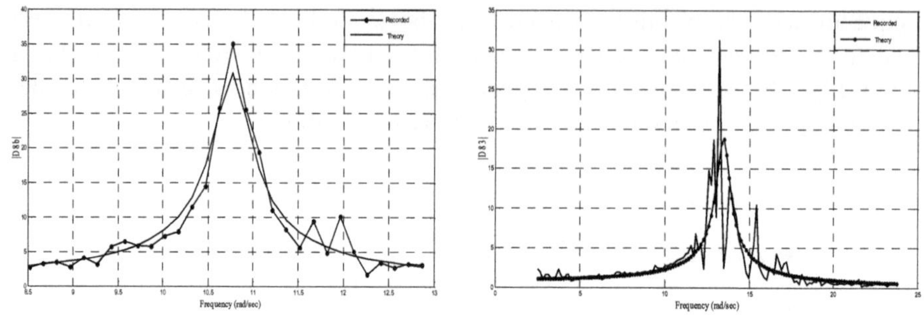

Fig. 3a,b GFRFs at the 8th floor with respect to basement motion (Fig. 3a, left) and the 3rd floor motion (Fig. 3b, right) obtained based on seismic recordings and theory in Eq. (14)

Table 1 Identified shear wave speeds and damping and their comparison with results from others (one number from Snieder and Safak (2006) using the uniform shear-beam model and the same earthquake but averaged over all D_{jb}, $j=1\text{-}10$ with 11-set recordings) and (two numbers from Table 11.1.1 in Chopra (1995) with Lytle and San Fan Fernando earthquake recordings respectively).

	Height (H_r) and location (z_R) in *m*	Identified 1st modal frequency in *rad/s*	Shear wave speed in *m/s*	Damping
$z_r = z_b$ $z = z_3$	48.2, 13.4	10.77 (12.08,10.13)	330 (322)	0.0305 (0.0244) (0.029,0.064)
$z_r = z_b$ $z = z_8$	48.2, 34.8	10.77 (12.08,10.13)	330 (322)	0.0187 (0.0244) (0.029,0.064)
$z_r = z_3, z = z_8$	34.1, 21.4	13.16	292	0.0281

3.2 Stochastic Wave-Motion Model

The difference between earthquake recordings and any wave/vibration-based models such as uniform shear-beam model or its generalized one with floor masses is well observed, which is primarily attributed to the deterministic approach for modeling and analysis. This issue can however be addressed with a stochastic wave-motion model, in which building parameters and/or seismic input are treated as random variables/processes. While various statistical responses for the stochastic model can be found within the framework of probabilistic structural dynamics (e.g., Lin and Cai, 1995), this study presents some analyses with selected random system parameters in wave responses and stochastic seismic acceleration input at building bottom.

For regular building, column/wall properties are not changed significantly from one floor to the other, and can be assumed to be the same without loss of generality. For earthquake-excited building motion, the largest frequency of interest is typically less than $n\,\omega_0$ with $n<N$. Therefore, for random floor mass with small floor-to-mass ratio ($r_{mj}=m_{fj}/m_{cj}<<1$), Eqs. (7-8) become

$$T_{jj^-} = T_{j^-j} \approx e^{-i\omega h_{ej}/v_j}, \qquad h_{ej} \approx 0.5 r_{m_j} h_j \tag{18}$$

which suggests that up-going and down-going waves transmit through the jth floor without loss of amplitude. As far as transmission coefficients are concerned, the floor functions like an extended column portion with extra height h_{ej} and zero damping. Accordingly, transmission coefficients for a building segment with a column connected to a floor mass are equivalent to those in a pure column without floor, but with an increased column length (h_j+h_{ej}) and reduced damping factor ($\gamma_{ej}=\gamma_{cj}/(1+h_{ej}/h_j)$). Alternatively, they can also be viewed as the equivalent transmission coefficients in the same column length but with decreased velocity and reduced damping factor. Based on Eq. (16), the fundamental modal frequency of the building with floor masses, denoted as Ω_1, is decreased in comparison with those without floor masses (ω_1). This can be seen clearly in Fig. 4a with $r_m=0$ and 0.1, where $r_m=0$ is the uniform shear-beam building model. The corresponding response amplitude is increased due to the reduced damping. For higher-order mode motion, the higher-order modal frequencies Ω_j of the building with floor masses will be reduced proportionally and the corresponding amplitude will be increased in general. The mean μ and standard variation σ of modal frequency Ω_j can be found as

$$\mu_{\Omega_j} = \omega_j\{1-0.5\mu_{r_m}\}, \qquad \sigma^2_{\Omega_j} = 0.25\mu^2_{\Omega_j}\sigma^2_{r_m} \tag{19}$$

where μ_{rm} and σ_{rm} are the mean and standard deviation of random ratio of total floor mass versus total building column mass. It can be proved that if r_{mj} ($j=1,2...N$) is constant and floor height h_j is random, Eq. (19) remains the same except μ_{rm} and σ_{rm} replaced by μ_h and σ_h respectively. The other statistical responses such as mean and standard deviation of frequency-response amplitudes at corresponding modal frequencies can be found numerically based on Eqs. (16,19).

For large floor-to-column mass ratio or other random system parameters, the statistical analysis for GIRF/GFRF must be carried out numerically or with Monte Carlo simulation. While not presented here, this paper shows influences of some system parameters in frequency responses. In particular, Fig. 4a shows the influences of large floor-to-column mass ratio ($r_m=1$) in GFRF, revealing similar phenomena observed with small r_m before. Figure 4b indicates that modal frequencies are insensitive to the change of column impedance ratio r_I, while the corresponding amplitudes are reduced significantly with decreased r_I.

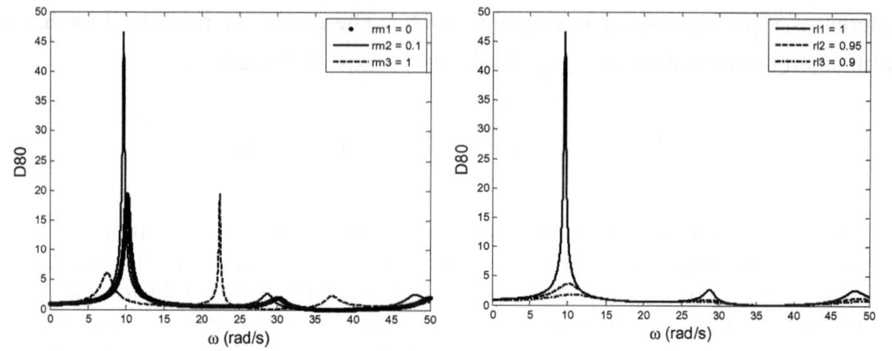

Fig. 4a,b GFRFs (D_{80}) at the 8th floor of a 11-story building with respect to bottom motion with $v_j=300$ *m/s*, $h_j=4.25$ *m*, $\gamma_{cj}=\gamma_{fj}=0.03$, $r_{lj}=1$ (only for Fig. 4a, left), $r_{mj}=0.1$ (only for Fig. 4b, right) and $r_{m11}=0.5r_{m10}$, for $j=1,2,\ldots,11$. Note that j (=1-3) of *rmj* and *rlj* in the legend indicates the case number, not the floor number.

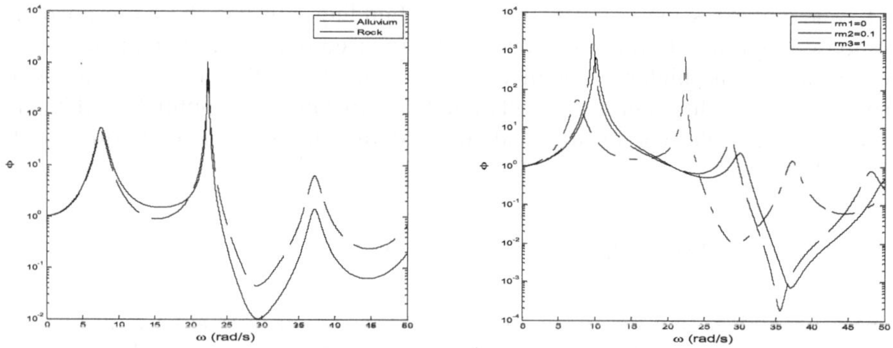

Fig. 5a,b Spectral density of acceleration at the 8th floor to ground acceleration spectral density with Kanai-Tajimi power spectra ($G_0=1$, site pre-dominant frequency ω_g, site damping ξ_g=0.34) with the same building parameters as Fig. 4a. Fig. 5a (left) ω_g=18.4 *rad/s* for alluvium and 27 *rad/s* for rock, and $r_{mj}=1$) and Fig. 5b (right) ω_g=18.4 *rad/s* for alluvium and $r_{mj}=0$, *0.1 and 1*).

For ground motion characterized by evolutionary stochastic process (Lin and Cai, 1995)

$$a(z_0,t)=\int_{-\infty}^{\infty} b(t,\omega)e^{i\omega t}dZ(\omega) \tag{20}$$

where Z is a stochastic process with orthogonal increment in frequency, and b is a deterministic function of both t and ω, the mean-square acceleration response with deterministic building parameters can be found as

$$E[a^2(z,t)]=\int_{-\infty}^{\infty} |b(t,\omega)|^2\,\Phi(\omega)d\omega, \qquad \Phi(\omega)=D_{R0}^2(\omega)G(\omega) \tag{21}$$

where E denotes ensemble average, and G and Φ are spectral densities of ground acceleration and response respectively. As a degenerated case with $b=1$ and $G=G_0(1+4\xi^2\omega^2/\omega_g^2)/[(1-\omega^2/\omega_g^2)^2+4\xi^2\omega^2/\omega_g^2]$ (Kanai-Tajimi model), Fig. 5a shows the spectral densities of acceleration at the 8th floor with seismic input at alluvium and rock sites. Since the rock pre-dominant frequency (ω_g=27 *rad/s*) is closer to the second modal frequency (ω_2 ~ 22 *rad/s*) than the alluvium one (18.4 *rad/s*), the peak with rock at the second modal frequency is larger than that with alluvium. This can also be seen with mean square accelerations 0.0076 and 0.026 m^2/s^4 for rock and alluvium respectively. Figure 5b shows the response spectral densities with different floor-to-column mass ratio, with corresponding mean square as 0.0188, 0.0102 and 0.0076 m^2/s^4 respectively for r_{mj}=0, 0.1 and 1.

4 Conclusions

This study proposes a wave-based approach to model and analyze seismic building motion. Alternative to vibration-based ones, this approach provides some perspective of seismic behaviors of building structures which traditional vibration-based approach does not show clearly.

References

[1] Chopra: Dynamics of Structures-Theory and Applications to Earthquake Engineering. Prentice-Hall, Inc., Englewood Cliffs (1995)

[2] Kohler, M.D., Heaton, T.H., Bradford, S.C.: Propagating waves in the steel, moment-frame Factor Building recorded during earthquakes. Bull. Seism. Soc. Am. 97(4), 1334–1345 (2007)

[3] Lin, Y.K., Cai, G.Q.: Probabilistic Structural Dynamics-Advanced Theory and Applications. McGraw-Hill, Inc., New York (1995)

[4] Safak, E.: Wave-propagation formulation of seismic response of multistory buildings. Journal of Structural Engineering, ASCE 125(4), 426–437 (1999)

[5] Snieder, R., Safak, E.: Extracting the building response using seismic interferometry: Theory and application to the Millikan library in Pasadena, California. Bull. Seism. Soc. Am. 96(2), 586–598 (2006)

[6] Todorovska, M.I., Ivanovic, S.S., Trifunac, M.D.: Wave propagation in a seven-story reinforced concrete building I: Theoretical models. Soil Dynamics and Earthquake Engineering 21, 211–223 (2001)

Author Index

MIX
Papier aus verantwortungsvollen Quellen
Paper from responsible sources
FSC® C105338

If you have any concerns about our products,
you can contact us on
ProductSafety@springernature.com

In case Publisher is established outside the EU,
the EU authorized representative is:
Springer Nature Customer Service Center GmbH
Europaplatz 3, 69115 Heidelberg, Germany

Printed by Libri Plureos GmbH
in Hamburg, Germany